Lamia Hadrich Belguith · Khaled Shaalan
Editors

Advancements in Machine Learning and Natural Language Processing: Innovations and Applications

3rd International Conference on Language Processing and Knowledge Management (LPKM'2024)

Springer

Editors
Lamia Hadrich Belguith
FSEGS
University of Sfax
Sfax, Tunisia

Khaled Shaalan
The British University In Dubai
Dubai, Dubai, United Arab Emirates

ISSN 2367-3370 ISSN 2367-3389 (electronic)
Lecture Notes in Networks and Systems
ISBN 978-3-031-85066-0 ISBN 978-3-031-85067-7 (eBook)
https://doi.org/10.1007/978-3-031-85067-7

This Springer imprint is published by the registered company Springer Nature Switzerland AG
The registered company address is: Gewerbestrasse 11, 6330 Cham, Switzerland

Lecture Notes in Networks and Systems 1303

The series "Lecture Notes in Networks and Systems" publishes the latest developments in Networks and Systems—quickly, informally and with high quality. Original research reported in proceedings and post-proceedings represents the core of LNNS.

Volumes published in LNNS embrace all aspects and subfields of, as well as new challenges in, Networks and Systems.

The series contains proceedings and edited volumes in systems and networks, spanning the areas of Cyber-Physical Systems, Autonomous Systems, Sensor Networks, Control Systems, Energy Systems, Automotive Systems, Biological Systems, Vehicular Networking and Connected Vehicles, Aerospace Systems, Automation, Manufacturing, Smart Grids, Nonlinear Systems, Power Systems, Robotics, Social Systems, Economic Systems and other. Of particular value to both the contributors and the readership are the short publication timeframe and the world-wide distribution and exposure which enable both a wide and rapid dissemination of research output.

The series covers the theory, applications, and perspectives on the state of the art and future developments relevant to systems and networks, decision making, control, complex processes and related areas, as embedded in the fields of interdisciplinary and applied sciences, engineering, computer science, physics, economics, social, and life sciences, as well as the paradigms and methodologies behind them.

Indexed by SCOPUS, EI Compendex, INSPEC, WTI Frankfurt eG, zbMATH, SCImago.

All books published in the series are submitted for consideration in Web of Science.

For proposals from Asia please contact Aninda Bose (aninda.bose@springer.com).

Contents

Enhancements in Machine Learning

Combining Syntactic and Semantic Information in Knowledge Extraction Pipeline

Martin Verrev(✉), Tanel Tammet, and Priit Järv

Tallinn University of Technology, Akadeemia tee 15a, Tallinn, Estonia
martin.verrev@taltech.ee

Abstract. This paper presents an experimental pipeline and an annotation scheme for enriching text with semantic information by leveraging implicit information extracted from Abstract Meaning Representation (AMR). AMR, as an unanchored semantic representation language, encodes a wide range of information that can be useful for downstream NLP tasks. This implicit knowledge encompasses frames, semantic roles, numeric value labels, concept grounding to Wikipedia entries, and fine-grained relationship types. The extracted information can improve the performance of existing logic-based end-to-end question-answering pipelines that rely on deterministic rule-driven semantic parsers for text-to-logic conversion based on Universal Dependencies (UD).

Keywords: semantic parsing · knowledge extraction · meaning representations

1 Introduction

One of the crucial tasks for constructing a knowledge base for commonsense question answering is to automate the process of extracting background knowledge from unstructured sources. While structured data sources like ConceptNet[1] and ontologies like WordNet[2] provide facts and simple rules, they contain a lot of un-parsed English phrases and also lack most of the common everyday knowledge - facts about the world that everyone is expected to know. To enrich these knowledge bases and ask questions using natural language, we need to perform semantic parsing of natural language phrases and sentences in a way compatible with the structured data sources.

To this end, we have developed an experimental symbolic question-answering pipeline in natural language[18][3]. The question-answering pipeline is a modular system consisting of four main components: (a) a semantic parser that translates the text into extended first-order logic by initially converting the text to

[1] https://conceptnet.io/.
[2] https://wordnet.princeton.edu/.
[3] https://github.com/tammet/nlpsolver/.

L. H. Belguith and K. Shaalan (Eds.): LPKM 2024, LNNS 1303, pp. 3–12, 2025.
https://doi.org/10.1007/978-3-031-85067-7_1

the Universal Dependencies graph using a Stanza parser and then converting the graph to first-order logic (FOL) formulae presented in JSON-LD-LOGIC – a language designed to represent FOL in JSON format [19]; (b) answer generation through deduction with our reasoner using first-order logic enhanced with numeric confidences [17] and exceptions [16]; (c) a commonsense knowledge base [6]; and (d) answer presentation in natural language, including the explanation.

This paper aims to describe an annotation scheme and experimental system for automatically annotating input texts with semantic knowledge obtained via AMR parsing. The annotated results are intended to be utilized by the semantic parsing component of the symbolic question-answering pipeline, which currently relies on syntax-derived rules for information extraction.

The goals for the proposed annotation scheme are: (a) to provide a lightweight and flexible annotation scheme that encodes information for further processing in an NLP pipeline; (b) to annotate the text with semantic information derived from AMR (c) to provide a method abstracting away from syntactic nuances, thus simplifying the handcrafted rules for UD $\rightarrow$ logic conversion in the existing experimental symbolic question-answering pipeline.

2 Background and Related Work

Combining syntactic and semantic information is an important task that can be performed either on model, notation, or output level. On the model level, the paper [13] describes integrating syntactic dependency information into the parser attention mechanism. On the notation level, extensions to AMR representation are being presented [3] AMR-DA is being used to boost the performance of NLP models by adding the information extracted from AMR to the NLP pipeline [14]. Model level annotation is also demonstrated by [7], where markers for plurality, annotations, quantifiers, and intensionality are directly added to the AMR graph.

Universal Dependencies (UD) [2] is a framework for consistent annotation of grammar. It solves the problem of different corpora having different tagsets and annotation schemes [21] by providing a universal scheme and category set suitable for parser development, cross-lingual support, and language parsing[5]. It combines Stanford Dependencies, Google universal part-of-speech tags, and Interset interlingua[4] for morphosyntactic tagsets – e.g., encoding additional morphological information to the syntactic form [9]. UD corpora consist of over 200 treebanks in over 100 languages.

Abstract Meaning Representation (AMR) is a notation based on propositional logic and neo-Davidsonian semantics. AMR encodes the semantics of a sentence into a directed graph. Nodes in the graph represent semantic terms in the sentence, and the edges identify the semantic relations between nodes. Concepts can be either English verbs, PropBank framesets, or specific keywords. AMR includes frame arguments based on Propbank conventions, general semantic relations, and relations for quantities, date entities, and lists. Annotations do

[4] https://ufal.mff.cuni.cz/interset.

not link between concept and original span – thus, input has to be aligned first [1]. AMR graph can be viewed as a conjunction of logical triples represented by $relation\,(variable, value)$, where a value can be either a constant or a variable. SMATCH (semantic match) score computes the F1-score for the maximum number of matched triples among all possible variable mappings between the generated and gold semantic graphs. We use the SMATCH score to evaluate the trained model for AMR parsing.

PropBank (Proposition Bank) is a corpus annotated with verbal propositions and their arguments plus a lexicon defining those argument roles per verb. [11] We use Propbank 3.4 [12] that contains not only frame files for verbs but also non-verbal predicates such as adjectives, prepositions, and multi-word expressions. In addition, the Propbank 3.4 contains unique AMR-specific abstract roles (*have-degree.91, have-experience.91, have-role.91* etc.)[5]. Those AMR-specific predicates are reifications for more general roles: e.g., we can infer that not just two entities have a relation but also more fine-grained information from that.

3 Comparing AMR and UD for Natural Language Understanding

The first step was to compare the level of natural language understanding for both UD and AMR representations. To that end, we used natural language inference (NLI). The NLI task predicts whether the hypothesis follows the premise given: the entailment label is either *true* if the hypothesis follows the premise, *false* if the hypothesis contradicts the premise, and *neutral* if the relationship between entries and premise cannot be established. We can think of NLI as a generalized question-answering metric where the premise is the context and the hypothesis is the question. Thus, we can use the same benchmarks for both UD and AMR to perform comparative analyses of both notations.

The following benchmarks were chosen: SICK[6], FraCaS[7], ProofWriter[8] and HANS[9]. A subset was chosen from each dataset: for FraCaS `dev` split was selected (70 samples); for SICK, the first 100 samples from the `validation` split were selected. Test cases converted for existing experimental question-answering pipeline were selected for HANS[10] (112 samples) and ProofWriter[11] (43 samples).

A conversion was performed for SICK and FraCaS tests to run on the existing experimental pipeline. The tests were normalized in two steps: first, by formulating a hypothesis as a question. Then, normalizing the labels across datasets. Punctuation was added for multi-sentence premises. To get AMR-based NLI

[5] https://github.com/amrisi/amr-guidelines/blob/master/amr.md.
[6] https://huggingface.co/datasets/sick.
[7] https://huggingface.co/datasets/pietrolesci/fracas.
[8] https://allenai.org/data/proofwriter.
[9] https://github.com/tommccoy1/hans.
[10] https://github.com/tammet/nlpsolver/blob/main/tests_hans.py.
[11] https://github.com/tammet/nlpsolver/blob/main/tests_allen.py.

Table 1. Accuracy scores for AMR and UD in NLI benchmarks

Dataset	UD	AMR
FraCaS	.46	.47
SICK	.68	.72
ProofWriter	.98	.05
HANS	1.0	.64

measures for existing HANS and ProofWriter test sets using the AMR4NLI [10][12] toolkit, the passages from original test sets for the experimental pipeline were split into a separate premise that could contain one or more sentences and a hypothesis. The test data, tools, and reports can be found at https://github.com/martinve/amr_ud_benchmarks (Table 1).

4 Annotation Format

Our main pipeline starts with parsing the UD graph. However, AMR can represent a broader range of semantic information than UD, such as events, entities, and relationships. AMR is also more modular than UD - each frame represents a specific type of semantic information, such as an event, an entity, or a relationship. In contrast, UD is a syntactic representation that aims to capture the grammatical structure of a sentence in a structured way.

Using AMR parse graphs provides us with the following additional information: (a) labeling actions and semantic roles; (b) adding context to numeric values; (c) reifying AMR-specific abstract roles; (d) fine-grained named entities; and (e) resolving verb ambiguity (Fig. 1).

```
Titanic {"role": "Patient", "wiki": "RMS_Titanic", "type": "ship", "cat": "product"} sank
↪ {"instance": "sink-01", "location": "Atlantic", "time": "1912"} in the Atlantic {"wiki":
↪ "Atlantic_Ocean", "type": "ocean", "cat": "location"} in 1912 {"instance":
↪ "date-entity", "year": "1912"}
```

Fig. 1. Example output for the sentence *"Titanic sank in the Atlantic in 1912."*

The annotator outputs original input, where each word token may optionally be followed by annotation data encoded as a JSON object containing key-value pairs. Thus, only text stripping between {} symbols is needed to recreate the original input.

The table below describes the semantics of possible keys for annotation objects. All the keys are optional. If no value is provided for a key, the key is omitted. If no values are declared for the annotation object, the object itself is removed (Table 2).

[12] https://github.com/flipz357/amr4nli.

Table 2. Semantics of annotation

Key	Semantics
instance	Reference to Propbank frame or date entity
role	Propbank core role for frame argument
wiki	Reference to linked entry on English Wikipedia, e.g. "`RMS_Titanic`"
type	Entity type, e.g. "`ship`"
vncls	Verbnet class ID, e.g. "`characterize-29.2-1`"
cat	Entity parent category e.g. "`product`"
ref	Coreference referent, e.g. "`John`" and "`he`"
polarity	Negative ("–") or positive ("+") polarity

5 Pipeline Description

The AMR annotation pipeline is a collection of scripts that takes text as input and returns text annotated with additional information extracted from AMR. The source code for the pipeline is published on Github[13] (Fig. 2).

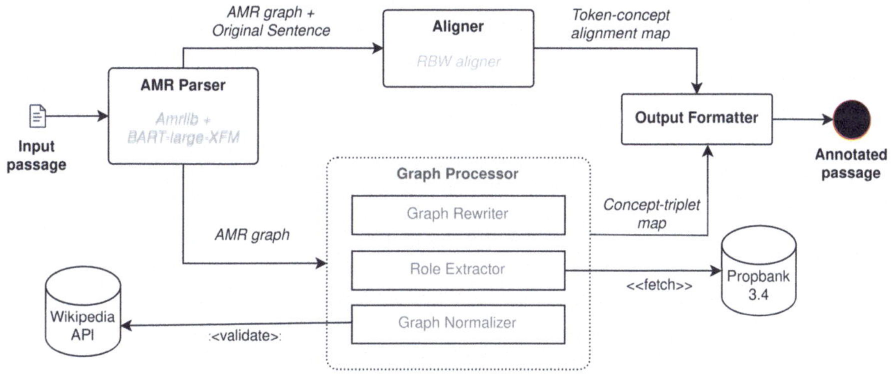

Fig. 2. High-level overview of the annotation pipleine.

It consists of the following primary components: (a) a *semantic parser* that constructs the AMR graph from input sentence; (b) a *graph processor* that simplifies the generated AMR graph, maps the roles to Propbank mappings, performs NER and groups graph concepts to a linear format; (c) an *aligner* that maps the processed AMR graph to input sentence; and (d) an *outputter* that filters redundant annotations and outputs the final sentence.

[13] https://github.com/martinve/amr-clausifier.

5.1 AMR Parser

During our initial study [20], a number of AMR parsers were identified. *Amrlib*[14] was chosen by applying the following criteria: interoperability - for the tool to be easily integrated into existing workflow; freshness - active development and being publicly available; lightness - given a choice between otherwise matching and equally performant parsers, the more lightweight one was picked; and publicly available models to be explored and extended.

A custom model was trained by finetuning BART-large [8] model with wikified LDC AMR 3.0[15] data. The trained model is available for download[16] and reaches a SMATCH score of 0.82. The model supports multi-sentence parsing, and for each input, we generate an AMR parse, a rooted, directed acrylic graph where nodes correspond to the concepts and edges define the relationships between those concepts.

5.2 Graph Processor

We are using Penman [4] toolkit[17] for AMR graph processing, and perform the following steps during the graph processing stage: (a) Normalizing the graph; (b) Mapping roles and named arguments; (c) extracting entity types, and (d) Grouping graph entities.

Graph normalization entails the following subtasks: rewriting named entities, cleaning up wiki roles, and unifying the quantifiers.

Named entity rewriting is performed to reduce the complexity of the initial AMR graph by rewriting the `:name` roles and merging the individual constituents of a named entity, defined by `:op1` to `:opN` in the AMR graph, with an underscore "_". This substitution simplifies the graph structure and reduces the number of *(subject, predicate, object)* triplets (Fig. 3).

```
(b / bear-02
   :ARG1 (p / person
      :wiki "Barack_Obama"
      :name (n / name
         :op1 "Barack"
         :op2 "Obama"))
   :location (s / state
      :wiki "Hawaii"
      :name (n2 / name
         :op1 "Hawaii")))
```

```
(b / bear-02
   :ARG1 (p / person
      :wiki "Barack_Obama"
      :name "Barack_Obama")
   :location (s / state
      :wiki "Hawaii"
      :name "Hawaii"))
```

Fig. 3. AMR graph for the sentence "Barack Obama was born in Hawaii." before and after name-rewriting.

[14] https://github.com/bjascob/amrlib.

[15] https://catalog.ldc.upenn.edu/LDC2020T02.

[16] https://cs.taltech.ee/staff/martin.verrev/model_parse_xfm_bart_large_wiki-v1_0_0.tgz.

[17] https://github.com/goodmami/penman.

Wikification is the task of linking the named entities to their respective Wikipedia entries. Different surface forms of a named entity are grounded in a standard form linked to the English Wikipedia. Named entities are often referred to differently, such as "US", "U.S.", "United States" or "United States of America". To annotate such named entities with a canonical form, we use `:wiki` role such as `:wiki"United States"`. During the **wiki cleanup** stage, we remove the empty or invalid `:wiki` roles. During our experiments, we encountered that the wikified model tends to hallucinate in case of incomplete information. In the case of non-empty Wikipedia roles, an API call to English Wikipedia is performed to verify if the page exists. Still, this approach does not take into account factual inconsistencies.

Quantifier unification is a step during the graph normalization stage to provide a unified representation of quantifiers in an AMR graph. To that end, we modify the graph the following way: Quantifiers in AMR are being represented with `:quant` for *"few"*, *"many"*, and *"some"*, but more general `:mod` (unspecified modifier) for *"all"* and *"any"*. We unify those predicates to use the `:quant` modifier.

Mapping Named Arguments. We match the named frame arguments in AMR - `:ARGn` with respective Propbank semantic roles. Our initial experiments were conducted with automatic substitution of generic Propbank roles as per Propbank specification[18]: `ARG0` ↦ `Agent`, `ARG1` ↦ `Patient`, etc. but as the argument roles are dependent on specific verbs, this resulted in incorrect substitutions.

To this end, we first identify possible Propbank frames for a given sentence, then determine if each frame is a Propbank core frame or AMR-specific keyword, and then fetch possible mappings for each keyword.

Let us consider the following example and get the Propbank mappings for a sentence "Barack Obama, 44th president of the United States of America, was born in Hawaii".

```
'h': {
  'roleset': 'have-org-role-91',
  'lemmas': {'m': 'have-role'},
  'roles': {
    ':ARG0': {'descr': 'office holder', 'role': 'Verb-specific'},
    ':ARG1': {'descr': 'organization', 'role': 'Verb-specific'},
    ':ARG2': {'descr': 'title of office held', 'role': 'Verb-specific'},
    ':ARG3': {'descr': 'description of responsibility', 'key': 'Verb-specific'}},
  }}
```

Fig. 4. AMR specific role mappings for the sentence "Barack Obama, 44th president of the United States of America, was born in Hawaii".

We retrieve the Propbank mappings via custom API integrated into our pipeline, as illustrated by Fig. 4. A result object is a dictionary where the key

[18] https://verbs.colorado.edu/propbank/EPB-Annotation-Guidelines.pdf.

maps to a variable in the AMR graph, "`roleset`" describes the roleset in PropBank, and "`roles`" describes all possible frame arguments for selected roleset.

Extracting Entity Types. We use the canonical list of NER tags from the AMR specification that fall into nine primary categories: individual (9 subclasses), organization (11 subclasses), location (30 subclasses), facility (20 subclasses), event (8 subclasses), product (12 subclasses), publication (5 subclasses), natural object (no subclasses) and miscellaneous (10 subclasses).[19]. That is more than 11 entity classes and seven value classes provided by standard Ontonotes annotations[20]. Let's consider the example sentence, "Titanic sank in the Atlantic in 1912." From the AMR parse graph, we can infer that Titanic is of type `ship` and Atlantic is of type `ocean`. Following this, we find the parent category for named entities. Thus, we can infer that Titanic belongs to the category `product` and the Atlantic belongs to the category `location`. Annotating the sentence with categories allows us to derive additional properties and default rules and behaviors for the entity under study if needed.

Aligner
AMR annotations do not include explicit mappings between elements of an AMR graph and the corresponding surface tokens of the text that evokes them. [15] Thus, the graph must be aligned to the surface form. In addition, the AMR contains abstract concepts, e.g. `have-org-role-91` that do not have a corresponding value in text. We are using a lightweight RBW Aligner[-] from *amrlib* package to align the graph with the text.

During the exploratory development phase, a preliminary evaluation of aligners was conducted. For the alignment basis, we chose Rule Based Word (RBW) aligner from *amrlib* package. RBW aligner is lightweight and easily integrated into the pipeline. The limitation of the aligner is that it does not align the spans but only words.

Combining the Results
As a result of the graph transformations, we have the following information: an original input text; a list of variable-concept mappings; a list of AMR triplets; a list of Propbank frames; and a list of alignments. First, we construct a list from the text $[(word_i, triplets)]$ where $word_i$ is the word in a sentence and $triplets$ is a list of optional triplets for that word. We use the variable-concept mappings to map the subject-predicate-object triplets to $word_i$. Finally, we encode the triplets as JSON and output the annotated sentence.

6 Conclusion

The current paper presented a system and an annotation scheme that can enrich a given input text with additional semantic and structural information

[19] https://github.com/amrisi/amr-guidelines/blob/master/amr.md#named-entities.
[20] https://huggingface.co/flair/ner-english-ontonotes-large.

from AMR that cannot be inferred via syntactic parsing alone, resulting in a more comprehensive capture of word knowledge. The pipeline output is a semi-structured text that significantly improves the efficiency of the syntactic parsing pipeline.

The pipeline described in this paper is available to download from https://github.com/martinve/amr-clausifier. The pipeline output can be used as a standalone text annotator or a pre-processor for downstream NLP tasks.

References

1. Chen, W.T., Palmer, M.: Unsupervised AMR-dependency parse alignment. In: Proceedings of the 15th Conference of the European Chapter of the Association for Computational Linguistics: Volume 1, Long Papers, pp. 558–567 (2017)
2. De Marneffe, M.C., Manning, C.D., Nivre, J., Zeman, D.: Universal dependencies. Comput. Linguist. **47**(2), 255–308 (2021)
3. Feng, L., Williamson, G., He, H., Choi, J.D.: Widely interpretable semantic representation: frameless meaning representation for broader applicability. CoRR abs/2309.06460 (2023). https://doi.org/10.48550/ARXIV.2309.06460
4. Goodman, M.W.: Penman: An open-source library and tool for AMR graphs. In: Proceedings of the 58th Annual Meeting of the Association for Computational Linguistics: System Demonstrations, pp. 312–319 (2020)
5. Haverinen, K., et al.: Building the essential resources for Finnish: the Turku dependency treebank. Lang. Resour. Eval. (2014)
6. Järv, P., Tammet, T., Verrev, M., Draheim., D.: Knowledge integration for commonsense reasoning with default logic. In: Proceedings of the 14th International Joint Conference on Knowledge Discovery, Knowledge Engineering and Knowledge Management - KEOD, pp. 148–155. INSTICC, SciTePress (2022). https://doi.org/10.5220/0011532200003335
7. Ji, Y., Williamson, G., Choi, J.D.: Automatic enrichment of abstract meaning representations. In: Proceedings of the 16th Lingusitic Annotation Workshop (LAW-XVI) within LREC2022, pp. 160–169 (2022)
8. Lewis, M., et al.: BART: Denoising sequence-to-sequence pre-training for natural language generation, translation, and comprehension. In: Jurafsky, D., Chai, J., Schluter, N., Tetreault, J. (eds.) Proceedings of the 58th Annual Meeting of the Association for Computational Linguistics. pp. 7871–7880. Association for Computational Linguistics, Online (2020). https://doi.org/10.18653/v1/2020.acl-main.703
9. Nivre, J., et al.: Universal Dependencies v2: an evergrowing multilingual treebank collection. In: Proceedings of the 12th Language Resources and Evaluation Conference. pp. 4034–4043. European Language Resources Association (2020). https://www.aclweb.org/anthology/2020.lrec-1.497
10. Opitz, J., Wein, S., Steen, J., Frank, A., Schneider, N.: AMR4NLI: interpretable and robust NLI measures from semantic graphs. CoRR abs/2306.00936 (2023). https://doi.org/10.48550/ARXIV.2306.00936
11. Palmer, M., Gildea, D., Kingsbury, P.: The proposition bank: an annotated corpus of semantic roles. Comput. Linguist. **31**(1), 71–106 (2005)

12. Pradhan, S., et al.: PropBank comes of age—larger, smarter, and more diverse. In: Nastase, V., Pavlick, E., Pilehvar, M.T., Camacho-Collados, J., Raganato, A. (eds.) Proceedings of the 11th Joint Conference on Lexical and Computational Semantics, pp. 278–288. Association for Computational Linguistics, Seattle (2022). https://doi.org/10.18653/v1/2022.starsem-1.24
13. Sataer, Y., Gao, Z., Fan, Y., Li, B., Gao, M., Shi, C.: Exploration and comparison of diverse approaches for integrating syntactic knowledge into AMR parsing. Appl. Intell. 1–21 (2023)
14. Shou, Z., Jiang, Y., Lin, F.: AMR-DA: data augmentation by abstract meaning representation. In: Muresan, S., Nakov, P., Villavicencio, A. (eds.) Findings of the Association for Computational Linguistics: ACL 2022, pp. 3082–3098. Association for Computational Linguistics, Dublin (2022). https://doi.org/10.18653/v1/2022.findings-acl.244, https://aclanthology.org/2022.findings-acl.244
15. Szubert, I., Lopez, A., Schneider, N.: A structured syntax-semantics interface for English-AMR alignment. In: Proceedings of the 2018 Conference of the North American Chapter of the Association for Computational Linguistics: Human Language Technologies, Volume 1 (Long Papers), pp. 1169–1180 (2018)
16. Tammet, T., Draheim, D., Järv, P.: GK: implementing full first order default logic for commonsense reasoning (system description). In: Blanchette, J., Kovács, L., Pattinson, D. (eds.) IJCAR 2022: Automated Reasoning. LNCS, vol. 13385, pp. 300–309. Springer (2022)
17. Tammet, T., Järv, P., Draheim, D.: Confidences for commonsense reasoning. In: Platzer A., S.G. (ed.) Automated Deduction – CADE 28. CADE 2021. LNCS, vol. 12699, pp. 507–524. Springer (2021)
18. Tammet, T., Järv, P., Verrev, M., Draheim, D.: An experimental pipeline for automated reasoning in natural language (short paper). In: International Conference on Automated Deduction, pp. 509–521. Springer (2023)
19. Tammet, T., Sutcliffe, G.: Combining JSON-LD with first order logic. In: 15th IEEE International Conference on Semantic Computing, ICSC 2021, Laguna Hills, CA, USA, 27–29 January 2021, pp. 256–261. IEEE (2021). https://doi.org/10.1109/ICSC50631.2021.00051
20. Verrev, M.: Evaluation of semantic parsing frameworks for automated knowledge base construction. In: International Conference on Intelligent Systems Design and Applications, pp. 554–563. Springer (2022)
21. Zeman, D.: Reusable Tagset conversion using Tagset drivers. In: Proceedings of the Sixth International Conference on Language Resources and Evaluation (LREC 2008). European Language Resources Association (ELRA) (2008)

Comparing Classical and Novel Techniques for Feature Selection: Toward a Machine Learning Method

Amal Ben Abdallah[1(✉)] and Maryam Elamine[2]

[1] LEG laboratory, Faculty of Economics and Management of Sfax, University of Sfax, Sfax, Tunisia
amal.benabdallah@fsegs.usf.tn

[2] MIRACL Laboratory, Faculty of Economics and Management of Sfax, Sfax, Tunisia

Abstract. The rapid proliferation of service providers across industries, notably in banking, has inundated customers with a plethora of options for financial services. Consequently, customer churn and engagement have emerged as paramount concerns for banks worldwide. By analyzing customer behavior, the research aims to explore the likelihood of churn. This paper proposes a method utilizing machine learning techniques, a subset of artificial intelligence, to predict customer churn in the banking sector. Our main focus was comparing classical feature selection techniques using SPSS's PCA and novel techniques using Python. Then to assess our theory, we trained Machine Learning models based on the features selected with each method and evaluated their effectiveness based on F1 score. The study employs various classifiers, including KNN, Naive Bayes and Random Forest, achieving an F1 score of 94.78 using the latter. This finding underscores the effectiveness of leveraging machine learning algorithms to address the pressing issue of customer churn in the banking industry. By identifying key predictors of churn and deploying robust predictive models, banks can proactively mitigate churn rates and enhance customer retention strategies, thereby fostering long-term profitability and sustainability in an increasingly competitive market landscape.

Keywords: customer churn · banks · machine learning · feature selection · classical techniques

1 Introduction

In today's dynamic and competitive market, numerous service providers are vying for consumers' attention. In this environment, service providers face the difficult task of understanding and adapting to changing consumer behaviors and growing expectations. The modern consumers who characterize the current generation are very different from those who came before them. They have high

L. H. Belguith and K. Shaalan (Eds.): LPKM 2024, LNNS 1303, pp. 13–22, 2025.
https://doi.org/10.1007/978-3-031-85067-7_2

standards and extensive requirements, especially when it comes to connectivity and personalized innovative solutions. Through access to a wider range of information deduction, they are equipped to evaluate new methods and technologies. This advanced knowledge changes the buying behavior and often leads to a phenomenon called "analysis analysis", in which consumers are fully analyze every aspect of the buying process to make an informed decision. Navigating this trend presents a significant challenge for contemporary service providers. To remain relevant and competitive, they must innovate and tailor their offerings to meet and exceed customer expectations, thus adding tangible value to their experiences. According to [12] underscore the imperative for corporations to prioritize consumer recognition, especially in light of escalating competitive pressures. They advocate for innovative marketing strategies aimed at meeting consumer expectations and fostering loyalty.

[5] echoes this sentiment, emphasizing the inadequacy of a blanket approach to sales in an increasingly competitive market environment. He suggests that organizations must strategically allocate marketing resources for maximum effectiveness. Leveraging technology, particularly data mining techniques [6], has emerged as a key tool for businesses seeking to maintain a competitive edge by extracting valuable marketing insights.

In today's rapidly evolving market landscape, customer retention is paramount as it is often more cost-effective than acquiring new customers [10]. However, delivering consistent, high-quality service while nurturing strong customer relationships presents a significant challenge. Addressing consumer needs, particularly mitigating customer churn, becomes a top priority for organizations. Customer churn occurs when clients disengage from a company or service, necessitating a continuous cycle of acquiring new clients through sales and marketing efforts.

Contrastingly, customer retention proves to be more economically efficient as it capitalizes on existing customer trust and loyalty. Therefore, the development of a predictive system capable of identifying potential churn early on becomes imperative for organizational sustainability. This paper aims to construct a framework utilizing machine learning techniques [20] to forecast client churn within the banking sector.

This research examines the application of PCA to credit approval issues, using data from a German, Spain and French Churns banks as a case study. The findings suggest that in order to improve the decision-making process when providing consumer loans, it is useful to remove variables from the credit scoring model that have the least amount of influence. SPSS and Machine Learning were the tool used to apply the PCA approach. The paper is organized as follows: an introduction to PCA, apart on its use in the banking, Description of Data and Model and a conclusion that includes a case study.

The remainder of this paper is structured as follows: Sect. 2 Related Work, Sect. 3 Methodology and Results and Sect. 4 concludes with policy suggestions and perspectives.

2 Related Work

Many works have currently been offered with a variety of strategies and models to address the issue of client turnover. Presently, numerous studies have introduced various methods and models aimed at addressing the challenge of customer churn.

[9] delve into developing a predictive model for customer churn in the electronic banking sector using data mining. By leveraging information technology to gather data from organizational databases, the study applies the decision tree method to construct a model integrating this knowledge. The results obtained allow for the identification of characteristics of customers who have terminated their contracts. In conclusion, banking managers can leverage these findings to anticipate future terminations and devise specific strategies for customers sharing similarities with those who have terminated their contracts. They should propose strategies for customers whose characteristics are becoming increasingly similar to those of terminating clients.

Recently, [14] made predictions about credit card customer churn using machine learning methods, including Random Forest, linear regression, and KNN, on a dataset comprising over 10,000 entries and 21 features. By adjusting hyperparameters and evaluating models based on ROC curve and AUC, as well as confusion matrix, these authors concluded that the Random Forest method performed the best, with accuracy reaching 96.25%. Additionally, the total transaction amount over the past 12 months, total number of transactions over the past 12 months, and total revolving balance are the three most important features significantly impacting customer churn prediction. This indicates that the more frequently customers use their credit cards, the less likely they are to churn, and by utilizing this model, banking managers can proactively take measures to combat customer churn.

In the same year, [14] studied the prediction of customer churn in the banking sector using exclusive customer data from a large Brazilian bank and exploring various supervised machine learning algorithms. The results indicate that the random forests method outperforms several other techniques such as decision tree, k-nearest neighbor, elastic networks, logistic regression, and SVM. The inquiry highlights that customers with closer relationships with the bank, owning a wider variety of products, and borrowing more have a lower propensity to close their current accounts. Furthermore, a back-of-the-envelope estimation suggests that the model has the potential to predict losses of up to 10% of the operating income reported by the top Brazilian banks in 2019, thus emphasizing its significant economic impact. The results reinforce the importance of investing in cross-selling and incentive strategies for existing customers, with positive effects on customer retention.

Moreover, [14] present a conceptual framework aimed at discovering confusion features that are correlated with independent variables and causally related to these dependent variables impacting churn by combining different algorithms, including SMOTE, ensemble ANN, and Bayesian networks, to address churn prediction problems on massive and multidimensional financial data, typically generated in financial institutions due to the use of interval-based features in customer relationship management systems, the effects of the curse and blessing of dimensionality are evaluated using the Recursive Feature Elimination method to overcome the problem of high-dimensional feature space. Additionally, causal discovery is conducted to find possible interpretation methods to describe the probabilities of causes leading to customer churn. Evaluation metrics on validation data confirm that the random forest and our ensemble ANN model, with an accuracy of 86%, outperform other approaches. Furthermore, causal analysis results confirm that certain independent causal variables representing the contribution level of the super guarantee, account growth, and account balance amount have been identified as causal confusion variables leading to customer churn with a high degree of certainty.

More recently, [2] created a model for the banking sector using artificial neural networks (ANN) capable of predicting whether the customer will terminate their contract. This prediction is based on input characteristics and the independent variable of the training dataset, and hyperparameters are adjusted during model training using the forward propagation algorithm and cross-validation techniques, enabling the model to perform well in terms of precision and accuracy rates. The results obtained show that the suggested model has an accuracy of 86% for predicting customer attrition. Compared to the results of the logistic regression model, ANN models are more effective in predicting customer churn in the banking industry. In fact, their study offers crucial insights into how to use machine learning approaches to increase customer retention and reduce customer churn, and banks can use this model to spot customers likely to terminate their contracts and take proactive measures to retain them.

[15] delves into the integration of AI and ML in financial forecasting in the US, exploring advancements and challenges from 2010 to 2024. Through systematic literature review and content analysis, it uncovers how AI and ML enhance accuracy, and efficiency, while addressing ethical and regulatory concerns. Findings reveal a revolution in financial forecasting, driven by deep learning, reinforcement learning, and hybrid models, greatly improving market trend analysis and asset price predictions.

For large machine learning models applied to non-traditional data, [3] advocate for leveraging pre-trained models and transfer learning techniques for applying large machine learning models, which are prevalent in economic applications. However, this approach faces challenges such as the need for substantial data and computational resources, as well as the complexity of implementation, particularly for economists lacking technical expertise [18].

To address these challenges, transfer learning, which involves adapting pre-trained models to similar economic tasks, emerges as a promising solution. For instance, standard ensemble learning models, such as random forests, are effective for handling panel data with strong collinearity or nonlinearity, but they require customization for specific tasks [1]. Customized models, like interpretable attention networks, have demonstrated superior performance compared to off-the-shelf models in various economic applications, such as fraud detection in insurance claims [7].

The domain of machine learning involves developing algorithms that enable computers to learn from and make predictions or decisions based on data. It encompasses a range of techniques including supervised, unsupervised, and reinforcement learning, and is applied across various fields such as finance, healthcare, and autonomous systems. Recent advancements have seen significant progress in deep learning, natural language processing, and the integration of multimodal data. Machine learning models continuously evolve to improve accuracy, efficiency, and adaptability to complex real-world tasks. [19] introduce the ZLaP technique, which improves zero-shot classification by leveraging geodesic distances within a graph-structured dataset that includes both image and text nodes, resulting in enhanced classification accuracy across various datasets. Another work by [16] uncovers a novel linear property in transformer decoders and introduces a cosine-similarity-based regularization technique, which enhances performance on benchmarks such as Tiny Stories and SuperGLUE. Recently, [21] introduce AutoCodeRover, a tool that leverages large language models to automate the resolution of GitHub issues, thereby significantly reducing the manual effort needed for program maintenance and improvement.

These studies reflect the evolving landscape of machine learning applications in economics, aiming to address the unique challenges and opportunities presented by non-traditional data sources.

3 Data Model and Results

Our dataset contains 10,000 records[1] that were used in this investigation for the case of German, French, and Spanish banks. Data were organized in a table in SPSS[2] and Python. Upon closer inspection, the dataset is not well balanced; 8,500 customers have exited (i.e., changed to another service) and 1,627 have not. As many authors have experimented with ML algorithms, specifically:

Naive Bayes: This model is known for its simplicity, efficiency, and effectiveness in handling large datasets with high-dimensional features. It's particularly useful when dealing with text classification or sentiment analysis tasks due to its probabilistic nature and assumption of feature independence [13,17].

Random Forest: Random Forest is an ensemble learning method based on decision trees, known for its robustness and ability to handle both classification and regression tasks. It's highly scalable and can handle large datasets with high dimensionality [4,11].

KNN (K-Nearest Neighbors): KNN is a simple yet powerful non-parametric method used for both classification and regression tasks. It works by storing all

[1] Data is available via this link https://www.kaggle.com/code/kmalit/bankcustomer-churn-prediction/data.kaggle.com/code/kmalit/bank-customer-churnprediction/data.

[2] SPSS (Statistical Package for the Social Sciences), http://www.spss.com/, accessed 2023.

available cases and classifying new cases based on a similarity measure (e.g., distance functions) [4,11].

Every bank usually has a complex procedure whereby each new customer undergoes a know your customer (KYC) examination. The onboarding procedure involves several crucial stages that guarantee the accuracy and completeness of the bank data acquired. As a result, all of the required consumer data collected is genuine and easily accessible. A distinct customer ID and related surname serve as identifiers for each individual customer. Customer information such as credit ratings, age, employment history, balance, number of items, and projected salary are all included in the dataset.

According to [8], the goal of PCA is to select fewer principal components that explain the most important variation in the data. PCA can help simplify data analysis, improve visualization, and make it easier to identify trends and relationships between factors by reducing the dimensionality of the dataset. The mathematical expression of PCA background by [8] for reduced dimensionality is as follows:

$$\mathrm{PC}_K = a_{k1}x_1 + a_{k2}x_2 + \cdots + a_{kp}x_p$$

where a_{k1} is the loading or weight of variable x_1 on principal component PC_K, and x_p is the pth variable in the data matrix X. The number of principal components used in the analysis, k, determines the reduced dimensionality of the dataset.

SPSS provides an easy-to-use interface along with a wealth of features, such as simple data pretreatment, clear result interpretation, and powerful statistical analysis capabilities. Compared to other software, PCA is preferred because of its wide user base and integration with other IBM products, which further increase its usability.

The data processing was carried out using the following tools: SPSS and machine learning. A flowchart of methods is presented in Fig. 1.

Table 1 presents the set of 8 features selected by SPSS in our attempt to reduce the dimension of our data, keeping only pertinent variables for further analysis.

Table 1. Communalities

	Initials	Extraction
Dependent_count	1.000	0.428
Months_on_book	1.000	0.570
Total_Relationship_Count	1.000	0.372
Months_Inactive_12_mon	1.000	0.394
Contacts_Count_12_mon	1.000	0.238
Total_Revolving_Bal	1.000	0.564
Total_Trans_Amt	1.000	0.852
Total_Trans_Ct	1.000	0.803

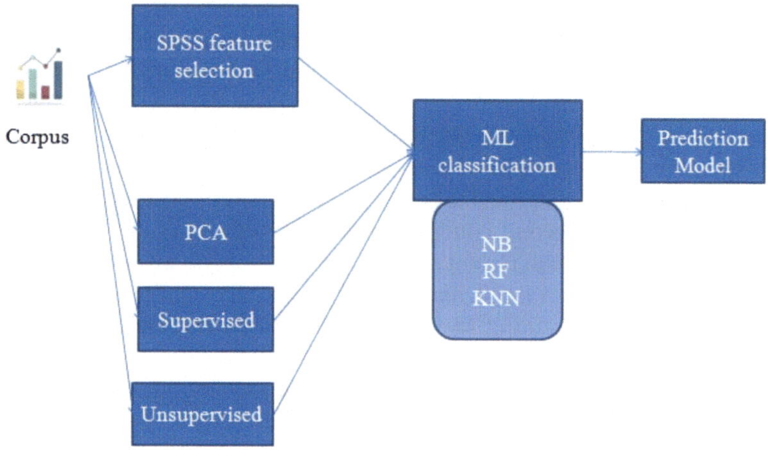

Fig. 1. Methodology framework.

We choose varimax, the most popular factor rotation technique in SPSS. Its goal is to lessen the variance of values that are not taken into account when composing the component. Varimax reduces the component's complexity. In the context of the original factorial solution (without rotation), the columns Extraction Sums of Squared Loadings include eigenvalues, an explanatory and cumulative version for each of the three components.

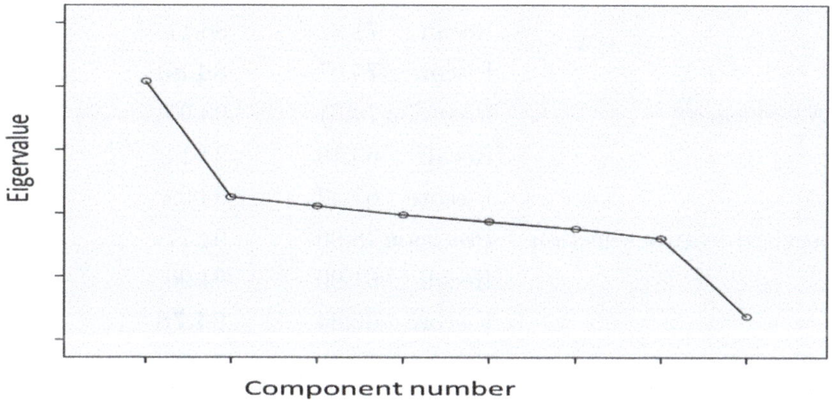

Fig. 2. SPSS Scree Plot

The eigenvalues for each of the primary components are graphically displayed in the Scree Plot (see Fig. 2). This figure determines the optimal number of components to retain in PCA by identifying the point at which the eigenvalues level off, indicating diminishing returns in explained variance.

In our attempt to experiment with Machine Learning (ML) techniques for feature selection and classification, we evaluated our data with three ML algorithms, namely: Naïve Bayes (NB), Random Forest (RF), and K-Nearest Neighbor (KNN). For KNN, we tested the algorithm with $K = 3$ and $K = 5$; in all our experiments, $K = 5$ achieved better results. Thus, the results presented in our work are with the second configuration.

Following the dimension reduction using PCA in SPSS, we selected the chosen variables, split our data, trained, and evaluated the ML algorithms. We dedicated 70% of our corpus to training the ML models, while the remaining 30% were reserved for testing.

We also experimented with feature reduction techniques through Python Programming Language. Using the Sklearn library, we performed three feature selection methods: PCA, supervised, and unsupervised feature selection.

For each method, we created three ML models (NB, RF, KNN) and observed the obtained results. Generally, RF gave us the highest F1 score with all tested methods (i.e., SPSS selected features, PCA, supervised, and unsupervised). We managed to achieve an F1 score of 94.78 with RF using the supervised method for feature selection. Table 2 summarizes the results of our classification with NB, RF, and KNN.

Table 2. Classification results

Algorithm	Method	Naïve Bayes	Random Forest	KNN
Statistic (SPSS)	Precision	81.81	84.51	83.61
	Recall	71.87	86.11	85.49
	F-score	75.07	**84.83**	83.98
PCA - Sklearn	Precision	74.51	93.97	84.50
	Recall	63.18	94.14	85.85
	F-score	67.42	93.99	84.91
Supervised feature selection	Precision	78.90	94.77	88.83
	Recall	61.90	94.90	89.34
	F-score	66.84	**94.78**	89.01
Unsupervised feature selection	Precision	70.02	75.51	73.09
	Recall	83.68	83.22	81.54
	F-score	76.24	76.77	76.15

4 Conclusion

Multivariate data analysis techniques like PCA are utilized in various domains to glean relevant information from complex data sets. Additionally, PCA offers a method for finding patterns in data and articulating it in a way that emphasizes its similarities and distinctions. Quantifying the significance of each dimension for characterizing a data set's variability is one benefit of employing PCA. This technique minimizes the number of dimensions while maintaining a high level of information. In this work, we discussed how PCA might be used in the banking industry to lower the dimension of data pertaining to credit problems. We employed 19 variables to describe churn bank data in our case study; however, using principal component analysis (PCA), we were only able to identify 3 factors that together account for more than 50% of the data. Furthermore, we achieved an F1 score of 94.78 with RF using the supervised method for feature selection. Future research will concentrate on creating a credit scoring system for the banking industry using PCA. Moreover, we will investigate the possibility of data balancing using data augmentation or under/over sampling. Deep Learning algorithms are also to be considered. Additionally, we will explore techniques for addressing data imbalance, such as data augmentation, as well as under and over sampling methods. We will also examine the potential integration of Deep Learning algorithms into our analysis.

References

1. Athey, S., Imbens, G.W.: Machine learning methods that economists should know about. Annu. Rev. Econ. **11**, 685–725 (2019)
2. Baby, B., Dawod, Z., Sharif, S., Elmedany, W.: Customer churn prediction model using artificial neural network: a case study in banking. In: 2023: International Conference on Innovation and Intelligence for Informatics, Computing, and Technologies (3ICT 2023), pp. 154–161. IEEE (2024)
3. Basu, I., Mukherjee, R.: Applications of machine learning techniques in macroeconomic analysis–a review. In: 2023 7th International Conference on Electronics, Materials Engineering & Nano-Technology (IEMENTech), pp. 1–4. IEEE (2023)
4. Breiman, L.: Random forests. Mach. Learn. **45**(1), 5–32 (2001)
5. Canning, G., Jr.: Do a value analysis of your customer base. Ind. Mark. Manage. **11**(2), 89–93 (1982)
6. Chen, P.: A random walk or color chaos on the stock market? Time-frequency analysis of S&P indexes. Stud. Nonlinear Dyn. Econom. **1**(2) (1996)
7. Farbmacher, H., Huber, M., Lafférs, L., Langen, H., Spindler, M.: Causal mediation analysis with double machine learning. Economet. J. **25**(2), 277–300 (2022)
8. Jolliffe, I.T., Cadima, J.: Principal component analysis: a review and recent developments. Philos. Trans. Roy. Soc. A: Math. Phys. Eng. Sci. **374**(2065), 20150202 (2016)
9. Keramati, M., Durand, A., Girardeau, P., Gutkin, B., Ahmed, S.: Cocaine addiction as a homeostatic reinforcement learning disorder. Psychol. Rev. **124**(2), 130 (2017)

10. Kim, M.K., Park, M.C., Jeong, D.H.: The effects of customer satisfaction and switching barrier on customer loyalty in Korean mobile telecommunication services. Telecommun. Policy **28**(2), 145–159 (2004)
11. Liaw, A., Wiener, M.: Classification and regression by random forest. R News **2**(3), 18–22 (2002)
12. Liu, D.R., Shih, Y.Y.: Integrating AHP and data mining for product recommendation based on customer lifetime value. Inf. Manage. **42**(3), 387–400 (2005)
13. McCallum, A., Nigam, K.: A comparison of event models for naive bayes text classification. AAAI-98 Workshop on Learning for Text Categorization **41**(42), 752 (1998)
14. Miao, X., Wang, H.: Customer churn prediction on credit card services using random forest method, pp. 649–656 (2022)
15. Olubusola, O., Mhlongo, N., Daraojimba, D., Ajayi-Nifise, A., Falaiye, T.: Machine learning in financial forecasting: a us review: exploring the advancements, challenges, and implications of AI-driven predictions in financial markets. World J. Adv. Res. Rev. (2024)
16. Razzhigaev, A., et al.: Your transformer is secretly linear. arXiv preprint arXiv:2405.12250 (2024)
17. Rennie, J.D., Shih, L., Teevan, J., Karger, D.R.: Tackling the poor assumptions of naive bayes classifiers. In: Proceedings of the Twentieth International Conference on Machine Learning (ICML-03), pp. 616–623 (2003)
18. Shen, C., Chen, X., Laloy, E.: Broadening the use of machine learning in hydrology. Front. Water **3**, 681023 (2021)
19. Stojnić, V., Laskar, Z., Tolias, G.: Training ensembles with inliers and outliers for semi-supervised active learning. In: Proceedings of the IEEE/CVF Winter Conference on Applications of Computer Vision, pp. 260–269 (2024)
20. Swamidason, I., Tatiparthi, S., Arul Xavier, V., Devadass, C.: Exploration of diverse intelligent approaches in speech recognition systems. Int. J. Speech Technol. **26**(1), 1–10 (2023)
21. Zhang, J., Zhu, M., Liu, F.: Find who is doing social good: using machine learning to predict corporate social responsibility performance. Oper. Manag. Res. **17**(1), 253–266 (2024)

Machine Learning Serving the School Orientation Process

Walid Ben Fradj[1(✉)], Mohamed Turki[2], and Faiez Gargouri[1]

[1] MIRACL Laboratory, ISIMS, University of Sfax, P.O. Box 242, 3021 Sfax, Tunisia
walid.benfradj@isimsf.u-sfax.tn, faiez.gargouri@usf.tn
[2] University of Hail, P.O. Box 2440, Hail 81451, Kingdom of Saudi Arabia
moh.alturki@uoh.edu.sa

Abstract. Regardless of the spatial or temporal context, school orientation plays a pivotal role in determining the future of a high school student. Guiding the student to make the right choice in orientation requires significant effort from multiple stakeholders, including parents, friends, administration, teachers, and academic counselors. In many cases, the high school student may disregard the advice provided by these stakeholders, even though they lack sufficient knowledge to reconcile their desires with their abilities. Consequently, the final decision regarding orientation may be based on the student's conviction, a random choice, imposed by parents, or an administrative decision. In this article, we propose a novel approach based on machine learning to predict the appropriate orientation choice for a new student based on their qualifications, results, and professional personality. In this context, we followed the standardized CRISP-DM methodology, widely used for data mining and data analysis projects. The data used were collected from students who have already undergone the orientation process. Subsequently, these data were carefully processed. During the modeling phase, we compared five algorithms, namely Decision Tree, KNN, Random Forest, SVM, and Logistic Regression. In the evaluation stage, we obtained promising results. Finally, according to our research, this approach represents an innovation in the field of secondary education, specifically in school orientation.

Keywords: School Orientation · Holland's Test · Professional Personality · Machine Learning · CRISP-DM

1 General Introduction

School orientation is central to shaping a student's future. This process involves allocating students according to the rates and procedures established by the Tunisian Ministry of Education. In the first year of secondary school, the goal is to anticipate and effectively guide students in choosing their future studies. Often, students are unaware of the opportunities available to them, necessitating a comparison between their real-world aspirations and their personal abilities. With technological advancements and the revolution in artificial intelligence, the

L. H. Belguith and K. Shaalan (Eds.): LPKM 2024, LNNS 1303, pp. 23–33, 2025.
https://doi.org/10.1007/978-3-031-85067-7_3

creation of an intelligent computer system to assist students in making informed choices is increasingly necessary. Today, Data Science has become a significant topic, especially in education. School orientation is a challenging issue being explored and addressed through Machine Learning models, making education and Data Science increasingly complementary disciplines. In this context, we present an artificial intelligence approach based on machine learning that predicts the best educational path for a student according to their desired choices, dominant professional personality type, and academic performance across various subjects.

The second section of this article presents school orientation in Tunisia. The third section discusses the analysis of a student's professional personality based on Holland's theory. In the fourth section, we introduce a new approach based on the CRISP-DM procedure and machine learning to assist students in school orientation. We conclude with a summary and future perspectives.

2 Related Works

According to [1], machine learning provides a significant advantage in predicting performance and organizing the field of education. [2] further elaborate that in the educational environment, the application of machine learning takes various forms to enhance the teaching and learning process, including predicting student performance, career paths, and academic orientation.

[3] proposed collaborative recommendation systems to assist stakeholders in the academic orientation field, supporting counselors in guiding high school students regarding their academic direction. The system utilizes student grades as input data to suggest academic possibilities by providing multiple recommendations.

[4] propose a recommendation system based on data mining techniques to aid students in making appropriate decisions regarding their academic journey. This system offers support for students to better choose the number of courses to enroll in based on the experience of previous students with similar academic achievements.

In [5], the authors present an intelligent school guidance system aimed at reducing academic orientation failure and creating an environment conducive to successful adaptation of school guidance based on students' cognitive trends and soft skills using the Internet of Things. The system offers an innovative solution to guide students in smart cities to choose a conscious and promising career path.

In [6], the authors focus on the application of Chatbot as an online guidance agent to support undergraduate and graduate students in making decisions regarding academic and career paths. The proposed approach concentrates on students' personalities to define the predominant personality type and obtain a suitable career path.

The work by [7] aims to improve high school students' orientation towards suitable fields and sectors using machine learning algorithms. The authors developed an intelligent recommendation system based on machine learning methods such as linear regression, decision trees, and random forest. The algorithms used in the article predict high school students' orientation based on the analysis of students' core curriculum data.

In [8], researchers address the automation of the student orientation process by leveraging Big Data technologies and machine learning algorithms. They evaluated four classification algorithms (SVM, Random Forest, Tree, Naive Bayes, and Neural Network) based on student grades and the number of absences. The researchers employed the WEKA tool to compare classification algorithms and concluded that the Naive Bayes classifier was the most accurate among those tested. The article highlights the importance of using Big Data to analyze large datasets quickly.

In the study by [9], the authors present a system capable of predicting students' personality types based on Holland's theory, which contributes to supporting their academic orientation. To achieve this, they employed classification algorithms such as decision trees, logistic regression, KNN, and SVM. These algorithms were trained on data collected from a sample of high school student registrations.

3 School Orientation

In Tunisia, the four years of secondary education are open to all holders of the basic education diploma. The goal for all students is to enter university or enter the workforce at the end of this education. It is divided into two phases: a general academic phase of one year and a specialized phase of three years. In the academic phase, all students except those in the sports stream undergo a common foundation year. At the end of this year, students are directed towards one of the four following streams: Arts, Sciences, Information Technology, and Economics-Services. At the end of the second year, students are again directed towards one of the six following sections: Arts, Economics and Management, Computer Science, Mathematics, Experimental Sciences, and Technical Sciences.

The role of the guidance counselor is to distinguish between students capable of successfully continuing their studies and those who are not. However, before reaching this selection stage, the counselor starts by providing information to students about the different sections. They are expected to know and describe all channels of educational mobility and institutional mechanisms that enable success in further studies. The guidance manual outlines the various steps and procedures to be applied to guide students under optimal conditions. After informing students, meetings are held with parents. Subsequently, the school administration issues grades for core subjects to help students make well-informed choices. Each student is required to fill out a preference form and submit it to the administration or electronically. Students will be informed of the results following the preliminary orientation council meeting. Collaboration with parents, teachers, and the guidance counselor assists the student in making the orientation decision. However, in practice, many students do not give the subject the attention it deserves, and they may later become victims of poor career choices. Learning from the experiences of students who have overcome this stage, especially those with a similar profile to the student in question, is crucial.

At the end of the first year of secondary education, the student must choose one of the following sections: Arts, Sciences, Economics-Services, or Information Technology. The criteria for choosing vary from one section to another.

3.1 Qualifications for Different Sections

- **Science:** Interest in mathematics, physical sciences, and technology; ability to observe, analyze, and think abstractly; passion for experiments and laboratory applications; scientific reasoning. Key subjects: Life and Earth Sciences, Mathematics, Physical Sciences, Technology.
- **Economics-Services:** Initiative, interest in economic phenomena, mathematical proficiency, mastery of French and English languages, interest in history and geography, organizational ability. Key subjects: History, Geography, French, English, Mathematics.
- **Information Technology:** Logical thinking, interest in new technologies, proficiency in mathematics and physical sciences, liking for computing, organizational ability. Key subjects: Mathematics, Computer Science, Physical Sciences, Technology.
- **Arts:** Love for literature and languages, love for reading, interest in the arts, interest in relationships and communication between individuals, writing skills, ability to express and edit. Key subjects: Arabic, French, English, History, Geography.

4 Analysis of the Student's Professional Personality

Before choosing a career path, students should ask themselves the following questions: In which school subjects am I excellent? Which school subjects do I enjoy? What are my preferences and personal characteristics? What are my academic ambitions? Which professions appeal to me? Do my abilities allow me to pursue studies in the desired field?

The Tunisian Ministry of Education employs a questionnaire consisting of 30 questions based on Holland's theory to determine the student's professional personality type. There are five questions for each personality type (Realistic, Investigative, Artistic, Social, Enterprising, and Conventional). The response to each question is one of the following options: "No", "A little", "Sometimes", "Yes", or "A lot", corresponding to values (1, 2, 3, 4, 5). Calculating the sum of values for each personality type results in six outcomes, with the highest representing the dominant personality type of the student in question.

The Holland test (or RIASEC Test or Hexagon RIASEC) refers to a theory of careers and vocational choices based on personality types. It was initially developed by the American psychologist John L. Holland. The idea is that if people are aware of their personality type, they can choose professions that align better with their personalities and make them happy. According to John Holland's theory, most people fall into one of six personality types: Realistic, Investigative, Artistic, Social, Enterprising, and Conventional.

Realists (R) tend to engage in activities requiring physical strength, skills, and coordinated movements. They avoid situations demanding verbal skills and social interaction, preferring to deal with machines, equipment, plants, and animals. They are characterized by a dislike for assistance and educational activities. They enjoy learning physical aspects and can organize their affairs independently.

Investigators (I) enjoy mystery-solving activities. They are interested in researching causes and relationships. They have unconventional values and attitudes, avoiding social interactions and dealings with others. They like collecting and analyzing information, finding and interpreting facts. They prefer thinking of solutions to problems and tend to value organization and understanding over authority.

Artists (A) avoid situations requiring physical skills. They prefer indirect relationships with others and enjoy addressing environmental problems through self-expression. They possess a strong capacity for emotional expression and avoid interacting with others. They are characterized by emotional intuition, spontaneity, creativity, and imagination.

Socials (S) have high social skills and fundamental human and religious values. They prefer teaching, social services, and psychological counseling. They avoid problems requiring mental problem-solving or new skills. They possess good verbal and communication skills with others.

Entrepreneurs (E) have a great ability to communicate their ideas and opinions. They can persuade and organize events and activities. They excel in verbal skills and prefer working within social contexts. They are enthusiastic, self-confident, and care about strength and social status, tending to undertake unusual and risky activities.

Conventionals (C) adhere to laws, rules, and regulations. They prefer working with those in power and avoid situations requiring personal relationships and physical skills. They can accomplish a lot of work in a short time and work well within a specific system. They are characterized by being organized, precise, number-oriented, and methodical.

The Holland test consists of the following 30 questions:

- TH1 (I) Do you tend towards activities that require thinking, analysis, and proof?
- TH2 (R) Do you tend towards hands-on activities, such as repairing broken devices?
- TH3 (E) Do you tend towards activities that involve taking on responsibilities, such as organizing trips?
- TH4 (A) Do you tend towards artistic activities, such as drawing, music, or acting?
- TH5 (E) Do you tend towards commercial activities?
- TH6 (I) Do you seek to continuously develop and expand your knowledge?
- TH7 (R) Do you tend towards activities that require physical effort, such as sports or strenuous work?

TH8 (C) Do you tend towards office-related professions, such as secretary or clerk?
TH9 (A) Do you tend towards professions involving artistic innovation?
TH10 (S) Do you tend towards activities that involve caring for others' problems and attempting to solve them?
TH11 (C) Do you prefer completing specific tasks that don't require extra effort or creativity?
TH12 (S) Do you tend towards professions that allow you to provide humanitarian services to others?
TH13 (R) Do you want to resemble a great inventor of machines?
TH14 (E) Do you want to resemble a businessperson?
TH15 (I) Do you want to resemble a researcher in a scientific or literary field?
TH16 (A) Do you want to resemble someone known for their artistic activity?
TH17 (S) Do you want to resemble a personality known for their love of helping others?
TH18 (C) Do you want to resemble someone who works in an office?
TH19 (S) Do you believe you have enough qualities to perform a job involving assistance or guidance to others?
TH20 (R) Do you believe you have enough qualities to perform a job requiring manual skill?
TH21 (E) Do you believe you have enough qualities to lead a group project where you are the leader?
TH22 (I) Do you believe you have enough qualities to perform a job requiring advanced knowledge usage?
TH23 (C) Do you believe you have enough qualities to perform a job requiring organization and attention to detail?
TH24 (A) Do you believe you have enough qualities to perform an artistic job?
TH25 (I) Do you tend towards acquiring knowledge and achieving excellence in your studies?
TH26 (R) Do you tend towards drawing maps and coloring them accurately during geography lessons?
TH27 (A) Do you tend towards art or music education sessions in your studies?
TH28 (C) When you need one of your textbooks, do you find it in its place without hassle?
TH29 (E) Do you tend towards taking on the responsibility of leadership or initiative within the classroom?
TH30 (S) Do you tend towards group work in your studies, driven by your desire to benefit those around you?

5 New Approach to School Orientation

In this section, we apply the six phases of the CRISP-DM [12] process to create an intelligent application to assist first-year secondary school students in making the right career choices. CRISP-DM (Cross Industry Standard Process for Data Mining) [10] is a field-tested method for guiding data exploration work.

It remains the only method proven effective for all data science projects [11]. The process includes descriptions of typical project phases, tasks within each phase, and an explanation of the relationships between these tasks. It provides an overview of the data exploration lifecycle. CRISP-DM divides the data mining process into six main steps: Business understanding, Data understanding, Data preparation, Modeling, Evaluation, and Deployment.

5.1 Business Understanding

Making the wrong career choice is among the major factors contributing to educational failure and dropout rates. The orientation phase for a first-year secondary school student is crucial for the continuation of their educational journey. The objective of this work is to assist the student in making an informed decision and choosing the right career path. Today, machine learning integrated into almost every field can provide a solution to address a significant issue in the academic sphere. The idea involves creating a learning sample based on data from students who have already undergone the orientation phase in the first year of secondary education. This data includes the student's aspirations, results in various subjects in the first year of secondary education, the type of professional personality, and the outcome achieved in the second year of secondary education. This outcome will be the class to predict for the new student in the first year. Possible values for this class are $(0, 1, 2, 3)$ corresponding respectively to ("failure", "success with a decline in results", "success with stable results", "success with improved results").

5.2 Data Understanding

Data collection was conducted through two similar forms, one in print and the other online, created using Google Forms. The first form was manually filled out at Eljem-Mahdia Secondary School in Tunisia, while the second form was shared in Facebook groups. Both forms were completed by students in the third and fourth years of secondary education. The data collection process spanned 40 days, resulting in a raw database containing 672 records with 264 online and 408 in-person submissions. To ensure a balanced database, 640 records were used, evenly distributed among the four sections: Sciences, Arts, Economics-Services, and Information Technology. The raw database is structured as (640, 46). Table 1 provides a description of the various attributes in the dataset.

5.3 Data Preparation

During the data preparation stage, we used the OpenRefine tool for data processing. We started by checking for the existence of empty rows and cells in the dataset "Orientationbrute.csv". Then we followed this processing procedure:

- Encode the values in the "decid" column: "L'élève seulement" with 1, "L'élève et les parents" with 2, "Les parents seulement" with 3, and "L'administration" with 4.

Table 1. Attribute Description

Attribute	Description
Sexe	Gender: Male or Female
Niveau	Grade: 3rd or 4th
Choix	Section: Sciences, Arts, Economics-Services, or Information Technology
Decid	Decision Maker for Orientation
Satisf	Satisfaction with the Choice: Yes or No
Qualif	Choice Alignment with Qualifications: Yes or No
Narb	Average in Arabic: Value between 0 and 20
Nfr	Average in French: Value between 0 and 20
Nang	Average in English: Value between 0 and 20
Nhist_geo	Average in History-Geography: Value between 0 and 20
Nmath	Average in Mathematics: Value between 0 and 20
Nphy	Average in Physical Sciences: Value between 0 and 20
Nsvt	Average in Life Sciences: Value between 0 and 20
Ntech	Average in Technology: Value between 0 and 20
MoyG1	Overall Average in the First Year of Secondary Education
TH1–TH30	30 Questions (Holland Test: Student's Professional Personality)
RESF	Outcome in the Second Year of Secondary Education

- Encode the values in the "choix" column: "Sciences" with 1, "Economie-Services" with 2, "Technologie de l'informatique" with 3, and "Lettres" with 4.
- Delete the "Satisf" column as all students in the sample are satisfied with their choices.
- Delete the "Qualif" column as all students have stated that the choice is in line with qualifications.
- Delete the "niveau" attribute as it does not influence the output class.
- Encode the responses to the Holland test questions: "No" with 1, "A little" with 2, "Sometimes" with 3, "Yes" with 4, and "A lot" with 5.
- Add attributes "Réaliste", "Investigateur", "Artiste", "Social", "Entrepreneur", and "Conventionnel" such as: $Realistic = PH2 + PH7 + PH13+PH20+PH26$; $Investigative = PH1+PH6+PH15+PH22+PH25$; $Artistic = PH4+PH9+PH16+PH24+PH27$; $Social = PH10+PH12+PH17+PH19+PH30$; $Enterprising = PH3+PH5+PH14+PH21+PH29$; and $Conventional = PH8 + PH11 + PH18 + PH23 + PH28$.
- Add an attribute "max" equal to the maximum of the values of attributes (Realistic, Investigative, Artistic, Social, Enterprising, and Conventional).
- Add an attribute "pers" designating the type of the student's professional personality such as: $pers = 1 if max = Realistic$; $pers = 2 if max =$

Investigative; $pers = 3ifmax = Artistic$; $pers = 4ifmax = Social$; $pers = 5ifmax = Enterprising$; and $pers = 6ifmax = Conventional$.

- Delete attributes PH1 to PH30 and "max" after obtaining the "pers" attribute.
- Convert "RESF" representing the output class into a numerical value such as: $RESF = 0$ if the student has failed, $RESF = 1$ if the student has succeeded but with a decrease in results; RESF = 2 if the student has succeeded with stable results; and $RESF = 3$ if the student has succeeded with improved results.

5.4 Modeling

In our case, the data is labeled so it is a supervised learning task. The prediction is of the qualitative variable *"RESF"* (0: failure; 1: success with a decrease in results; 2: success with stable results; and 3: success with improved results). Therefore, we are dealing with a classification problem. The data on which we will perform the learning are recorded in a CSV file where the format is $(640, 14)$, all attributes are of integer type and no missing values.

According to the correlation between the different variables, we deduce that the type of professional personality, grades in different subjects, and especially the overall average in the first year of secondary education are well correlated with the *RESF* class, which represents the final outcome to be predicted. In the next step, we decompose the dataset into explanatory variables X and the target variable y with: X = ['sexe', 'choix1', 'decid', 'Narb', 'Nfr', 'Nang', 'Nhis_geo', 'Nmath', 'Nphy', 'Nsvt', 'Ntech', 'MoyG1', 'pers'] and y = 'RESF'. The Train/Test split phase brings us back to the construction of the training set (80%) and the test set (20%). Finally, the trained models are: Decision Tree, KNN, Random Forest, SVM, and Logistic Regression.

5.5 Evaluation

Based on the confusion matrix (See Table 2), the evaluation of the different models used is presented in Table 3. We conclude that the best model in terms of accuracy is the SVM (Accuracy = 96%).

Table 2. Matrix Confusion

0	22	1	0	0
1	7	48	1	0
2	0	0	27	1
3	0	0	3	18
	0	1	2	3

Table 3. Model evaluation

Model	Evaluation Metrics			
	Precision	Recall	F1-Score	Accuracy
Decision Tree	0.890	0.910	0.895	0.90
KNN	0.795	0.810	0.795	0.79
Random Forest	0.887	0.907	0.89	0.90
SVM	0.955	0.97	0.962	**0.96**
Logistic Regression	0.650	0.670	0.655	0.65

5.6 Deployment

The created system consists of two complementary applications. The first prompts the student to answer the questions of the Holland test to obtain a code indicating the dominant type of the student's professional personality: "1" for "Realistic", "2" for "Investigative", "3" for "Artistic", "4" for "Social", "5" for "Enterprising", and "6" for "Conventional". In the second application, the student is asked to complete the following attributes: gender, career choice, decision-maker for this choice, grades in various core subjects studied in the first year, overall average in the first year, and the code of the professional personality obtained from the first application. The prediction yields one of the following outcomes: "Failure", "Success with a decrease in results", "Success with stable results", or "Success with improvement in results".

6 Conclusion

In order to address the issue of students falling behind due to poor career choices, we have proposed a novel approach that involves implementing a machine learning solution to assist first-year high school students in making crucial decisions regarding career paths. The project required significant efforts in both the theoretical and technical aspects, resulting in a solution that meets our needs and combines efficiency, performance, and reliability. To successfully carry out this work and achieve all the intended goals, we navigated through critical yet rewarding stages. Throughout this report, we detailed the various phases of the project life cycle. We conducted a theoretical study defining some basic concepts to clarify our choices. Subsequently, we focused on understanding the business problem and preparing the data. Then we moved on to the modeling phase. During this process, following the CRISP-DM procedure, we had the opportunity to employ various techniques such as data cleaning and preparation using the powerful tool OpenRefine, programming with the Python language, utilizing libraries such as sklearn, pandas, numpy, scipy, matplotlib, seaborn, and streamlit, and practicing machine learning. The approach presented is certainly open to several improvements to enhance its performance. Among these enhancements, expanding the database, including the study of career choices in the second year of high school, automating data preparation, and testing other classical and deep learning algorithms to determine the most effective one are potential avenues for improvement.

References

1. Velakanti, G., Mathur, A.: WITHDRAWN: machine learning approach to find the abilities in a candidate for steady employment in engineering field: a literature survey. Mater. Today: Proc. (2020). https://doi.org/10.1016/j.matpr.2020.10.585.
2. Asthana, P., Hazela, B.: Applications of machine learning in improving learning environment. In: Multimedia Big Data Computing for IoT Applications: Concepts, Paradigms and Solutions, pp. 417–433 (2020)

3. Castellano, E.J., Martinez, L.U.I.S., Sánchez, P.J.: OrieB, a linguistic CRS for supporting decision making in academic orientation. In: Computational Intelligence in Decision and Control, pp. 841–846 (2008)
4. Vialardi, C., Bravo, J., Shafti, L., Ortigosa, A.: Recommendation in higher education using data mining techniques. In: International Working Group on Educational Data Mining (2009)
5. El Mrabet, H., Moussa, A.A.: Smart school guidance and vocational guidance system through the internet of things. In: Proceedings of the 2nd International Conference on Networking, Information Systems & Security, pp. 1–5 (2019)
6. Zahour, O., Eddaoui, A., Ouchra, H., Hourrane, O.: A system for educational and vocational guidance in Morocco: chatbot E-orientation. Procedia Comput. Sci. **175**, 554–559 (2020)
7. Tarik, A., Farhaoui, Y.: Recommender system for orientation student. In: Big Data and Networks Technologies 3, pp. 367–370. Springer (2020)
8. Ouatik, F., Erritali, M., Ouatik, F., Jourhmane, M.: Students' orientation using machine learning and big data (2021)
9. El Mrabet, H., Ait Moussa, A.: A framework for predicting academic orientation using supervised machine learning. J. Ambient. Intell. Humaniz. Comput. **14**(12), 16539–16549 (2023)
10. Rivo, E., de la Fuente, J., Rivo, Á., García-Fontán, E., Cañizares, M.Á., Gil, P.: Cross-Industry Standard Process for data mining is applicable to the lung cancer surgery domain, improving decision making as well as knowledge and quality management. Clin. Transl. Oncol. **14**, 73–79 (2012)
11. Gaultier, D.: Méthode CRISP: la clé de la réussite en data science. Business & Decision (2023). https://fr.blog.businessdecision.com/methode-crisp-la-cle-de-la-reussite-en-data-science/
12. Chapman, P.: CRISP-DM 1.0: step-by-step data mining guide (2000). https://www.semanticscholar.org/paper/CRISP-DM-1.0%3A-Step-by-step-data-mining-guide-Chapman/54bad20bbc7938991bf34f86dde0babfbd2d5a72. Accessed 23 May 2024

Improving Arabic Fake News Detection Across Context-Aware Attention Deep Model Based on Natural Language Processing

Enas Tariq Khudair[1], Onsa Lazzez[1,2](✉), Tarek M. Hamdani[1,3], Ahmed T. Sadiq[4], and Adel M. Alimi[1,5]

[1] REGIM-Lab.: REsearch Groups in Intelligent Machines, University of Sfax, National Engineering School of Sfax (ENIS), BP 1173, 3038 Sfax, Tunisia
onsa.lazzez@enis.tn

[2] Industry 4.0 Research Lab Polytech-Sfax (IPSAS), Avenue 5 August Rue Said Aboubaker, 3002 Sfax, Tunisia

[3] Higher Institute of Computer Science Mahdia (ISIMa), Computer Science Department, University of Monastir, Monastir, Tunisia

[4] Department of Electrical and Electronic Engineering Science, University of Technology-Iraq, Baghdad, Iraq

[5] Faculty of Engineering and the Built Environment, University of Johannesburg, Johannesburg, South Africa

Abstract. With the enormous growth of social data in recent years, fake news detection has gained increasing research attention and has been widely explored in various languages. Arabic language nature imposes various challenges, such as the complicated morphological structure and the limited resources. Therefore, the current state-of-the-art methods for fake news detection remain to be enhanced, which inspired us to explore the application of the emerging deep-learning architecture to Arabic text classification. In this paper, we present an ensemble model that enhances the attention mechanism to detect fake news in Arabic sentences. An attention mechanism unit is incorporated to highlight the critical information from the contextual feature vectors. The context-related vectors generated by the attention mechanism layers are then concatenated and passed into a classifier to predict the final label. The experimental results show that the attention mechanism improves the model's performance while yielding 99.8% in accuracy.

Keywords: Arabic Language · Social Network · Fake News · Arabic Fake News Detection

1 Introduction

Recently, Social Network Sites (SNSs), such as Instagram, Twitter, and Facebook, have emerged as attractive communication platforms [1]. SNSs have

L. H. Belguith and K. Shaalan (Eds.): LPKM 2024, LNNS 1303, pp. 34–43, 2025.
https://doi.org/10.1007/978-3-031-85067-7_4

opened venues for internet subscribers to interactively express and share their news in different life aspects. Consequently, the volume and diversity of the published content have massively increased.

Fake news detection uses advanced approaches, such as computational linguistics or natural language processing, to analyze people's shared content towards specific events or topics. Fake news detection is concerned with computationally recognizing whether shared content is fake or real [3]. Currently, the application of fake news detection has brought significant advances in observing public opinions regarding political movements, product preferences, and customer satisfaction. In addition, fake news detection has been widely investigated recently for ubiquitous languages on social media, such as English [4].

In particular, Arabic has very complex and rich morphological structures and several intentional and derivation forms. For instance, words' position and the sentence type, whether nominal or verbal, can influence the transitional meaning of the entire sentence. Moreover, a single word can be used to express various emotional conditions. Consequently, Arabic expressions are subjective to domains and contexts. In addition, variations of the suffix and prefix have a negative impact on sentence representation [5]. Negating words and transliterated and misspelled words change the overall meaning of the sentence. Most online users use a combination of colloquial and standard Arabic, in addition to sentiment lexicons and available corpora shortage.

Inspired by the human's attention while reading, a new era of deep neural networks has recently emerged. The attention mechanism allows focusing on the most discriminating components of the sentence, i.e., the context-related words. The remainder of this paper is organized as follows: Sect. 2 reviews the related works. Section 3 explores the proposed method. Section 4 presents the dataset and model settings. Section 5 demonstrates and discusses the experimental results, and finally, Sect. 6 concludes the paper and presents future research.

2 Related Works: Arabic Fake News Detection Methods

The swift transport of information has been aided by the emergence of social networks, such as Facebook and Twitter. Thus, there are concerns about whether the information is reliable. In particular, anybody with a social media account can broadcast anything without worrying about the truth or the trustworthiness of a source.

In recent years, fake news analysis has gained remarkable attention, and its applications have spread to all possible domains [9]. Therefore, many systems are conducted on the English language since it is a dominant language of science, besides other Indo-European languages [10].

Moreover, the Arabic language is a native language for more than 168 million speakers using the Internet. According to the Internet World Stat 2016 ranking, the Arabic language recorded a fast growth rate of 6091.9 in the number of Internet users. However, only a few researchers have considered Arabic SA compared to other languages. Regarding SA, the Arabic language has several challenges. The dialectal varieties and morphological complexities of the Arabic language require heavy pre-processing and advanced dictionaries (lexicons) more than those used in other languages.

As stated in [1], Arabic text has many different inflectional and derivation forms. In addition, the position of the words within a sentence and the type of sentence, whether nominal or verbal, may change the transitional meanings of words, which makes Arabic opinions highly subjective to context domains, where words can be found in different polarity classes for different contexts. The Arabic language is diversified in terms of word suffixing, prefixing, and affixing (one three-letter root can generate different words in each case), which can affect the dimensionality of words and sentences.

Few studies have been conducted to detect fake news in the Arabic language using various techniques such as deep contextualized embedding models [1], textual analysis [2], attention-enriched Mini-BERT fake news analyzer [3] and deep learning [4].

Thus, the researches on Fake News detection in Arabic social media are few and require more attention and effort to achieve optimal results.

As stated in [5] Arabic text has many different inflectional and derivation forms, the position of the words within a sentence and the type of sentence whether it's nominal or verbal may change the transitional meanings of words.

This make Arabic opinions highly subjective to context domains, words can be found in different polarity classes for different contexts. In addition, Arabic language is diversified in term of word suffixing, prefixing, and affixing, which can affect in the dimensionality of words and sentences. In the following section we grouped the recent work on the Arabic Fake news detection.

The authors in [7] developed a deep model to automatically detect Arabic fake News. They used convolutional neural network as deep neural network to classify fake and real news claims in order to solve the problem from the fact checking perspective. They performed experiments on an Arabic balanced corpus to build the deep model and detect the Arabic Fake news. They achieved an accuracy of 91%.

In [8] the authors constructed a large and diverse Arabic fake news dataset and they developed a transformer-based classifiers to identify fake news. They are based on based on deep contextualized embedding models and they achieved an accuracy of 98%.

In [9], the authors detect the Arabic fake news by a revolutionary combined BERT architecture. They applied their proposed architecture on two Fake News datasets: Covid-19 fakes, and Satirical and they obtained an accuracy of 88% respectively.

The authors in [2] proposed an improved learning model for detecting Arabic fake news in Twitter. They use the TF-IDF method, to convert the words into features in order to determine the features that have a high rank, the fuzzy model to determine the relevant features for the user and the random forest algorithm. They obtained an accuracy of 89.5%.

Table 1. Summary of Arabic Fake News detection methods applied in 2022–2023

Ref	Dataset	Methods	Results
[7]	Different datasets	CNN and ML	91%
[8]	Kaggle dataset	Transformer	98%
[9]	Covid-19 fake	BERT	88%
[2]	Twitter data	Fuzzy logic and TF-IDF	89.5%
[6]	Twitter data	POS and BERT	89.2%
[8]	BBC-Arabic, CNN -Arabic sites	NLP and ML	98.6%
[3]	Qatar news site	NLP and DL	79.7%
[4]	Arabic Twitter data of rumors and nonrumors	Topic-based Features and ML	97.2%

In [1], the authors indicated the lack of available data for Arabic fake news detection. They used the true stories that are abundantly available online, using a part of speech tagger (POS). The authors used various BERT-based models and achieved a good performance using Arabert. They evaluated their models according to accuracy and F-score, which were recorded at 89.23% and 89.25% respectively.

The authors in [6] proposed various exploratory analyses to identify the linguistic features of Arabic fake news but with sarcastic content. Their main work was to show that although news items with sarcastic content might be trustworthy and verifiable, the text has distinguishing features on the lexica-grammatical level. They construct several machine learning models to identify sarcastic fake news websites and achieved accuracy up to 98.6% using two Arabic datasets collected from BBC-Arabic and CNN-Arabic.

Authors in [3], applied the low dimensionality statistical embedding to detect the untrustworthy sources. They used this method on various Arabic language corpora including the Arabic credibility corpus and two other corpora they created from Qatar News. They obtained an accuracy of 79.7%.

In [4], Arabic fake news in tweets were detected using a gradient boosting approach. The authors used Twitter data and topic-based features method. They employed the classical ML methods for classification. On 60% of the training data, the eXtreme Gradient Boost (XG-Boost) had 97.18% accuracy in classifying the data from fake news A summary of these studies is shown in Table 1.

3 The Proposed System for Arabic Fake News Detection

Recently, Bidirectional Encoder Representations from Transformers (BERT) have demonstrated a highly effective performance at text understanding, such models have presented a superior performance on many text analysis fields [12].

For our fake news classification task, we chose three recently developed pretrained Arabic contextualized embedding models based on Transformer architecture, which are AraBERT, ArBERT and MarBERT [13]. The pretrained models have already been trained on a large amount of Arabic data. As a result, these

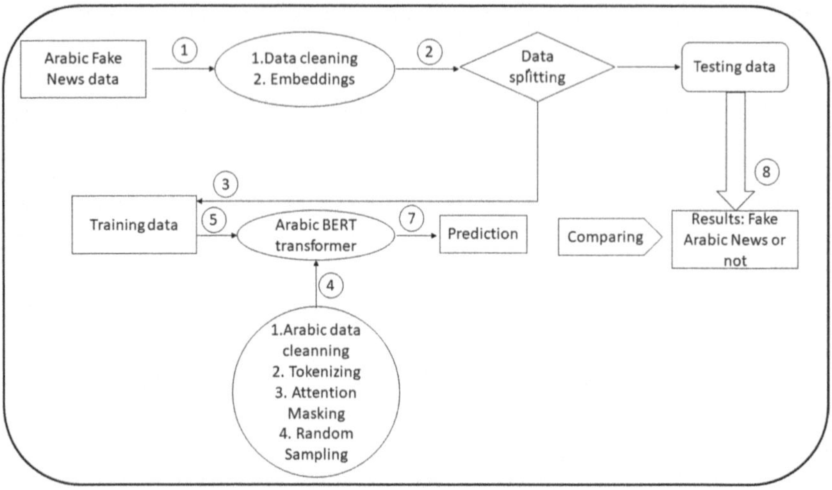

Fig. 1. Arabic Fake News detection system

models are familiar with the Arabic language, letters, vocabulary and stop words, as well as how words are used in a sentence.

In this study, the Arabic-language-based transformer model was used to detect fake news on Arabic news. At the beginning, a step of preprocessing was used to remove useless words, and then Arabic BERT model-acquired data-based preprocessing was employed. After preparing unique input tokens and attention masks, training was employed using the pretrained Arabic BERT method (AraBERT, ArBERT and MarBERT). The three data splits were performed on the data, and tested on a Arabic BERT classification model (Fig. 1).

3.1 Preprocessing Text

The non-availability of writing rules in social media applications makes some, if not most, social users tend to use an informal writing style.

Text preprocessing is essential to prepare the social textual data before feeding them into the feature extraction and classification stage.

In fact, textual data were processed first, where the Arabic text was freed with URLs, the white spacing, the '@' character, and other useless signs, and then these data were processed using mandatory BERT preprocessing methods.

In these steps, the tokenizer from AraBERT, ArBERT and MarBERT was loaded into the workspace. The reason for using this version of BERT is that it is an Arabic-sentence-based trained model.

The model in the proposed study was trained on 11M parameters with 256 hidden layers, 4 attention heads, and 4 hidden layers. This Arabic version was mainly trained on Google tensor processing units (TPUs) that were designed to train large data tensors. Therefore, simple frequency or other features may not fulfil the sentiments in Arabic text. After BERT preprocessing, encoding was applied, as illustrated in Fig. 2.

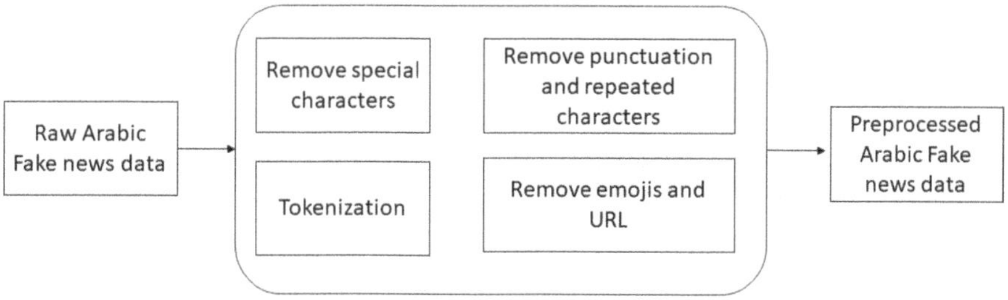

Fig. 2. Arabic text preprocessing

3.2 Arabic Models for Textual Data Analysis

In this work, we illustrate one of the most famous NLP tasks. In particular, NLP is a field of artificial intelligence that deals with human text and language. NLP can be applied to many tasks, including translation from one language to another and textual data classification.

Recently, transformers, illustrated in Fig. 3, have been introduced and applied at an increasing rate. Transformers are deep learning models released in 2017 and are mainly used in NLP tasks. Later in this section, we will present the BERT-based architecture models, including 12 attention layers, each of which has 12 attention heads and 768 hidden dimensions, used in this paper, such as AraBert, ArBert, and MarBert. The following three models were chosen to conduct our experiments based on their ability to perform NLP tasks for Arabic text or multilingual text based on the reviewed literature work [8].

The Arabert model, provided by AUB University, is a BERT architecture based on the Arabic language. In this model, the authors tried to achieve the same success that BERT achieved for the English language for most NLP tasks. It was released in March 2020, and it has already achieved optimal performance on most assigned Arabic NLP tasks [4]. Various studies on this Arabert architecture explain the training methodology in detail. It is worth mentioning that the model has been pre-trained on a large dataset consisting of 70 million sentences, corresponding to 24 GB of text [3].

Moreover, we also used two more powerful transformer-based language models for the Arabic language, which are ARBERT and MARBERT. These two models were tested on five tasks for the Arabic language, and each produced positive results [1,10].

4 Experimental Results

In this section, we describe our fake news classification model. First, we describe the fine-tuning and optimization that we used to tune our models to obtain the highest performance. Next, we list the evaluation criteria that we used to assess our models during both the training and the testing phases. We describe the outcomes and results of our experiment using the translated dataset from Kaggle. Finally, we discuss the general results and sum up our findings.

4.1 Arabic Fake News Dataset

For our experiment, we include a fake news detection dataset from Kaggle to build our classification model. The data is originally written in English. However, since we are developing Arabic text-based fake news detection, we have to translate the dataset into Arabic. Originally, the dataset consisted of two columns: Text and Classes. We translated the text to Arabic using Python code and the package "google-trans", after which we exported it into an xlsx format. The data possess some Russian and German letters that we could not translate using our English-to-Arabic translator code. Therefore, we first eliminated them as the first step in the preprocessing stage. Secondly, we removed the remaining English words that could not be translated into Arabic, such as names and terms (ex: Donald Trump, Jennifer, etc.). In fact, the Kaggle dataset originally contained 20,000 instances, but after we preprocessed the data, we ended up with 16,000 instances in the data pool. But after we preprocessed the data, we ended up with 16,000 instances in the data pool. We removed punctuation and unnecessary stopping words, as well as English and Arabic numbers. Finally, during the training process, we tokenized each sentence using the pretrained model tokenizer. The dataset used in this approach was split into 90/10, 80/20, and 70/30 parts. In these splits, the first arguments, 90, 80, and 70, show the training data ratio from the total data, whereas the second figures, 10, 20, and 30, show the testing data ratio. We removed punctuation and unnecessary stopping words, as well as English and Arabic numbers. Finally, during the training process, we tokenized each sentence using the pre-trained model tokenizer.

4.2 Experimental Fine-Tuning

A BERT model is a pre-trained model that is already constructed and trained on a huge amount of data. However, when we apply it to a specific task, we have to fine-tune the parameters. During the fine-tuning process, we can adjust various parameters, such as the optimizer, the learning rate, the number of epochs, and the dropout value. The parameter values we used are as follows:

- Optimizer: We tried more than one optimizer as part of the fine-tuning process. Specifically, we tried the SGD optimizer, the ADAM, and finally the ADAMW.
- Learning rate: As we get many well-known values from previous fake news detection codes for English text, we tried 0.001, 0.0001, and finally 1e-5.
- Number of epochs: We tried many numbers of epochs from 1 to 100. We applied the stop accuracy method that allowed us to stop the training when we reached the highest accuracy so as not to waste time and storage.
- Dropout value: We tried 0.1, 0.25, and 0.5, with each value producing a slightly different result.
- In the last step, we divided our data randomly into 80 testing data.

The fine-tuned parameters used in the transfer-learning-based classification of Arabic news are presented in Table 2.

Table 2. Fine-tuned parameters of Arabic fake news detection Arabic Bert model

Parameters	Values
Batch size	20
Adam optimizer learning rate	0.00005
Adam optimizer Epsilon	10-8
Hidden size	256
Maximal length	280
Epochs	100

We construct our fake news classification system using our translated dataset. As we fine-tuned the system, we tried many configurations and parameters to get the best possible results. Table 3 illustrates all the trials we conducted along with the corresponding results for all the models, and hence, the best values are visible for each variable. The pretrained Arabic BERT (AraBERT, MarBERT and ArBERT) classifier showed the simple deep-learning classifier results obtained with forward and backward propagation [8,9]. The classifier prediction results for the testing data on various splits are shown in Table 4.

Table 3. Model performance translated data with fine tuning steps

Model	Accuracy (%)	Precision (%)	Recall (%)	F-Score (%)
ArBERT	84.2	82.1	88.4	85.1
MarBERT	81.2	77.0	90.2	83.0
AraBERT	99.8	99.7	99.8	99.8

As illustrated in Table 5, recent studies on the Arabic language containing rumor and nonrumor text that was employed by user-based features, and ML classifiers. The XG-boost method classification performed the best, with 97.1% accuracy.

Table 4. Arabic Fake News Classification results of Arabic Bert-based Transfer Learning Classifiers using various data splits

Model	Splits	Accuracy (%)	Precision (%)	Recall (%)	F1-Score (%)
MarBERT	70/30	69.1	69.5	84.7	75.2
	80/20	79.6	76.3	89.0	82.3
	90/10	81.2	77.0	90.2	83.0
ArBERT	70/30	72.8	89.6	82.9	77.2
	80/20	82.6	81.4	87.2	84.4
	90/10	84.2	82.1	88.4	85.1
AraBERT	70/30	88.4	89.6	94.3	91.9
	80/20	97.5	99.1	98.6	99.1
	90/10	99.8	99.7	99.8	99.8

Table 5. Comparison of fake news data using various machine learning and DL techniques

References	Methods	Accuracy
[?]	ML Classifer	97.1%
[?]	BERT and other ML methods	77%
Proposed approach	AraBERT, MarBERT, ArBERT	99.8%, 81.2%, 84.2%

5 Conclusion

In this study, we have introduced a novel deep-learning approach for Arabic fake news representation and classification by adopting the transformer architecture. Due to the self-attention mechanism, the transformer makes it easy to automatically extract local features and reduce the computation complexity. Besides, BERT is skilled in learning contextual characteristics due to its ability to learn syntax features of linguistics. This model approaches are evaluated using a translated dataset of Arabic fake news. The evaluation results showed that the proposed approach outperforms the baseline method and is relatively competitive compared to other related work.

Acknowledgements. The research leading to these results has received funding from the Ministry of Higher Education and Scientific Research of Tunisia under the grant agreement number LR11ES48.

References

1. Kim, J., et al.: Retracted: extracting user interests on Facebook. Int. J. Distrib. Sens. Netw. **10**(6), 146967 (2014)
2. Birjali, M., Beni-Hssane, A., Erritali, M.: Machine learning and semantic sentiment analysis based algorithms for suicide sentiment prediction in social networks. Procedia Comput. Sci. **113**, 65–72 (2017)
3. Alowaidi, S., Saleh, M., Abulnaja, O.: Semantic sentiment analysis of Arabic texts. Int. J. Adv. Comput. Sci. Appl. **8**(2) (2017)
4. Shu, K., et al.: Fake news detection on social media: a data mining perspective. ACM SIGKDD Explor. Newsl. **19**(1), 22–36 (2017)
5. Monti, F., Frasca, F., Eynard, D., Mannion, D., Bronstein, M.M.: Fake news detection on social media using geometric deep learning. arXiv preprint arXiv:1902.06673 (2019)
6. Sahoo, S.R., Gupta, B.B.: Multiple features based approach for automatic fake news detection on social networks using deep learning. Appl. Soft Comput. **100**, 106983 (2021)
7. Yang, S., et al.: Unsupervised fake news detection on social media: a generative approach. In: Proceedings of the AAAI Conference on Artificial Intelligence, vol. 33, no. 01 (2019)
8. Abd Elaziz, M., Dahou, A., Orabi, D.A., Alshathri, S., Soliman, E.M., Ewees, A.A.: A hybrid multitask learning framework with a fire hawk optimizer for Arabic fake news detection. Mathematics **11**, 258 (2023)
9. Mishra, S., Shukla, P., Agarwal, R.: Analyzing machine learning enabled fake news detection techniques for diversified datasets. Wirel. Commun. Mob. Comput. **2022**, 1–18 (2022)
10. Nirav Shah, M., Ganatra, A.: A systematic literature review and existing challenges toward fake news detection models. Soc. Netw. Anal. Min. **12**(1), 168 (2022)
11. Bsoul, M.A. Qusef, A., Abu-Soud, S.: Building an optimal dataset for Arabic fake news detection. Procedia Comput. Sci. **201**, 665–672 (2022). ISSN 1877-0509
12. Burnard, P.: Teaching the analysis of textual data: an experiential approach. Nurse Educ. Today **16**(4), 278–281 (1996)
13. Virmani, C., Pillai, A., Juneja, D.: Extracting information from social network using NLP. Int. J. Comput. Intell. Res. **13**(4), 621–630 (2017)
14. Bakir, V., McStay, A.: Fake news and the economy of emotions: problems, causes, solutions. Digit. Journal. **6**(2), 154–175 (2018)

Towards a Maude-Based Approach for Formal Modeling Deep Neural Networks

Angham Boukhari[1(✉)], Aïcha Choutri[2], Faiza Belala[2], and Ahmed Hadj Kacem[1]

[1] ReDCAD Laboratory, ENIS, University of Sfax, Sfax, Tunisia
anghamboukhari@fsegs.u-sfax.tn, ahmed.hadjkacem@fsegs.usf.tn
[2] LIRE Laboratory, Abdelhamid Mehri Constantine 2 University, Constantine, Algeria
{aicha.choutri,faiza.belala}@univ-constantine2.dz

Abstract. Rapid evolution of Deep Learning (DL) and its underlying architecture, Deep Neural Networks (DNNs), presents both significant advances and ethical challenges. As DNNs become increasingly prevalent across various sectors, ensuring their reliability, interpretability, and robustness becomes paramount. However, this also poses considerable challenges, particularly in understanding their internal workings and ensuring that their behavior complies with ethical and legal standards. The objective of this work is to propose a new approach to respond to these challenges by integrating formal methods, notably exploiting the Maude language. Our method emphasizes the creation of multi-view models to capture different aspects of DNNs, thus facilitating comprehensive understanding and analysis. Through the application of Model Driven Engineering (MDE) principles and Maude language features, our approach enables the formalization of DNNs structures, initialization mechanisms, and processing. We demonstrate the utility of our approach by presenting a meta-model for DNNs and providing a detailed formal specification in Maude. Their dynamic behavior is described with rewriting rules to model operational dynamics. Our results indicate that the proposed approach provides a solid basis for understanding and modeling the behavior of DNNs, thereby contributing to their reliability and robustness in critical applications.

Keywords: Deep neural networks · deep learning · performance · meta-modeling · Maude strategy · rewriting logic

1 Introduction

Deep Learning (DL) is a major branch of machine learning (ML) managed by deep neural networks (DNNs) [18]. Its rapid development has led to advances in areas such as computer vision, speech recognition, automatic translation, and

L. H. Belguith and K. Shaalan (Eds.): LPKM 2024, LNNS 1303, pp. 44–58, 2025.
https://doi.org/10.1007/978-3-031-85067-7_5

many others, making AI the most significant field in computer science [14]. Indeed, without DL and mathematical expertise, it is very difficult to build a well-adapted learning model. However, the rapid evolution of DL, and by extension DNNs, also raises ethical challenges related to the diversification of existing DNNs and their methods of architectural configuration and learning, particularly regarding the prior modeling of different perspectives of the DNNs, the guarantee of interoperability of the produced models, and the automation of verifications. These ethical concerns have become crucial as AI applications continue to spread across various sectors of society, raising complex questions about governance, transparency, and regulation of these emerging technologies. Although various types of deep neural networks exist and have achieved notable success, they still remain complex to create and use [14]. Indeed, the performance of DL models can vary greatly depending on the type of hardware used [4]. Furthermore, data scientists encounter difficulties in improving the performance of DNNs by evaluating algorithms, adjusting parameters, and selecting the best model to ensure accurate and reliable results [15]. The integration of formal methods in this process can play a crucial role by providing rigorous mathematical frameworks for the design, analysis, and verification of DL models, thereby contributing to strengthening the reliability, interpretability, and robustness of AI systems. Our study addresses this issue and proposes a novel approach based on the use of partial models from different viewpoints in the design and analysis of DNNs, as well as the application of the Maude language of strategies for executing and verifying the produced formal models. We employ methods and tools from Model-Driven Engineering (MDE) [17] as well as the formal language Maude and its tools [10].

Our approach contributes in the following ways:

- Metamodeling, which involves designing multi-view models to represent specific aspects of the DNNs. This approach focuses on simplifying and accurately representing the composition of partial models to form a global model of the DNNs. By using consistent and well-defined metamodels, we ensure that critical system properties and behaviors are preserved during each step of the composition process. For instance, when composing partial models representing different layers of a DNNs, our metamodel ensures that the data flow and transformations between layers are accurately maintained. This involves defining clear interfaces and interactions between components, ensuring that the overall behavior of the DNNs is consistent with its intended design. Additionally, validation checks are incorporated at each step to verify that properties such as data integrity, consistency, and expected behavior are preserved throughout the modeling process.
- The use of formal methods to describe DNNs. This method is applied to formalize the structure, initialization, and processing of DNNs in Maude in order to prove their behavior. This step allows for exploring and understanding the functioning of DNNs in depth, which can lead to significant improvements in their efficiency and reliability. Furthermore, the Maude language of strategies,

through its practical system, will enable the execution of the produced formal models.

Through this study, we aspire to illustrate how this specification can effectively handle DNNs at every stage, thereby ensuring their consistency and reliability. Additionally, we plan to expand our approach to accommodate more sophisticated types of neural networks, addressing the growing demands for complexity and performance in DL systems.

The rest of this paper is organized as follows: Sect. 2 presents the fundamental concepts of DNNs, MDE, and the Maude language, which form the foundation of our research. Section 3 reviews related work to place ours in a broader context and highlight its relevance. Section 4 is then dedicated to presenting our meta-model for DNNs. Section 5 focuses on presenting our Maude-based semantics for DNNs. Subsequently, Sect. 6 presents discussions on the obtained results and perspectives. Finally, Sect. 7 includes the conclusion and discussion of future work.

2 Background

2.1 Deep Neural Networks (DNNs)

DNNs are advanced algorithms of artificial neural networks. They are computational models consisting of many simple processing units (akin to neurons) that work in parallel and are arranged in interconnected layers [6]. Simple neural networks consist of an input layer and an output layer; when more layers are stacked, the networks are called deep [16,18]. A DNN learns to perform particular tasks through training, during which the strength of connections between units is learned. Subsequently, the trained DNN is used to perform the same task on novel inputs [6]. DNNs are employed for addressing complex problems involving voluminous data, making them particularly powerful for tasks such as computer vision, speech recognition, automatic translation, and many others.

2.2 Model Driven Engineering (MDE)

Model Driven Engineering (MDE) is a software development methodology that emphasizes the creation and manipulation of models throughout the software development lifecycle [21,23]. At its core, MDE revolves around the concept of domain models, which are abstract representations of the problem-specific domain being addressed by the software system [13]. These domain models capture essential aspects of the problem space, including its entities, relationships, constraints, and behaviors. One of the key objectives of MDE is to raise the level of abstraction in software development by shifting the focus from low-level implementation details to higher-level domain concepts [19]. By doing so, MDE aims to improve software productivity, quality, and maintainability by enabling automatic code generation and model transformations [13,23]. This approach allows developers to express system requirements and designs in a more intuitive and platform-independent manner, facilitating better communication between stakeholders and promoting reusability across different projects [23]. The MDE process typically involves several stages, including domain modeling, platform-independent modeling, platform-specific modeling, code generation, and model transformation [21]. Each stage focuses on different aspects of the software development process, from conceptualization to implementation, while leveraging domain-specific languages (DSLs) and modeling tools to facilitate model creation, manipulation, and analysis. Overall, MDE offers numerous benefits, such as improved software maintainability, enhanced productivity, reduced development costs, and increased flexibility to accommodate changes in requirements [19]. It has found applications in various domains, including but not limited to, software architecture, system design, software testing, and software evolution [13,21].

2.3 Maude Language

Maude is a high-performance language and system supporting both equational and rewriting logic computation for a wide range of applications, including the development of theorem proving tools, language prototyping, executable specification, and analysis of concurrent and distributed systems, as well as logical framework applications in which other logics are represented, translated, and executed [8]. Rewriting logic is a logic of concurrent change that can naturally deal with state and concurrent computations. It serves as a general semantic framework for providing executable semantics to various languages and models of concurrency, particularly supporting concurrent object-oriented computation. Moreover, rewriting logic's characteristics as a logical framework make it suitable for representing and executing many other logics [20]. Maude comes in two flavors: Core Maude and FULL Maude.

Core Maude implemented in C++ and providing all of Maude's basic functionality [9], supports module hierarchies with functional and system modules. Additionally, it provides the META-LEVEL module, which likely serves as a foundational module for meta-level operations or reasoning about the system [8].

On the other hand, FULL Maude is an extension of Maude, written in Maude, that endows Maude with a powerful and extensible module algebra. This algebra allows Maude modules to be combined to build more complex modules, supports parameterized modules that can be instantiated with mappings called views, and enables moving up and down the reflective tower using Full Maude commands [9].

In our proposed approach, we utilize **Full Maude** to model and formalize the structure and behavior of DNNs. This choice is based on Full Maude's powerful and extensible module algebra, which allows us to combine modules to build more complex and parameterized models, providing the flexibility needed for our comprehensive approach.

3 Related Work

In recent years, there has been a growing interest and momentum in the integration of DNNs with formal methods. This section delves into some works in this field, emphasizing the application of Formal Methods. These endeavors utilize DNNs formalization to specify and verify models.

For instance, authors in [11] present innovative methodologies that harness formal methods to model, verify, and simulate the properties of neuronal behaviors across various neural network architectures, effectively bridging the gap between neuroscience and computer science. In their research, they initially introduce the synchronous language Lustre, which encodes the anticipated behavior of the circuit in a precise and modular manner. This language operates entirely automatically, obviating the need for manual proof generation. It leverages the "kind2" model checker, rooted in neural network modeling, to verify and specify property satisfaction. The second approach involves the use of the Coq Proof Assistant, enabling the formulation of mathematical definitions, algorithms, and theorems. This tool ensures the validity of properties in a generalized context using Coq, affirming the truth of desired behaviors irrespective of series length and parameters. The verification tool employed in the second approach is "theorem proving", entailing formal mathematical logic reasoning applied to the validation of manually crafted systems. Their approach does not specifically focus on DNNs. Additionally, they utilize formal languages such as Lustre and Coq, which are based on a declarative, data-flow-oriented approach and on propositional and predicate logic. However, they do not address rewrite logic, which is essential in our case.

In [22], the authors propose a novel approach for formalizing deep neural networks using rewriting logic and the Maude language. They have developed a self-learning algorithm, known as "Backpropagation", which enables the neural network to train autonomously. This approach specifies multilayer perceptrons using two different methods: vector representation and concurrent object-oriented programming. The main goal is to demonstrate the feasibility and applicability of using rewriting logic and Maude for modeling and analyzing neural networks in a practical context. However, this formalization may be complex for non-experts in the field, as it requires a deep understanding of the underlying concepts. Additionally, the article does not address modeling in a way that is accessible to non-experts, which could limit its usefulness for those seeking to understand neural networks without prior expertise in the field.

Authors in [25] explore formal verification techniques tailored to neural networks. They focused on top-performing tools presented in the context of VNN-COMP 2022 to create a model for controlling and monitoring LHC equipment. However, this article solely concentrates on verifying networks within a specific domain and does not address the verification of their structure and behavior in a general sense. In the next section, we will briefly present and explain our earlier research contributions. By reviewing these previous contributions, we can better illustrate the evolutionary path that has guided us to our current approach.

4 A Metamodel for DNNs

Our contribution introduces a model for DNNs, conceptualizing them as software systems. This model, named "MM-DNN", developed via Model-Driven Engineering (MDE), distinguishes the structure and behavior of DNNs. It is built using Model-Driven Architecture (MDA) [24] and the declarative language called Query/View/Transformation (QVT) [1]. MM-DNN aims to provide a common language for defining, manipulating, and transforming all types of DNNs seamlessly into other languages, with a focus on MAUDE. Developed using tools like ECORE language [3] and the Eclipse IDE [12], the model delineates two aspects: structure (Layers, Neurons, Connections) and behavior (Initialization, Data processing, Learning). This model utilizes meta-classes and relationships to formally represent DNNs, with the DNN class as the core and various subclasses representing different types of DNNs [5].

4.1 Structural Part

The Structural Part of MM-DNN, highlighted in blue (refer to figure in [5] Subsect. 3.2), encompasses the three fundamental elements of a DNN and their connections. A DNN instance, belonging to a specific subclass (e.g., FeedForward or RNN), comprises a single InputLayer, one or more HiddenLayers, and a single OutputLayer. These elements are interconnected hierarchically through instances of Neuron subclasses (InputNeuron, HiddenNeuron, OutputNeuron) and Link subclasses (IHLink, HHLink, HOLink) [5], following a connection topology detailed in Subsect. 4.1.2.

Basic Elements. The basic elements Layer, Neuron, and Link are represented by the classes **Layer**, **Neuron**, and **Link** respectively. The **Layer** class includes an attribute **nblayer** denoting the layer number and is specialized into InputLayer, HiddenLayer, and OutputLayer subclasses, corresponding to the input, hidden, and output layers. Neuron class embodies individual computing units with attributes such as activation value (**x**), threshold (**t**), and state (**st**). Its subclasses (**InputNeuron**, **HiddenNeuron**, **OutputNeuron**) represent neuron categories. The **Link** class models connections between neurons, storing weights and traversal state. Subclasses **IHLink**, **HHLink**, and **HOLink** represent specific link types [5].

Connection Topology. Connection topology is defined by associations between Neuron subclasses and their respective cardinalities, adhering to the DNN's hierarchical structure. In MM-DNN, rules governing connection topology are deduced from Structural Part: InputNeurons in an InputLayer are connected to HiddenNeurons in the corresponding HiddenLayer via IHLink instances. HiddenNeurons in a HiddenLayer are interconnected either via HHLink instances to the next HiddenLayer or, if it's the last hidden layer, via HOLink instances to OutputNeurons in the OutputLayer [5].

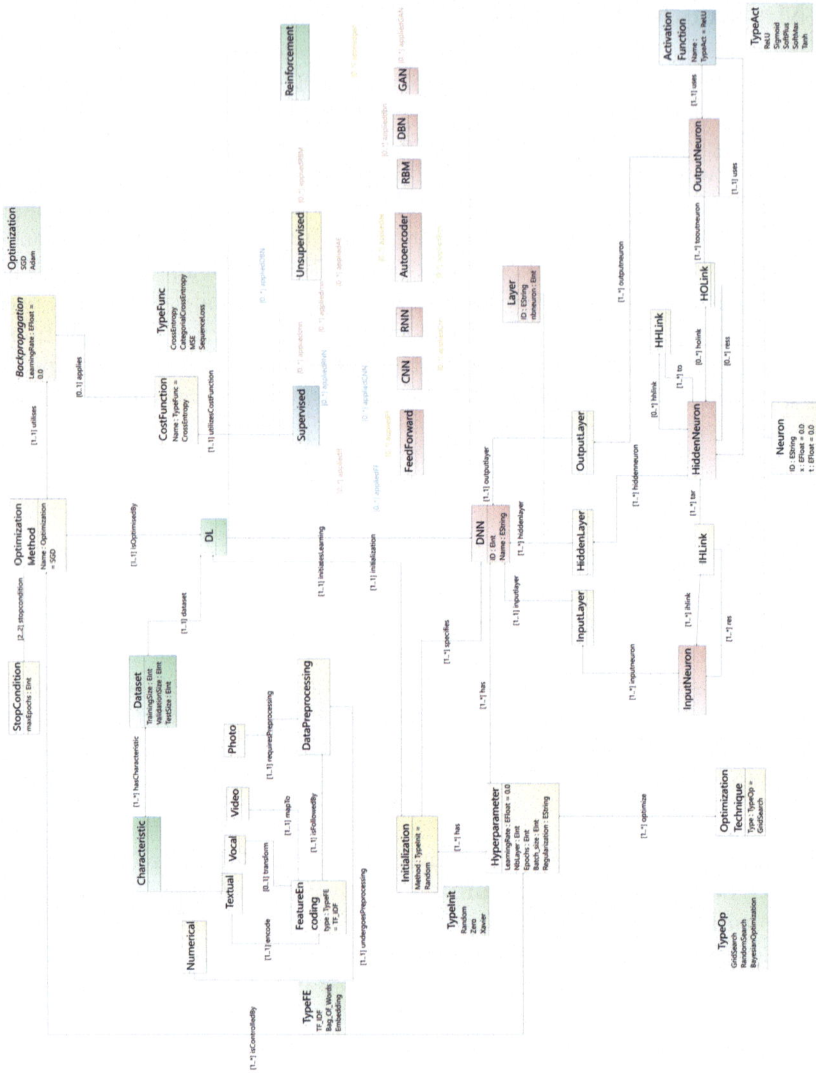

Fig. 1. Meta-Model MM-DNN

4.2 Behavioral Part

In this part, we explore the behavior of our meta-model as outlined in our paper [5]. It covers three key areas: parameter initialization, data processing and propagation, and learning. We've enhanced this section to include detailed insights into DNNs behavior, as depicted in Fig. 1.

Modeling Initialization. In the **DNN** class, we define various initialization methods through the **Initialization** class. These methods include **Random**, which initializes weights with small random values, **Zero**, initializing all weights to zero, and **Xavier**, tailored for certain activations. Hyperparameters like **LearningRate**, **NbLayer**, **Epochs**, **Batch-Size**, and **Regularization** are also defined in the **Hyperparameter** class. Optimization techniques such as **GridSearch**, **RandomSearch**, and **BayesianOptimization** are used to optimize these hyperparameters. The **Initialization** class is a fundamental component of the **DL** (Deep Learning) class, maintaining a composition relationship with it.

Modeling Data Processing and Propagation. The first crucial step in the operation of a neural network, particularly in MM-DNN, is **'Data Preparation'**, managed by the **'Dataset'** class. It includes attributes such as **'TrainingSize'**, **'ValidationSize'**, and **'TestSize'**. Data in various formats (numeric, textual, vocal, video, photo) is processed by the **'Characteristic'** class within **'Dataset'** and converted using the **'FeatureEncoding'** class, employing techniques such as 'TF-IDF', 'Bag-of-Words', and 'Embedding'. After preprocessing by the **'DataPreprocessing'** class, the data enter the propagation process, managed by subclasses of 'Neuron' and association relationships defined in the structural part of MM-DNN. The **'ActivationFunction'** class, highlighted in blue in Fig. 1, plays a crucial role in transforming the data at each step.

Modeling Learning. Learning in MM-DNN involves several key steps:

- **Initialization:** Initializing the model is vital. The 'Initialization' class, integrated into the 'DL' class, handles this task, contributing to both model initialization and learning.
- **Data Preparation:** This step involves cleaning, transforming, and normalizing data to prepare them for model training.

After model initialization and data preparation, learning begins. This includes various types such as **Supervised**, **Unsupervised**, and **Reinforcement** learning, applicable to different types of neural networks like 'FeedForward', 'CNN', 'RNN', etc.

In supervised learning, a cost function (e.g., 'CrossEntropy', 'MSE') measures the difference between the model output and the expected output for optimization. Unsupervised and reinforcement learning use evaluation functions for performance measurement.

Backpropagation computes the direction to adjust network parameters to minimize prediction error, while optimizers like 'SGD' or 'ADAM' perform these adjustments.

Finally, once the stop condition is met, the learning process ends, otherwise, it iterates for further model refinement.

4.3 Illustrative Example

To explain the process of MM-DNN instantiation and facilitate its understanding, we illustrate it by considering the Feedforward type of DNNs. It involves first instantiating the Structural part of MM-DNN, and then the Behavioral part to define an example model of the chosen type.

Structural Part. In this section, we present the instantiation of MM-DNN in an automated manner using the Sirius tool [2]. This representation is specific to the Feedforward type of DNNs and is visualized in Fig. 2. This illustration allows:

1. Transforming the abstract MM-DNN into a visually intuitive graph, facilitating the understanding and manipulation of DNNs concepts.
2. Clearly establishing conformity between elements at different levels. This conformity is illustrated by:
   - Representing model layers with rectangles to visualize the network structure.
   - Illustrating neurons with circles to denote individual computing units.
   - Indicating the active or inactive state of the neuron through circle filling or coloring.

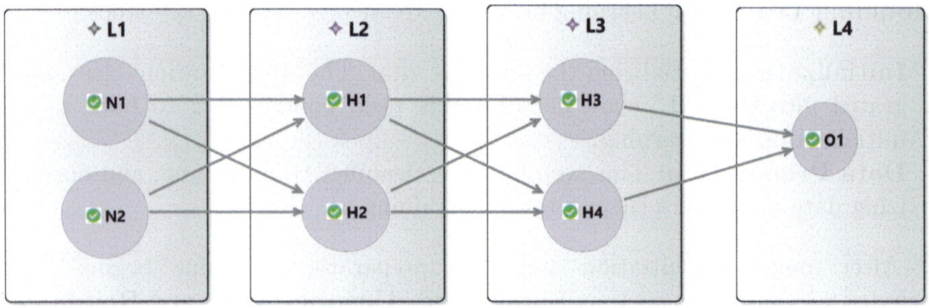

Fig. 2. Illustration of Feedforward DNNs Instantiation

Behavioral Part. To complete the Feedforward model in MM-DNN, we propose using UML Activity Diagrams to illustrate the logical flow of its functionalities: Initialization, Data processing and propagation, and Learning. The diagram (Fig. 3) depicts the general operation process of a Feedforward DNNs. It starts with initializing the network's parameters, including weight initialization methods (random, zero, or Xavier) and configuring hyperparameters like learning rate, number of layers, and epochs. After setting up the network, data propagation begins upon receiving input data. The learning algorithm is automatically activated, and the diagram concludes with output production after data propagation.

By employing this UML Activity Diagram, we aim to provide a clear representation of the Feedforward DNNs instantiation process, aiding understanding and implementation.

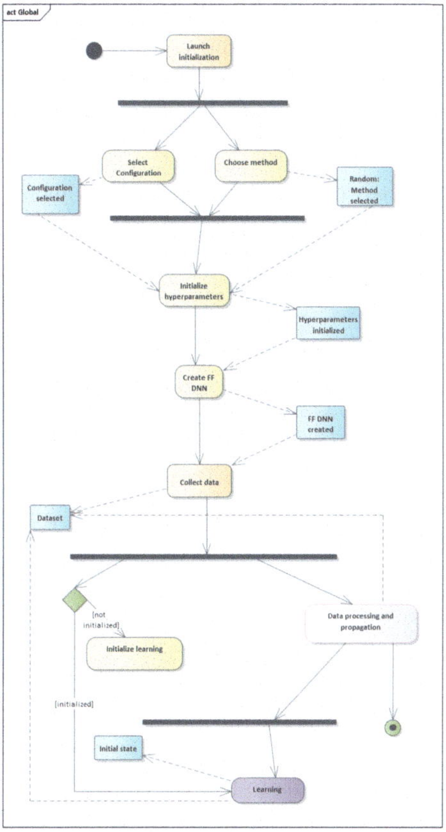

Fig. 3. Behavioral Flow Diagram

5 Maude-Based Semantics for DNNs

In this section, we delve into the formal semantics of DNNs using the Maude language. Our focus will be on Feedforward type of DNNs, and we will explore how Maude can be utilized to provide a rigorous mathematical foundation for understanding their behavior.

5.1 Mapping Rules

The formalization of DNNs in Maude represents a fundamental step in our approach, essential for precise modeling and formal specifications of DNNs. This phase revolves around the use of a model transformation tool, XSLT (Extensible Stylesheet Language Transformation) [7], to convert DNNs models, expressed in XML according to a meta-model, into Maude specifications. Our initial step involves establishing mapping rules that bridge the structural and behavioral aspects of DNNs with their formal definition in Maude. These rules will serve as the basis for translating the visual and conceptual representations of DNNs into a precise and executable formal specification. Table 1 details how each key element of the feedforward DNNs model is mapped and represented in Maude.

Table 1. Mapping between FF DNNs and Its Formalization in Maude

	FeedForward DNNs	Mapping Rules (between FF DNNs and MAUDE)
FF Structural		`Class FeedForward ∣ ID : Nat, Name : String, IL : InputLayer, OL : OutputLayer, HL : HiddenLayer` `rl [Activation-Ne] : Activate ( < Ne : Neuron ∣ ID : S, x : X1, t : X2, b : X3, st : 1>, F) .`

FF Behavioral	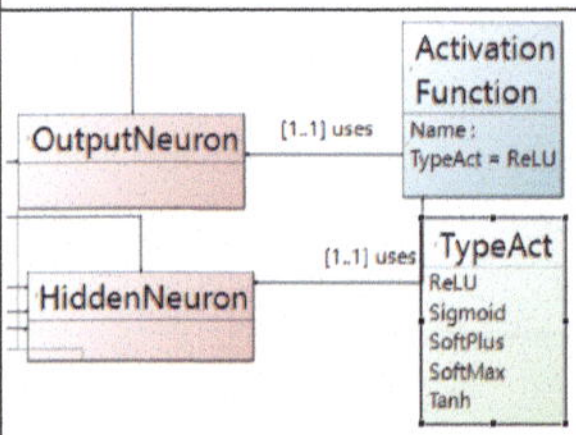	`=> < Ne : Neuron ∣ ID : S, x : F(X1, X2, X3), t : X2, b : X3, st : 1> .`
		
		`rl [Propagation-Init] : Propagation ( < AF : FeedForward ∣ ID : 1, Name : 'FF, IL: IL1, HL : HL1, OL : OL1 > , F)`
		`=> < AF : FeedForward ∣ ID : 1, Name : 'FF, IL: Activate(IL1,F), HL : HL1, OL : OL1 > .`

5.2 Static Specification of DNNs

Next, we provide a specification of a four-layer feedforward DNNs in Maude. We specify the neurons in each of the layers, as well as the weights and biases of all the links. Our representation of perceptrons in Maude revolves around the definition of the `FeedForward` class:

```
Class FeedForward | ID : Nat, Name : String, IL : InputLayer,
OL : OutputLayer, HL : HiddenLayer .
```

and their configurations with:

```
ops initialFeedForward : -> Configuration .
```

To specify a layer, we define it as follows:

```
eq L1 = < ID : 'L1', nbneuron : 2, IN : NN1; NN2 >
```

We specify it as an operation with its identifier and number of neurons.

To represent a neuron:

```
eq NN1 = < ID : 'NN1', x : 2.0, t : 0.5, b : 0.1, st : active >
```

Each neuron instantiation is represented with its ID, current value (`x`), threshold (`t`), bias (`b`), and an attribute `st` to indicate if the neuron is already activated or not.

```
eq Link1 = < weight : 0.5, st : active >
```

Each link instantiation is represented with its weight and status, either active or not.

This static specification provides a detailed description of the structure and configuration of the feedforward DNNs in Maude, enabling precise modeling and analysis.

5.3 Dynamic Specification of DNNs

We extend our formalization to cover the dynamic behavior of feedforward DNNs. This dynamic specification focuses on how input data evolves through each network layer and how activation functions shape the output.

In our formalization, we define rewriting rules to describe the transformations applied to input data as it moves through the network layers. These rules encompass mathematical operations such as matrix multiplications, bias additions, and activation function applications. To activate each layer, we employ a set of rewriting rules.

The rule [Activation-La] governs the activation of neurons within a layer using a specific function:

```
rl [Activation-La] : Activate (<La : Layer  | ID : S, nbneuron : J ,
    IN : IN1; IN2 >, F) => <La : Layer  | ID : S, nbneuron : J , IN
    : Activate (IN1, F); Activate(IN2, F) > .
```

For individual neuron activation, the rule [Activation-Ne] is used:

```
rl [Activation-Ne] : Activate ( < Ne : Neuron |  ID : S, x : X1, t :
    X2, b : X3, st : 1>, F)
=>  < Ne : Neuron |  ID : S, x : F(X1, X2, X3), t : X2, b : X3, st :
    1>  .
```

The rule [Link-Activation] formalizes weight adjustment for active links:

```
rl [Link-Activation] : Activate ( < Link : Link | weight : W, st :
    active>, F)
=>  < Link : Link | weight : F(W), st : active>  .
```

Additionally, we define rules for data propagation through network layers. [Propagation-Init] describes propagation through the input layer, [Propagation-Hid] through hidden layers, and [Propagation-Fin] through output layers. They are described in Maude as follows:

```
rl [Propagation-Init] : Propagation( < AF : FeedForward | ID : 1,
    Name : 'FF, IL: IL1, HL : HL1, OL : OL1 > , F)
 => < AF : FeedForward | ID : 1, Name : 'FF, IL: Activate(IL1,F), HL
     : HL1, OL : OL1 >  .
rl [Propagation-Hid] : Propagation( < AF : FeedForward | ID : 1,
    Name : 'FF, IL: IL1, HL : Activate(HL1, F), OL : OL1 > , F)
 => < AF : FeedForward | ID : 1, Name : 'FF, IL: IL1, HL :
     Activate(HL1, F), OL : OL1 >  .
rl [Propagation-Fin] : Propagation( < AF : FeedForward | ID : 1,
    Name : 'FF, IL: IL1, HL : HL1, OL : Activate(OL1, F) > , F)
 => < AF : FeedForward | ID : 1, Name : 'FF, IL: IL1, HL : HL1, OL :
     Activate(OL1, F) >  .
```

6 Results Discussion and Perspectives

Our Maude-based approach offers a robust framework for representing both the structure and behavior of intricate DNNs. By establishing a broad metamodel for DNNs, we create a flexible and adaptable model that captures the essence of these networks. This metamodel encompasses various classes like 'DNN', 'FeedForward', 'RNN', 'CNN', each defining specific network characteristics. Relationships between these classes illustrate the interactions among different parts of a DNN, simplifying the modeling of complex architectures.

Transforming this general metamodel into Maude specifications tailored to feedforward neural networks results in a precise and executable representation of DNNs. The static specification provides detailed insights into a DNN's architecture, covering layers, neurons, weights, biases, and activation functions. Such granularity enables thorough modeling and analysis of diverse DNNs types. Meanwhile, the dynamic specification captures operational dynamics, including input data propagation, neuron activation, and weight adjustments during learning. These formal rewriting rules serve as a foundation for understanding and reasoning about DNNs behavior across various scenarios.

Maude stands out as a formal language for DNNs specification due to its expressive framework and solid semantics based on rewriting. Its modularity and extensibility align well with evolving research, allowing seamless integration of new concepts and techniques. Our method offers several advantages, including a formal foundation for understanding and analyzing DNNs, precise representation of network aspects, and flexibility for exploring new ideas.

However, our approach also faces limitations, such as managing complexity in large networks and potential gaps in representing specific learning algorithms or advanced architectures. Moving forward, we aim to explore formal verification techniques to assess DNNs model accuracy and reliability. This includes verifying critical properties like convergence, stability, and robustness, as well as employing model checking algorithms to automatically analyze DNNs models against specifications. These efforts can enhance our understanding and engineering of neural network systems, driving advancements in artificial intelligence and machine learning.

7 Conclusion

The motivation for the proposed approach to modeling and formalizing the structure and behavior of DNNs, as presented in this paper, arises from the pressing need for rigorous methodologies to address the growing complexity and ethical concerns surrounding artificial intelligence systems, particularly DNNs. With the rapid advancement of AI technologies and their increasing integration into various sectors of society, ensuring the reliability, interpretability, and robustness of DNNs has become paramount. Our method, which combines MDE principles with the formal language Maude, offers a systematic framework for designing DNNs models. By leveraging metamodeling techniques and formal methods, we provide a structured approach to capturing the essential properties and behaviors of DNNs, thus contributing to the development of more transparent, accountable, and trustworthy AI systems. Looking ahead, our work opens new possibilities for applying formal verification and analysis to improve the consistency, reliability, and security of DNNs, thereby advancing the field of AI and its ethical implications. We also intend to extend our approach to support more advanced types of neural networks to meet the increasing complexity and performance needs of DL systems.

Acknowledgment. This work was partially supported by the LABEX-TA project MeFoGL: "Méthode Formelles pour le Génie Logiciel".

References

1. Arrassen, I., Esbai, R., Meziane, A., Erramdani, M.: QVT transformation by modeling: from UML model to md model. In: 2012 6th International Conference on Sciences of Electronics, Technologies of Information and Telecommunications (SETIT), pp. 86–91. IEEE (2012)
2. Bedini, F., Maschotta, R., Zimmermann, A.: A generative approach for creating eclipse sirius editors for generic systems. In: 2021 IEEE International Systems Conference (SysCon), pp. 1–8 (2021). https://doi.org/10.1109/SysCon48628.2021.9447062
3. Budinsky, F.: Eclipse Modeling Framework: A Developer's Guide. Addison-Wesley Professional (2004)
4. Chaber, P., Ławryńczuk, M.: Pruning of recurrent neural models: an optimal brain damage approach. Nonlinear Dyn. **92**(2), 763–780 (2018). https://doi.org/10.1007/s11071-018-4089-1
5. Choutri, A., Boukhari, A., Belala, F., Kacem, A.H.: Vers l'amélioration des caractéristiques des réseaux de neurones profonds: une approche de méta-modélisation (2023)
6. Cichy, R.M., Kaiser, D.: Deep neural networks as scientific models. Trends Cogn. Sci. **23**(4), 305–317 (2019)

7. Clark, J., et al.: Xsl transformations (xslt). World Wide Web Consortium (W3C) **103** (1999). http://www.w3.org/TR/xslt
8. Clavel, M., et al.: The maude system. In: International Conference on Rewriting Techniques and Applications, pp. 240–243. Springer (1999)
9. Clavel, M., et al.: Maude 2.0 manual (2003). http://maude.cs.uiuc.edu
10. Clavel, M., Durán, F., Hendrix, J., Lucas, S., Meseguer, J., Ölveczky, P.: The maude formal tool environment. In: Mossakowski, T., Montanari, U., Haveraaen, M. (eds.) CALCO 2007. LNCS, vol. 4624, pp. 173–178. Springer, Heidelberg (2007). https://doi.org/10.1007/978-3-540-73859-6_12
11. De Maria, E., et al.: On the use of formal methods to model and verify neuronal archetypes. Front. Comput. Sci. **16**, 28 (2021)
12. Eclipse, I.: Eclipse ide, pp. 1–20 (2009). www.eclipse.org
13. Estévez, E., García, A.S., García, J.G., Ortega, J.G.: A novel model-based approach to support development cycle of robotic arm applications. IFAC Proc. Vol. **47**(3), 3465–3470 (2014). https://doi.org/10.3182/20140824-6-ZA-1003.02163, https://www.sciencedirect.com/science/article/pii/S1474667016421415
14. Goodfellow, I., Bengio, Y., Courville, A.: Deep Learning. MIT Press (2016)
15. Hartmann, T., Moawad, A., Schockaert, C., Fouquet, F., Le Traon, Y.: Meta-modelling meta-learning. In: 2019 ACM/IEEE 22nd International Conference on Model Driven Engineering Languages and Systems (MODELS), pp. 300–305. IEEE (2019)
16. Hinton, G.E.: Learning multiple layers of representation. Trends Cogn. Sci. **11**(10), 428–434 (2007). https://doi.org/10.1016/j.tics.2007.09.004
17. Kent, S.: Model driven engineering. In: International Conference on Integrated Formal Methods, pp. 286–298. Springer (2002)
18. LeCun, Y., Bengio, Y., Hinton, G.: Deep learning. Nature **521**(7553), 436–444 (2015)
19. Ma, Y., Hao, Y., Qian, Y., Chen, M.: Cloud-assisted humanoid robotics for affective interaction. In: 2016 2nd International Conference on Control, Automation and Robotics (ICCAR), pp. 15–19. IEEE (2016)
20. Maude, S.: Maude overview - the maude system. https://maude.cs.illinois.edu/w/index.php/Maude_Overview
21. Mens, T., Gorp, P.V.: A taxonomy of model transformation. Electron. Notes Theor Comput. Sci. **152**, 125–142 (2006). https://doi.org/10.1016/j.entcs.2005.10.021
22. Santos-García, G., Palomino, M., Verdejo, A.: Rewriting logic using strategies for neural networks: an implementation in maude. In: International Symposium on Distributed Computing and Artificial Intelligence 2008 (DCAI 2008), pp. 424–433. Springer (2009)
23. Silva, E.D.A., Valentin, E., Carvalho, J.R.H., Barreto, R.D.S.: A survey of model driven engineering in robotics. J. Comput. Lang. **62**, 101021 (2021).https://doi.org/10.1016/j.cola.2020.101021, https://www.sciencedirect.com/science/article/pii/S2590118420300812
24. Soley, R., et al.: Model driven architecture. OMG White Paper **308**(308), 5 (2000)
25. Sommart, T.: Formal verification of neural networks. Technical report (2023)

Anti-pattern Based IoRT-Aware Business Process Structure Verification Approach

Najla Fattouch[1(✉)], Imen Ben Lahmar[2], and Khouloud Boukadi[1]

[1] FSEG Sfax, MIRACL Laboratory, Sfax University, Sfax, Tunisia
{najla.fattouch,khouloud.boukadi}@fsegs.usf.tn
[2] ISIM Sfax, ReDCAD Laboratory, Sfax University, Sfax, Tunisia
imen.benlahmar@isims.usf.tn

Abstract. The IoRT-aware Business Process is a new business type. It embedded IoT and robot technologies to automate classic BP. To effectively execute an IoRT-aware BP, it is significant to ensure that is accurately modeled. Referring to the literature, various modeling languages can be used to model an IoRT-aware BP. The Business Process Model and Notation (BPMN) is among the most attractive and descriptive languages. However, the latter lacks a formal aspect, potentially leading to inconsistencies and ambiguities. Therefore, to effectively check an IoRT-aware BP modeled using BPMN, it seems relevant to cope with a consistent formalism (e.g., Petri Net (PN), Discrete Event System Specification (DEVS), etc.), simplifying the process structure verification. From this perspective, we propose an approach that deals with DEVS formalism to check IoRT-aware BP structure (e.g., deadlock, loop, dead task, etc.). To evaluate the proposed approach, we consider a case study of IoRT-aware BP in agriculture by checking its process model structure. The findings show the effectiveness of our approach in capturing structure errors.

Keywords: Structure Verification · Anti-pattern · DEVS · IoRT-aware BP

1 Introduction

In recent years, the IoRT-aware Business Process has emerged through integrating the Internet of Robotic Things (IoRT) technology within the classic Business Process (BP). This integration allows business managers to automate their processes, where tasks are accomplished using IoT and robot devices rather than humans [8].

According to the literature, heterogeneous modeling languages have been developed to model IoRT-aware BP. The Business Process Model and Notation (BPMN) provides an attractive standard for modeling processes [7]. It is distinguished by its straightforward and comprehensible notations. Additionally, BPMN can be effectively used to design automated processes incorporating the

L. H. Belguith and K. Shaalan (Eds.): LPKM 2024, LNNS 1303, pp. 59–68, 2025.
https://doi.org/10.1007/978-3-031-85067-7_6

newest technologies, such as the IoRT. However, the BPMN defines some limits, for example, the lacking a formal aspect, which might cause inconsistencies, ambiguities, and incompleteness in the developed processes [11].

To bridge the BPMN gaps, it is relevant to adopt an expressive formalism to ensure accurate modeling and verification of BP.

Back to the literature, diverse formalisms (e.g., Petri Net, Discrete Event System Specification (DEVS), etc.) are developed to deal with processes modeled using the BPMN language. During this work, we cope with DEVS formalism to simulate an IoRT-aware BP. It has several advantages, which make it one of the most attractive formalisms. It is distinguished by its capacity to articulate the system's development by outlining its behaviors via transition functions, model interactions, state variables, etc.

During this work, we focus on the checking of an IoRT-aware BP model structure using DEVS formalism. Consequently, verifying the structure of the process model (e.g., deadlock, loops, timer events, etc.) aims to preempt any potential issues within it and ensure its flawless execution.

Back to the literature, numerous approaches cope with the DEVS to model the BP, for instance, [3,5], and [6]. However, these approaches do not address the checking of the process structure.

To bridge the existing approaches limit, we aim during this work to check the structure of IoRT-aware BP, where we consider a set of anti-patterns. Indeed, addressing the anti-patterns enables the detection and prevention of undesirable constructs within the process model [13].

To demonstrate the proposed approach's effectiveness, we consider an IoRT-aware BP model in agriculture. The outcomes demonstrate the effectiveness of our suggestion in detecting structure errors.

The remainder of this paper is structured as follows: Sect. 2 gives an overview of the approaches that tackle BP checking. Section 3 details the proposed approach. Section 4 addressed the approach experimentation. Finally, Sect. 5 outlines the main axes of our work and highlights its future directions.

2 Related Work

Numerous studies address BP's verification behaviors. This section provides a comprehensive overview of the approaches published in this field. This section gives an overview of the recently published approaches.

Among the recent approaches, we cite [15], where authors deal with checking a BP structure. During their proposal, the authors considered the BPMN when designing their BPs. The authors address the Petri Net(PN) formalism to achieve the formal verification goal of the process. Nonetheless, this latter lacks a hierarchical structure that impedes the modeling and checking of complex processes. Moreover, the proposed approach ignores some process structure details, such as checking task names, overlapping events, etc.

Another approach presented in [11] that copes with the BP formal verification. It considers the RECATNets, which are defined as high-level algebraic nets based on a combination of the PN and abstract data types. Nevertheless, despite their relevance to capture process errors, the suggested approach does not consider the anti-patterns.

In the BP checking context, the authors in [4] propose an approach to checking the process management data. However, the proposed approach is limited to the data dimension without considering the process structure. Moreover, the authors tackle neither a formal language to accomplish their proposal nor an automated BP.

In [17], the authors focus on examining the temporal behaviors of the traditional BP using Petri Nets (PN). However, it is imperative to initially validate the structure of the BP model before delving into assessing its temporal characteristics. Flaws in the process model structure could lead to erroneous process temporal behaviors.

Another approach published in [9], where authors deal with the modeling and verification of automated BP that integrated the IoRT technology. However, the proposed verification does not address the process structure.

Some other attractive approaches, for instance, [1,2], and [10] deal with the BP modeling using the BPMN standard. While these approaches have made notable contributions, there is still room for further exploration, such as ensuring the accurate execution of the automated BP through checking its structure.

To close the current approach gaps, we propose in this work an approach that deals with the checking of an IoRT-aware BP. The proposed approach aims mainly to guarantee the correct process's internal structure. Our proposal is based, on one hand, on the DEVS formalism regarding its advantages such as its hierarchical structure. On the other hand, it addresses a set of anti-patterns.

3 IoRT-Aware Business Process Structure Verification

In the BP model, it is significant to check, whether its structure is correct along with meeting a set of anti-patterns. A process with a correct structure is defined as a model structure consistent with the real world [13]. Therefore, verifying the structure of the process model aims to mitigate potential issues and ensure its accurate execution.

The checking of a process structure model can be accomplished by investing in a set of relevant patterns and anti-patterns. A pattern is a solution for a modeling issue. In contrast, an anti-pattern is a model that intends to capture process modeling issues [13].

During this work, we address checking the IoRT-aware BP structure considering DEVS formalism (see Fig. 1). Notably, the transformation of IoRT-aware BP into DEVS is elaborated in our work [6], which has been accepted but has not yet been published.

To achieve our process-checking goal, we address two anti-pattern categories: syntax and control-flow errors.

Syntax errors cope with the process parts (e.g., gateway, task, etc.) that are wrongly modeled. Referring to the literature, we select a set of heterogeneous syntax anti-patterns. In what follows, we enumerate those are used in our work:

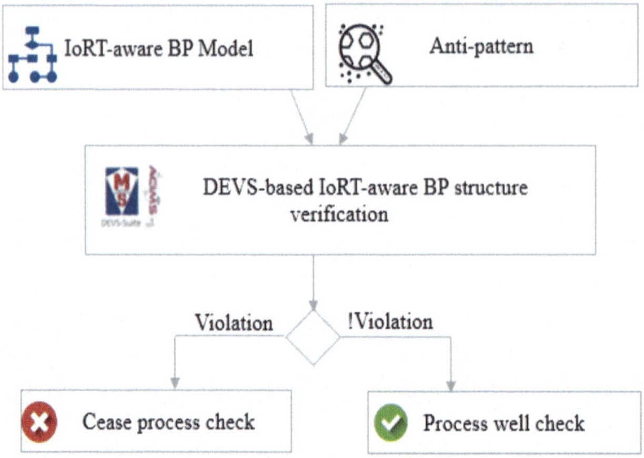

Fig. 1. Overview for anti-patterns based process structure checking approach.

- An IoRT-aware BP lacks a start event: Designates the process without a start event to launch its workflow. However, in Business Process Management (BPM), the process should define an event to trigger the process workflow and tasks. The lack of a start event can potentially hinder the process management.
- An IoRT-aware BP lacks an end event: Gives an IoRT-aware BP without an end event to finish its execution. This event determines when the process achieves its goal. Dealing with a process without at least one end event can result an ambiguity regarding the successful accomplishment of tasks. Furthermore, it may lead to a mismanagement of process resources.
- Sequence flow crosses pool boundaries: It refers to the interaction between pools using sequence flow. Indeed, pools depict heterogeneous entities that should interact using only message flow. Therefore, considering sequence flow to connect two pools leads to an inconsistency with the process model.
- Divergency AND/ OR gateways with more than one terminate event: Sets forward the use of more than one end event for a divergence AND/ OR Gateway.
- Common tasks name: Depicts the IoRT-ware BP tasks that define the same names. However, within the process, each task should possess a unique name to distinguish it from others.

The control-flow anti-patterns category defines problems of the process model's control flow [12]. In what follows, we give an overview of the considered control-flow anti-patterns:

- Deadlock: Designates two tasks waiting for each other to proceed. Regarding this situation, no task can complete its execution. Therefore, the process will be blocked.
- Dead Component: Refers to the IoRT-aware BP components (e.g., sensor, actuator, event, etc.) without any input and output flows.

- Infinite loop: Depicts one or more tasks are carried out perpetually, where there is no end condition.
- Issue on timer event: Referred timer events that are not appropriately matched with tasks needed. Consequently, dealing with this anti-pattern leads to errors within the process model.
- Overlapping Events: Depicts a set of events that are intended to be launched under similar circumstances, without any obvious priories or distinctions. Thus induces conflicts within the process flow.

To accomplish our checking goal, we suggest associating each process component with a Structure Control Model (SCM) (See Fig. 2). The SCM (Colored in blue) is modeled using the DEVS atomic model. The latter depicts the discrete event system behaviors and it presents how the system reacts to a receiving event.

The SCM serves to check the structure of an IoRT-aware BP model. It defines an input port *in* and an output port *out* to connect with the controlled process component model (e.g., task, event, etc.).

In the IoRT-aware BP checking structure based on heterogeneous anti-patterns, we address numerous process components. Each considered component is modeled using DEVS atomic model (see Fig. 2). Each component (colored in gray) defines an *in* and *out* ports to ensure the interaction with its predecessors and successors. Moreover, we add an *inCS* (in Control Structure) and *outCS* (out Control Structure), which connect it with its associated SCM.

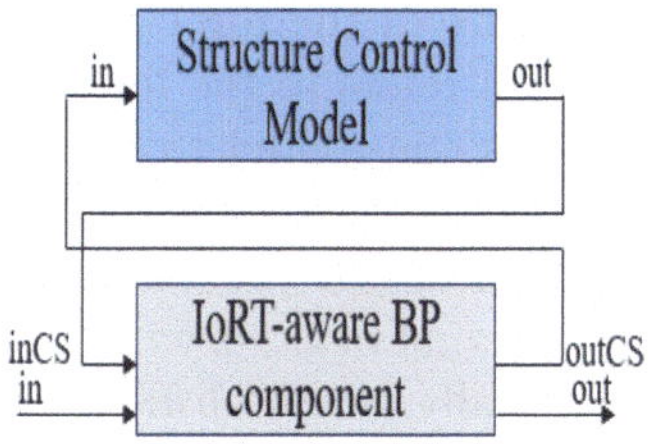

Fig. 2. Structure Control Model associated with an IoRT-aware BP component.

It's significant to highlight that due to the limited space, we are unable to provide further details for the formal description of the proposed approach based on DEVS.

4 Experimentation

To evaluate our proposal we consider IoRT-aware BP models in the agriculture field. The process models are developed using an extended BPMN 2.0 modeler plugin [9].

During our work, we cope with an IoRT-aware BP model for the harvesting of the tomatoes. Figure 3 shows the considered IoRT-aware BP model. The latter defines two sensors which are *Capture Rainfall* and *Capture Temperature* to retrieve rainfall and temperature data respectively. The sensors' output is stored using *Storage Rainfall Data* and *Storage Temperature Data* tasks. Considering the stored data, the *Make Harvesting Tomatoes Decision* robot script is launched to make the tomato harvest decision. Therefore, the process stops its execution whether there is no harvesting need. Otherwise, the robot *Start Harvesting Tomatoes* initiates its action.

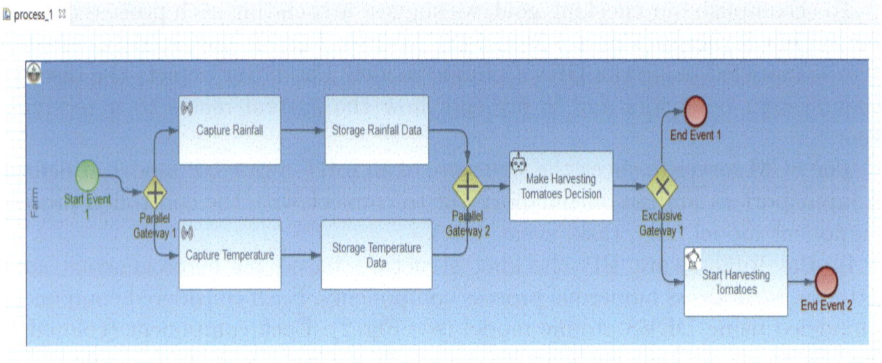

Fig. 3. Example of an IoRT-aware BP model.

To formally check the structure of the IoRT-aware BP, we simulate it using the DEVS-Suite Tool. Simulation refers to the imitation of real-world processes using some developed tools. It is considered to predict the impact of applying one or more changes on the process performance [14].

Figure 4 shows the simulation of our processes using the DEVS-Suite tool, where each process component is associated with an SCM (Structure Control Model). As shown in Fig. 4, initially, all the IoRT-aware BP components are in the *Waiting* state, and the control models are in the *Wait_Check* state, which indicates that are biding to launch the verification action.

Using the IoRT-aware BP model (see Fig. 3), we cope with a first scenario to check the process structure. This scenario gives a process correctly modeled (see Fig. 5). The process components are correctly checked where they change their state to *Terminate*. Moreover, for each component (e.g., task, event, etc.), the SCM alters its state to *well_check* which designates that the component is structured verified.

Moreover, we deal with a second scenario (Fig. 6), where we define the *Start Harvesting Tomatoes* as a dead task without input and output flows. Moreover, we do not define an end event for the process.

Table 1 provides additional insights into the examined scenarios. It indicates that the defined structural control models did not detect violations in the first scenario. However, during the second scenario, the control models identified some violations, such as the dead task and the end event. This table depicts the

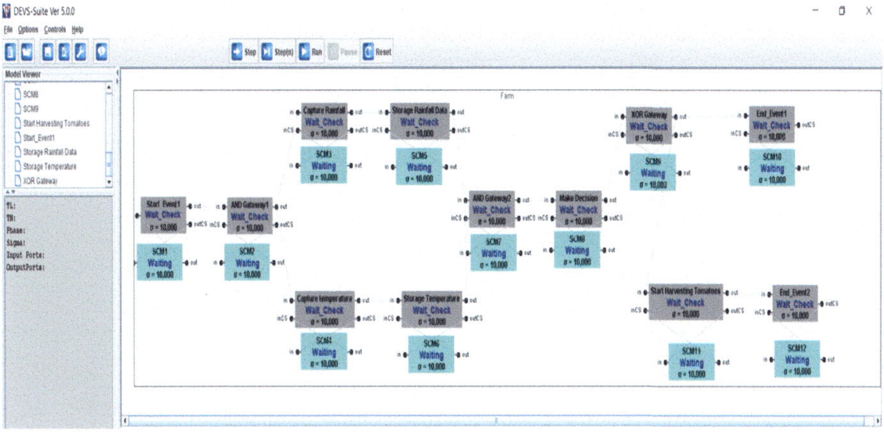

Fig. 4. Simulation of an IoRT-aware BP model using DEVS-Suite tool.

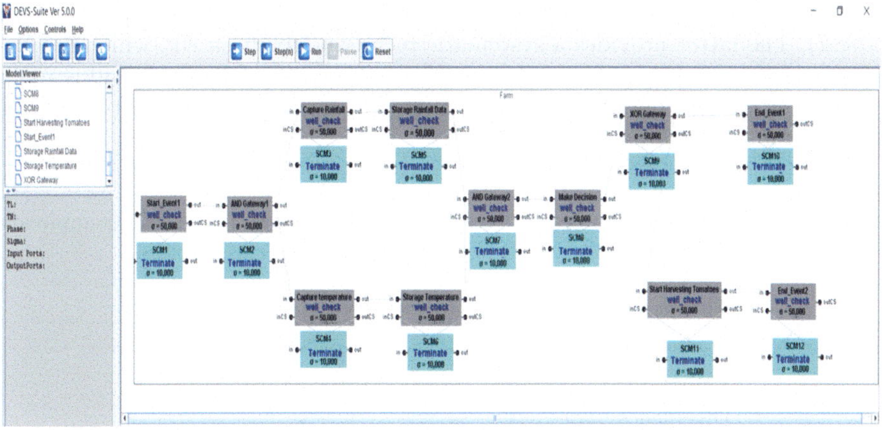

Fig. 5. Process structure well checked using DEVS formalism (First scenario).

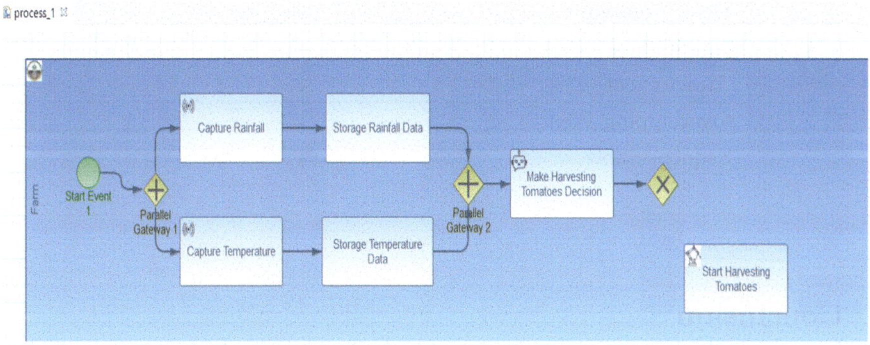

Fig. 6. An IoRT-aware BP model considered for the second scenario.

Table 1. Results of the first and second scenarios checking structure behaviors.

Anti-patterns	Fist scenario	Second scenario
Start event	✓	✓
End event	✓	Violation
Divergence Gateways	✓	Violation
Common tasks name	✓	✓
Deadlock	✓	✓
Dead Component	✓	✓
Infinite loop	✓	Violation

relevance of our proposed structure-checking approach in ensuring the integrity and compliance of the IoRT-aware BP.

To evaluate the efficacy of our verification approach, we compare it with the Block-Structure Collaborative Business Process (BS-CBP) approach (see Table 2). The latter provides a systematic technique to check BP, which is published in [16]. It copes with the anti-patterns to check a BP. Table 2 illustrates that our proposed approach addresses a greater number of anti-patterns compared to the BS-CBP.

Table 2. A comparative analysis of our anti-pattern based BP checking approach versus the BS-CBP one.

Anti-patterns	BS-CBP approach	Our approach
Start event	-	✓
End event	-	✓
Divergence Gateways	-	✓
Common tasks name	-	✓
Deadlock	✓	✓
Dead Component	✓	✓
Infinite loop	-	✓
Timer event	-	✓
Overlapping event	-	✓
Sequence flow	-	✓

5 Conclusion

The current research addresses the checking of an IoRT-aware BP structure using a set of fixed anti-patterns. The process checking goal is achieved considering the DEVS formalism. Therefore, each process component is associated with a DEVS atomic model called the Structure Control Model (SCM). To assess our proposal, we cope with a process in the agriculture field.

In future work, we plan to check the process spatial behaviors regarding the integrated robot devices.

References

1. Alyani, Z.: Extension of the functional dimension of BPMN based on MDA approach for sensitive business processes execution. In: The 6th International Conference on Information and Knowledge Systems, vol. 486, pp. 243–254 (2023)
2. Alyani, Z., Turki, M.: An MDA approach for extending functional dimension for sensitive business processes execution. In: The 22nd International Conference on Intelligent Systems Design and Applications, vol. 715, pp. 147–155 (2022)
3. Bazoun, H., Bouanan, Y., Zacharewicz, G., Ducq, Y., Boyé, H.: Business process simulation: transformation of BPMN 2.0 to DEVS models (WIP). In: The Symposium on Theory of Modeling and Simulation - DEVS Integrative M&S Symposium, p. 20 (2014)
4. Beerepoot, I., et al.: The biggest business process management problems to solve before we die. Comput. Ind. **146**, 103837 (2023)
5. Boukelkoul, S., Maamri, R.: Optimal model transformation of BPMN to DEVS. In: The International Conference of Computer Systems and Applications, pp. 1–8 (2015)
6. Fattouch, N., Ben Lahmar, I., Boukadi, K.: A model-based approach for the transformation and verification of an IoRT-aware business process. In: The International Conference on Intelligent Systems Design and Applications (2023)
7. Fattouch, N., Lahmar, I.B., Boukadi, K.: IoT-aware business process: comprehensive survey, discussion and challenges, pp. 100–105 (2020)
8. Fattouch, N., Lahmar, I.B., Boukadi, K.: A comprehensive architecture for an IoRT-aware business process outsourcing into fog and cloud computing. In: The Tunisian-Algerian Joint Conference on Applied Computing, vol. 3067, pp. 164–172 (2021)
9. Fattouch, N., Lahmar, I.B., Boukadi, K.: Towards a meta-modeling approach for an IoRT-aware business process. In: The International Conference on Modelling and Simulation, pp. 29–35 (2022)
10. Fattouch, N., Rekik, M., Wakrime, A.A., Boukadi, K.: A model-driven engineering approach for business process based SaaS services composition. In: The 16th International Conference on Computer Systems and Applications, pp. 1–8 (2019)
11. Kheldoun, A., Barkaoui, K., Ioualalen, M.: Formal verification of complex business processes based on high-level petri nets. Inf. Sci. **385**, 39–54 (2017)
12. Koschmider, A., Laue, R., Fellmann, M.: Business process model anti-patterns: a bibliography and taxonomy of published work. In: The European Conference on Information Systems - Information Systems for a Sharing Society (2019)
13. Lehmann, B., Alexander, P., Lichter, H., Hacks, S.: Towards the identification of process anti-patterns in enterprise architecture models. In: International Workshop on Quantitative Approaches to Software Quality co-located with 27th Asia-Pacific Software Engineering Conference, vol. 2767, pp. 47–54 (2020)
14. López-Pintado, O., Dumas, M., Berx, J.: Discovery, simulation, and optimization of business processes with differentiated resources. Inf. Syst. **120**, 102289 (2024)

15. Meyer, T.: A symmetric petri net model of generic publish-subscribe systems for verification and business process conformance checking. In: International Workshop on Petri Nets and Software Engineering, vol. 3430, pp. 88–109 (2023)
16. Roa, J., Chiotti, O., Villarreal, P.: Specification of behavioral anti-patterns for the verification of block-structured collaborative business processes. Inf. Softw. Technol. **75**, 148–170 (2016)
17. Zhang, M., Feng, F., Zhang, Z., Wen, J.: A new business process verification approach for e-commerce using petri nets. Int. J. Enterprise Inf. Syst. (IJEIS) **16**(1), 92–107 (2020)

Advanced NLP Techniques for Arabic Language

Opinion Analysis Based on a Sentiment Lexical Ontology and Deep Learning Models: Tunisian Dialect Case

Tahar Alimi(✉), Rahma Boujelben, and Lamia Hadrich Belguith

ANLP Research Group - MIRACL Laboratory, Faculty of Economic Sciences and Management of Sfax, Sfax, Tunisia
alimitahar2020@gmail.com, lamia.belguith@fsegs.usf.tn

Abstract. Social networks have become a major source for the expression and dissemination of information and news in various languages. Users can express their opinions and sentiments towards different topics through written comments in dialectal languages. The objective of this work is to build a multi-class sentiment analysis model for Tunisian dialect comments scraped from social media websites, based on a lexical ontology and deep learning models. First, we created the TDCOR_Train, a multi-domain Tunisian Arabic dialect dataset. Subsequently, we developed the TDSO lexical ontology, which will be projected to automatically annotate each comment as positive, negative, or neutral. Further-more, we trained four deep learning models, namely Convolutional Neural Networks (CNN), Long Short-Term Memory (LSTM), Bi-LSTM, and RCNN (RNN + CNN) on the annotated dataset. Both RCNN and Bi-LSTM produced the best results, achieving accuracy rates of 81.27 and 82.01. To assess the performance of our sentiment analysis method, we fine-tuned the bert-base-arabertv02 transfer transformer which returned an accuracy rate of 81.17. Moreover, all models exceeded the accuracy rate of 85.00 when evaluated on two test corpora, the first is TSAC_New, and the second is TSAC_New_TDSO, which is TSAC_New dataset auto-annotated by the TDSO lexical ontology.

Keywords: sentiment analysis · deep learning · ontology · annotation · polarity

1 Introduction

Social media, a support for popular revolutions, have evolved to become an important engine for acquiring information in various domains. They have increasingly expanded beyond personal use. People communicate, share content and express their opinions on the social media websites about a large number of subjects using different forms of languages. Today, it becomes important to automate the sentiment analysis task of this heterogeneous content. Generally, a point of view can take one of the three polarities, namely positive, negative and neutral.

L. H. Belguith and K. Shaalan (Eds.): LPKM 2024, LNNS 1303, pp. 71–80, 2025.
https://doi.org/10.1007/978-3-031-85067-7_7

In recent years, resources and works on multi-class opinion analysis based on the Tunisian dialect appeared. Despite existing works, the Arabic dialect and more particularly the TD requires more annotated resources to develop any opinion analysis system. Manual annotation of large corpora can take a long time and can generate errors due to the lack of concentration. To overcome this problem, we propose in this work the TDSO (Tunisian Dialect Sentiment Ontology) to automatically annotate a multi-domain dataset of comments scraped from social media networks. Then, we will test the effectiveness of this annotation through four deep learning models, namely CNN, LSTM, Bi-LSTM and RCNN (RNN + CNN). These models were evaluated on two test corpora, the first is TSAC_New which is a multi-class manually annotated dataset, and the second is TSAC_New_TDSO, which is the TSAC_New dataset auto-annotated by the TDSO lexical ontology. Moreover, the transfer transformer bert-base-arabertv02 of [3] reached a close accuracy rate which reflects the importance of TDSO.

2 Tunisian Dialect Specificities

Tunisian Dialect (TD) often called «Darija» or «Tounsi» is the main communication language in radio, television, blogs, social media publications, marketing web-sites... etc. It can be written using the Arabic alphabet, as it can be written using the Latin alphabet. In [2], a study of 1.2M comments extracted from social media networks including 16M words and 1M unique words showed that 53% f comments use Roman characters, 87% of which are written in «Tunizi», 34% use the Arabic alphabet and 13% use the code switching. TD includes very specific words which distinguish it from other Arabic dialects such as (« باهي – bahi», «Ok»), (« برشة – barcha», «A lot»), (« عالسلامة – asslama», «Hi»), (« يزي – yezzi», «Stop» or «It is enough»)... Thus, some non-Arabic phonemes can be used in TD such as « ڨ– g», « پ – p». ...

In the TD morphological plane, the category of the dual number of the Modern Standard Arabic (MSA) does not exist. For example, in TD we use the prefix « نـ » with the first singular person (« نكتب / niktib», «I write»), so we simultaneously use the same prefix « نـ » and the suffix « و » with the plural pronoun (« نكتبو / niktbou», «We write»).

3 Related Work

In recent decades, several studies on the Arabic opinion analysis have been published. In this section, we initiate our review by exploring different sentiment analysis approaches, specifically those addressing sentiment analysis in both well-resourced and low-resourced languages. Subsequently, we delve into research dedicated to sentiment analysis that makes use of ontological resources specifically proposed for the TD.

3.1 Sentiment Analysis Based on Machine Learning Approaches

Automatic approaches are based on machine learning models without any direct human intervention.

Sentiment Analysis Based on Classic Machine Learning Models: Machine learning offers for classification tasks three primary approaches: supervised learning, semi-supervised learning and unsupervised learning. [14] used a supervised machine learning to enrich sentiment analysis resources in Arabic dialects with a dataset of 5514 tweets in MSA/TD related to the 2017 national assembly elections. They conducted a binary multi-class classification. Their study proved that the Support Vector Machine outperformed the Naïve Bayes model.

In order to standardize the automatic processing of Tunisian dialect, [9] firstly built a model to identify TD and MSA from a dataset extracted from social media networks. They subsequently created a parallel dataset containing 64K sentences with 2K manually annotated sentences. Among various classifiers, the SVM algorithm achieved the highest binary classification accuracy rate at 71.09%, surpassing Naïve Bayes, Decision Trees and Random Forest.

Sentiment Analysis based on Deep Learning Models: In recent In recent years, several researchers focused on opinion analysis in the Tunisian dialect using deep learning models. One notable contribution comes from [10], who developed a model based on three deep learning models which are CNN, LSTM and Bi-LSTM. This approach has two levels of analysis: the sentence level and the aspect level. Their study relied on a dataset of Tunisian dialect comments scraped from the official Facebook pages of various supermarkets. At the sentence level, both LSTM and Bi-LSTM scored an accuracy rate of 87%. Also, through the combined use of Bi-LSTM and CNN, at the aspect level, the models demonstrate strong performance. The aspect categorization model achieved an accuracy rate of 62% and the sentiment classification achieved an even higher accuracy rate of 78%.

3.2 Sentiment Analysis Based on Ontologies

The MSA language faced significant challenges in the field of ontological processing. As a result, numerous domain-specific ontologies have emerged. Additionally, researchers have explored the use of lexical databases referred to as 'lexical ontologies'. The significance of linguistic ontologies has grown substantially across diverse domains, including the processing of massive datasets (Big Data) [6], information retrieval and extraction [1], and data integration [5]. The major obstacle to construct an Arabic ontology is the absence of a high-level standard ontology that can serve as a reference. In [13], the authors introduced the ASO (Arabic Sentiment Ontology) which focuses on the semantic relationships between sentiments. Some researches based on ontologies have been conducted on the Tunisian Dialect. In [7] the authors have contributed to treat the Tunisian dialect in a module of understanding of dialogue systems. They used domain ontology to cover the lexicon used in all station services, and a task ontology which groups together all the tasks that can be carried out in this domain, such as the information about the train time, tickets reservation, etc.

4 Dataset Collection and Data Preprocessing

4.1 Dataset Collection

We have opted to diversify the sources of our data by incorporating various social networks namely Facebook, Twitter, Youtube, Instagram, TikTok. We used Python scripts as well as various applications and plugins. We built seven separate corpora, each one focused on a distinct domain: politics, education, e-commerce, family, sports, Tunisian General Labor Union (TGLU), and daily discussions. These corpora were combined to create the multi-domain dataset, TD-COR_Train (Tunisian Dialect CORpus) of over 30,000 comments.

4.2 Data Preprocessing

We conducted preprocessing on the multi-domain dataset to prepare it for the next process. Preprocessing involves removing ambiguities, standardization, and normalization of the writing form of letters and words. So we proceeded to remove, diacritics, punctuation marks, URLs, and extra spaces between words. We also perform a substitution for all 'alif' characters (أ, إ) by representing them in a single format ا while removing the 'hamza' character. Additionally, we eliminated all occurrences of each character that exceed two.

Transliteration of Arabizi Comments: Arabizi comments were transliterated into Tunisian Arabic script to ensure a uniform dataset written in Arabic characters. As a reference, we used the correspondence table of [8] where Line 1 indicates the phonetic representation of the graphemes, Line 2 represents the Arabic characters, and Line 3 shows the corresponding Arabish characters (Fig. 1).

1	[ðˤ]	[a :]	[ʔ]	[b]	[θ]	[ʒ]	[ħ]	[x]	[d]	[ð]	[s]	[ʃ]	[sˤ]
2	ض	ة	ء	ب	ث	ج	ح	خ	د	ذ	س	ش	ص
3	dh th d	a e h	2	b p	th	j	7 h	5 kh	d	dh	s	ch, (sh)	s
1	[a][a :]	[tˤ]	[ðˤ]	[ʕ]	[ɣ]	[q]	[k]	[l]	[m]	[n]	[h]	[w][u :]	[j][i :]
2	ا ى	ط	ظ	ع	غ	ق	ك	ل	م	ن	ه	و	ي
3	a e é è	6 t	th dh	3 a	4 gh	9 q	k	l	m	n	8, h	ou, w	i, y

Fig. 1. Correspondence Arabizi/Arabic Tunisian dialect

Words Normalization: We eliminated all diacritics and useless units such as website links. For example, (« مَعْلُومَاتٌ مُهِمَّةٌ جِدًّا عَلَى الرَّابِطِ http://www.infos.inf» «Very important information in the link http://www.infos.inf») becomes «

« معلومات مهمة جدا على الرابط » («Very important information in the link»). Comments were preprocessed based upon removing extra spaces and punctuation marks and emoticons. For example, « باش نخرج توة !! » («I will go out now») becomes « باش نخرج توة » («I will go out now»).

POS Tagging: At this stage, a grammatical category must be assigned to each term, which is referred to as morphosyntactic tagging (Part Of Speech Tagging: POS or POS-tag or simply POS-T). We experimented with several tools, including Stanza, Stanford-POSTagger, Arabycia, and the TArC dataset. The tool that proved to be the most suitable for the Tunisian dialect is TTR (Tunisian TReebank) developed by [12]. It is a treebank used to analyze comments written in the Tunisian dialect. The TTR pipeline generates a tree where POS-Tags represent the nodes, and the words are the leaves (final classes of the tree). The Fig. 2 shows an example of morphosyntactic labeling based on TTR.

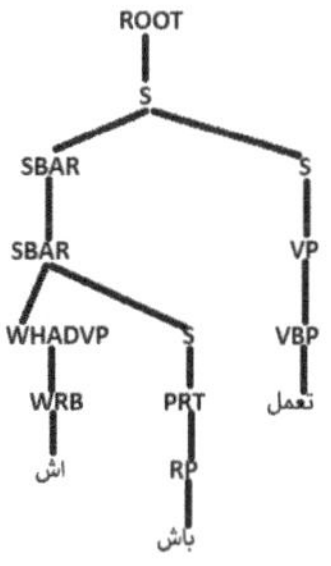

Fig. 2. An extract from TTR morphosyntactic labeling

Lemmatization: Lemmatization aims to reduce a word to its root form while retaining its meaning, unlike stemming. In this phase, we utilized the **TArC** corpus [8]. For each word found in TArC, we extracted its lemma; otherwise, we defined the word itself as its lemma. TArC successfully identified the lemmas for 37.39% of the total words within the TDCOR_train dataset.

5 Building TDSO Lexical Ontology and Data Annotation

5.1 TD Lexical Ontology

The Tunisian Dialect Sentiment Ontology (TDSO) is an automatic sentiment-based lexical ontology designed for Tunisian dialect words written in Arabic script. We constructed the TDSO lexical ontology using a dataset focused on politics, as it en-compasses the highest number of comments and words. The Fig. 3 illustrates the process of developing **TDSO** lexical ontology.

After the preprocessing phase, we determined the POS-Tag and lemma for each word, at the end of this preliminary step; each token/word/lemma should have tags and lems, as shown in the Table 1 below.

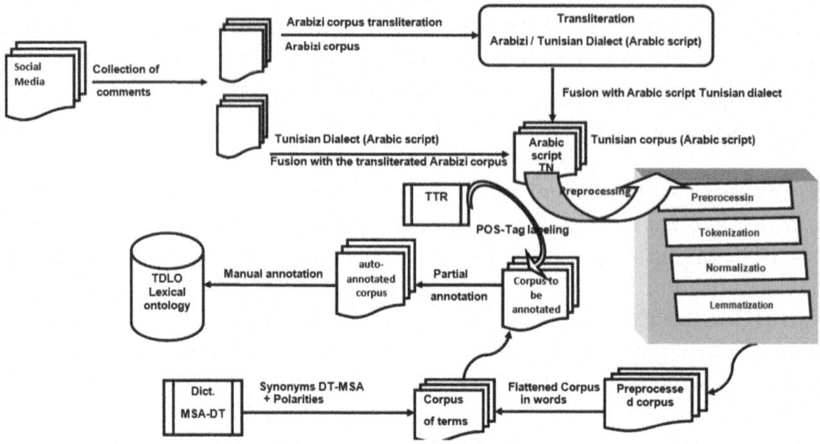

Fig. 3. Process of developing TDSO lexical ontology

Table 1. Dataset source of creation of TDSO

tokens	tags	lems
الاسلاميين	VBP	الاسلاميين
السياسة	DTNN	السياسة
اللي	IN	اللي
النهضة	NN	النهضة
او	CC	او

The next step consists on identifying for each word a list of MSA synonyms as shown in the Fig. 4. We have used the dictionaries proposed by [4] to attribute to each TD word, MSA synonyms according to the POS-Tag of each word.

Subsequently, we obtained a total of 21,162 unique words in Tunisian dialect. Within this word set, only a small percentage, 4.39% (928 out of 21,162 words), has one or more Modern Standard Arabic (MSA) synonyms. Words that are common between the TDSO lexical ontology and the dictionaries mentioned earlier retain their polarities, while in other words, polarity assignment requires manual classification as positive, negative, or neutral.

Thereafter, we generated partial graphs; each graph represents the potential relationships between pairs of nodes. It is important to note that the TDSO lexical ontology comprises five nodes: the term, its lemma, its POS-Tag, its polarity, and its list of synonyms in MSA. At the initial level, there are ten possible combinations of pairs of nodes, resulting in ten partial graphs. After obtaining partial graphs in the form of trees, we proceed to merge them, creating a graph that represents the entire data-frame. The fusion process occurs between any two graphs that share a common node.

In the following figure, an example presented in the purple color which illustrates a relationship between nodes in the final graph. In fact, the word راجل, Man which is neutral have the word رَجُلٌ, Man as a synonym in MSA. It has the same meaning as the plural word رجال Men and "NNS" as a POS-Tag (noun, common, plural).

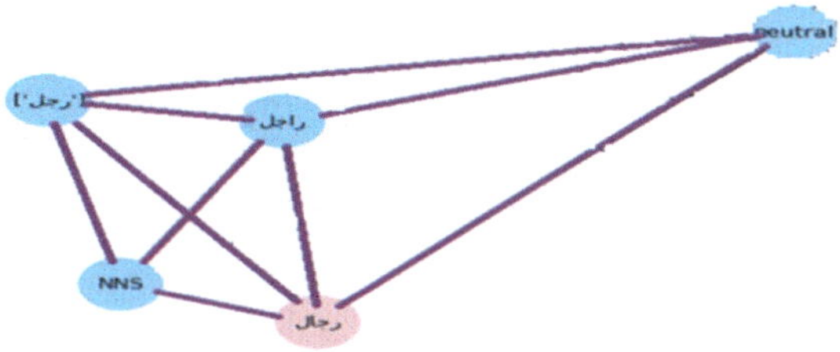

Fig. 4. An extract of TDSO final Graph

TDSO lexical ontology is the final graph resulting from the previous step converted into an XML, GML, GRAPHML, ... file whose all formats can be used as lexical bases (lexical ontologies). The Fig. 5 shows an extract from TDSO lexical ontology in XML format which contains 21,162 entries.

```xml
<row>
  <index>45</index>
  <tokens>موجودة</tokens>
  <tags>JJ</tags>
  <lems>موجود</lems>
  <syn_MSA>['موجود']</syn_MSA>
  <polarity>neutral</polarity>
</row>
<row>
  <index>46</index>
  <tokens>التراكن</tokens>
  <tags>NN</tags>
  <lems>التراكن</lems>
  <syn_MSA>[]</syn_MSA>
  <polarity>neg</polarity>
</row>
<row>
  <index>47</index>
  <tokens>و</tokens>
  <tags>CC</tags>
  <lems>و</lems>
  <syn_MSA>['و']</syn_MSA>
  <polarity>neutral</polarity>
</row>
<row>
```

Fig. 5. An extract of TDSO lexical ontology

5.2 Evaluation of TDSO Lexical Ontology

We utilized the TSAC bi-polar dataset of [11]), originally annotated into positive and negative classes with a distribution of 51.88% for the positive polarity and 48.12% for the negative polarity. To evaluate TDSO lexical ontology, we added 1K manually annotated neutral comments written in Tunisian dialect

(Arabic script) to the TSAC dataset to have a new multi-class dataset named TSAC_New. TSAC_New contains 117175 duplicated words (including smileys) of which 24864 unique words.

Given that TDSO contains 21,116 unique words, we were able to cover approximately 32.94% of the words with the TDSO lexical ontology, which has a significant impact on the annotation rate.

The similarity between the original TSAC_New comments annotation and the polarization via the TDSO lexical ontology is 8850 out of 18067 (47.49%). This rate is particularly significant considering that TSAC_New contains more unique words than TDSO, and TDSO only covered 32.94% of the words in TSAC_New. The true classification into positive, negative, and neutral polarity reached respectively the rates of 20.72%, 82.14% and 18.30%.

5.3 Annotation

Term Annotation: Identifying the polarities of words in any comment relies on the polarities assigned to the words in the TDSO lexical ontology. We observed that the automatic annotation of terms reached a rate of 47.46%. However, it is crucial to remember that we projected a domain-specific lexical ontology, originally created on the political dataset, into other multi-domain corpora. Some words may appear both in the ontology and the annotated dataset, which inherently influences the coverage rate of TDSO. Consequently, these closely overlapping rates hold significant importance.

Comment Annotation: This phase involves the automatic annotation of Tunisian dialect comments using the TDSO lexical ontology based on the polarities of the words within a comment. We classify a comment as neutral if it doesn't contain any positive or negative words. Otherwise, if a comment has more positive words than negative words, it is considered positive; otherwise, it is considered negative. This process was divided the multi-domain (TDCOR_Train) into 18.18% for positive, 40.88% for negative and 40.93% for neutral comments.

6 Sentiment Analysis Based on TDSO

6.1 Our Proposed Method

During the opinion analysis process, we chose to use approaches based on deep learning models. These models are based on artificial networks called neural networks. The advantage is to be able to use several layers (hidden and/or visible) which guarantees to improve results compared to traditional machine learning models.

6.2 Deep Learning Models

For deep learning models, we tried to progress from the simplest such as LSTM (Long Short-Term Memory) and CNN (Convolutional Neural Networks) to the most sophisticated namely Bi-LSTM and RCNN (RNN +CNN).

6.3 Experiments and Discussion

As previously mentioned, during the opinion analysis process, we tested four deep learning models: CNN, LSTM, Bi-LSTM, and RCNN. Each model was applied to all the corpora. The complete process of a model is divided into two parts: the first is the model compilation and training phase and the second is the prediction phase.

Learning Phase: All models were compiled using the categorical_crossentropy loss function since we are dealing with multi-class classification. After the compilation, each model must be trained to be able to classify the input test data.

Predicting Phase: The four deep learning models mentioned above were tested on the same manually annotated test dataset, TDCOR_Test, which contains 7K comments. The distribution of negative, neutral, and positive comments is as follows: 63.67% negative, 09.20% neutral, and 27.13% positive. The Bi-LSTM model achieved the highest accuracy rate of 82%.

Test Performance: To assess the performances of our approach, we made an evaluation in two ways. First, we fine-tuned the bert-base-arabertv02 transfer transformer which returned an accuracy rate of 81.17 after 10 epochs with a batch size of 32 during training and evaluation, and a weight_decay of 0.01. Second, we utilized two multi-class corpora; the first is the TSAC_New corpus mentioned above, and the second is TSAC_New_TDSO which is the TSAC_New corpus auto-annotated by projecting on it the TDSO lexical ontology. It is noteworthy that the two models, Bi-LSTM and RCNN, delivered the best results across all corpora, achieving accuracy rates of 86.36% and 86.53%, respectively, on TSAC_New and the accuracy rates of 86.21% and 85.89%, respectively, on TSAC_New_TDSO.

After conducting these tests, we can conclude that the training dataset annotated with TDSO proved to be beneficial for developing a sentiment analysis model. Consequently, we can rely on ontology for the automatic annotation of corpora and continue to expand its application to other Arabic dialects.

7 Conclusion and Perspectives

In this paper, we presented an approach of opinion analysis of Tunisian dialect texts. We focused on the detection of the polarities based on TDSO lexical ontology and deep neural networks. So, we built and preprocessed a multi-domain dataset by collecting 30,347 comments from social media networks. As the political dataset contains more vocabulary, we considered it for building TDSO sentiment ontology. Then, we projected it to annotate the whole TDCOR_Train dataset into three classes. To evaluate the effectiveness of this approach, we fine-tuned the bert-base-arabertv02 transfer transformer which reached a very close score. Moreover, we tested the four deep learning models (CNN, LSTM, Bi-LSTM, and RCNN) on two manually annotated corpora. We noted that all used models yielded very close results. This reflects, on the one hand, the importance and richness of the present domain lexical ontology. On the other hand, the close accuracy and polarization rates across various corpora affirm that the lexical ontology should not be limited to a particular domain but should cover a specific natural language. For future works, first, we must reinforce the TD-COR_Train corpus and the TDSO lexical ontology to cover various domains. Second, we need to work with other Arabic dialects.

References

1. Abderrahim, M.A., Chikh, M.A.: Using Arabic wordnet for semantic indexation in information retrieval system, p. 7 (2013)
2. Abidi, K.: La construction automatique de ressources multilingues à partir des réseaux sociaux: application aux données dialectales du maghreb, p. 138 (2019)
3. Antoum, W., Baly, F., Hajj, H.: Arabert: transformer-based model for Arabic language understanding, p. 07 (2020)
4. Boujelbane, R., Khemekhem, M.E., Belguith, L.H.: Mapping rules for building a Tunisian dialect lexicon and generating corpora, p. 10 (2013)
5. Coletta, R., Castanier, E., Valduriez, P., Frisch, C., Ngo, D., Bellahsene, Z.: Public data integration with websmatch. In: Proceedings of the First International Workshop on Open Data - WOD 2012, p. 5. ACM Press, Nantes, France (2012). https://doi.org/10.1145/2422604.2422606
6. Gayo, J.E.L., Kontokostas, D., Auer, S.: Multilingual linked open data patterns, p. 19 (2012)
7. Graja, M., Jaoua, M., Belguith, L.H.: Building ontologies to understand spoken Tunisian dialect. IJCSEA **1**(4), 23–32 (2011). https://doi.org/10.5121/ijcsea.2011.1403
8. Gugliotta, E., Dinarelli, M.: Tarc un corpus d'arabish tunisien, p. 9 (2020)
9. Kchaou, S., Boujelbane, R., FsiH, E., Belguith, L.H.: Standardization of dialect comments in social networks in view of sentiment analysis: case of Tunisian dialect, p. 8 (2022)
10. Masmoudi, A., Hamdi, J., Belguith, L.H.: Deep learning for sentiment analysis of Tunisian dialect. CyS **25**(1) (2021). https://doi.org/10.13053/cys-25-1-3472
11. Medhaffar, S., Bougares, F., Estève, Y., Hadrich-Belguith, L.: Sentiment analysis of Tunisian dialects: linguistic resources and experiments. In: Proceedings of the Third Arabic Natural Language Processing Workshop, pp. 55–61. Association for Computational Linguistics, Valencia, Spain (2017). https://doi.org/10.18653/v1/W17-1307
12. Mekki, A., Zribi, I., Ellouze, M., Belguith, L.H.: Treebank creation and parser generation for Tunisian social media text. In: 2020 IEEE/ACS 17th International Conference on Computer Systems and Applications (AICCSA), pp. 1–8. IEEE, Antalya, Turkey (2020). https://doi.org/10.1109/AICCSA50499.2020.9316462
13. Samir, T., Ibrahim, A.N.: Semantic sentiment analysis in Arabic social media.pdf (2017)
14. Sayadi, K., Liwicki, M., Ingold, R., Bui, M.: Tunisian dialect and modern standard Arabic dataset for sentiment analysis: Tunisian election context, p. 11 (2016)

Disfluent-to-Fluent Tunisian Dialect Speech Translation with Fine-Tuning Pre-trained Language Models

Emna Boughariou[1(✉)], Younés Bahou[2], and Lamia Hadrich Belguith[1]

[1] Sfax University, Sfax, Tunisia
emnaboughariou@gmail.com
[2] HA'IL University, Hail, Kingdom of Saudi Arabia

Abstract. In contrast to written texts and prepared speeches, conversational/spontaneous speech has a very high degree of freedom and includes a huge number of disfluencies. Detecting disfluencies using transformer-based models has advanced state-of-the-art performance. In this work, we aim to process disfluencies in the spontaneous tunisian dialect speech by generating fluent utterances from disfluent transcripts. We propose a transformer-based model by fine-tuning the pre-trained T5 language model. Using this model, we achieved an F-Measure score of 74,71% based on the evaluation data set part of DisCoTAT.

Keywords: Disfluencies · Tunisian dialect · spontaneous transcriptions · transformers · Fine-Tuning

1 Introduction

In this work, we investigate the field of processing the Tunisian Dialect (TD) spontaneous speech by proposing an effective approach for disfluencies removing from hand-transcribed utterances.

The TD is a subgroup of Arabic dialects usually identified with Maghreb Arabic. It is the official mother tongue spoken by all Tunisians, regardless of their origin and social affiliation. Nevertheless, the TD is not standardized or taught, and has no official status. Most Arabic native speakers are unable to produce a sustained spontaneous speech in Modern Standard Arabic (MSA); in unwritten situations where spoken MSA would normally be expected. Hence, most Tunisian speakers typically resort to frequent code-switching between their dialect and MSA in their daily life such as in talk shows on radio and TV channels [13].

Regarding this disturbance in spoken TD speech, disfluencies can occur frequently in all forms of spontaneous speech, whether casual discussions or formal arguments [21]. Disfluencies are disruptions to the regular flow of speech, typically occurring in spontaneous speech. These irregularities in the text can alter its syntactic and semantic structure, thereby adversely affecting the performance

L. H. Belguith and K. Shaalan (Eds.): LPKM 2024, LNNS 1303, pp. 81–90, 2025.
https://doi.org/10.1007/978-3-031-85067-7_8

of downstream NLP tasks, such as machine translation, question-answering, and summarization.

Most of the existing NLP tools, designed for the TD processing, are developed for well-formed fluent text without considerations of disfluencies. Therefore, in spite of their very high accuracy on fluent text, utilizing them for solutions on disfluent text is relatively less accurate.

The problems pertaining to disfluency processing in TD can be broadly summarized as (but not limited to): *a)* written language difficulties like enumeration, synonymy, voluntary repetitions, etc. *b)* spontaneous speech difficulties like pronunciation variability, out-of-vocabulary words, utterances segmentation, etc. *c)* TD specificity's like agglutination, reuse of borrowed vocabulary, syntactic dependencies, etc.

A preliminary version of this work was presented by Bougharìou et al. in [4] and [5]. Our current study aims to propose an alternative model for processing complex disfluencies in spontaneous spoken TD. The model is fine-tuned using the pre-trained T5 language model based on the transformer architecture with attention [16]. To the best of our knowledge, this is the first work that takes into account disfluency processing in TD using transformer-based models.

The rest of this paper is organized as follows. Section 2 reviews related work. Section 3 introduces the proposed method. Section 4 presents the details of the experiments. Section 5 presents and discusses the results of the evaluation. And, Sect. 6 concludes our work and points to interesting future research directions.

2 Related Work

Detecting disfluencies in spoken TD is rarely studied by researchers. Authors in [15] adopted a symbolic rule-based approach to delimit and correct disfluencies automatically using a set of rules and patterns, in a very restricted domain and with a limited TD vocabulary. Authors in [23] proposed transcription conventions to annotate disfluencies in TD corpora that are used later to annotate manually only incomplete words, repetitions and onomatopoeia words in the TD corpus STAC [22].

For other languages, several approaches have been proposed for handling speech disfluencies. These approaches are divided into four significant categories: *a)* noisy-channel methods, *b)* transition-based, *c)* sequence tagging, and *d)* translation-based. Noisy-channel models [9,12] detect disfluencies by computing the similarity between the reparandum and the repair. Transition-based approaches concurrently detect both the syntactic structure and the disfluencies of the input sequence [19,20]. These approaches require corresponding annotated training datasets that include both syntactic and disfluency annotations. Sequence tagging, which is the most common approach for disfluency detection, classifies each token of the input sequence as fluent or disfluent. This approach includes a wide range of different models varying from Hidden Markov Models [10], Conditional Random Fields (CRFs) [8], LSTMs [11], RNNs [21], to more recent Transformer-based architectures [7,17].

Our work falls into the translation-based methods approach. Disfluencies processing is converted to a Machine Translation (MT) task that translates a disfluent input (the source language) into a fluent sequence (the target language).

3 Method

According to [3], disfluencies in TD are classified into two main types: Simple Disfluencies (SDs) and Complex Disfluencies (CDs). SDs affect only one minimal token, they include syllabic elongations, speech words, word-fragments, and simple repetitions. CDs affect several tokens and may break the morphological flow of the utterance; they include complex repetitions, insertions, substitutions, and deletions. Table 1 and Table 2 present examples of SDs and CDs in TD.

Table 1. Examples of SDs in TD

SD type	Example
Syllabic elongations	صوووتك موش واضح [Swwwtk mw$ wADH] (Your voiiiice is not clear)
Speech words	يعني [y'ny] (meaning)
Word-fragments	وقتاش يبدا ال المؤتمر [wqtA$ ybdA Al AlmWtmr] (When the the-conference starts)
Simple repetitions	ثلاثة أمم ثلاثة وزراء [vlAvp Omm vlAvp wzrA'] (Three emm three ministers)

In [4], SDs are removed automatically using a rule-based approach using hand-crafted rules and on the basis of morpho-syntactic and semantic features. The purpose of this work is to present an alternative model for removing CD in the spontaneous spoken TD. Our model imitate the MT sequence-to-sequence (seq2seq) principle; it consists of translating a disfluent sequence as the source language to a fluent output as the target language. We propose to build this disfluent-to-fluent model on the basis of the transformer architecture. Indeed, the Transformer architecture was presented in 2017 to obtain high-performance MT systems [18]. Transformers use an attention mechanism without resorting to any recurring or conventional network. This technology allows the modelling of dependencies without considering their distances in the input or output sequences. Otherwise, the transformer architecture aims to solve seq2seq tasks while easily managing long-range dependencies. Thus, the attention mechanism

Table 2. Examples of CDs in TD

CD type	Example
Complex repetitions	علاش هذا التبذير علاش [ElA$ h*A Altb*yr ElA$] (Why this extravagance why)
Insertions	عندك مثال كان عندك مثال [Endk mvAl kAn Endk mvAl] (you have an example if you have an example)
Substitutions	موضوع يهم المسنين أهه سامحني الشباب [mwDwE yhm Almsnyn Ohh sAmHny Al$bAb] (A topic that interests the elderly euuh forgive me the youth people)
Deletions	تران نورمال إلي يخرج تو تران إلي يخرج تو [trAn nwrmAl Ily yxrj tw trAn Ily yxrj tw] (Normal train that leaves now train that leaves now)

allows focusing on parts of the input sequence while the model predicts the output sequence. It allows us to trace a correlation between the input and output words. For example, if the model predicts the word " فيروس [fayrus] (virus)", it would be very likely to find a high weight for the word " كورونا [kwruna] (corona)" in the input sequence.

However, designing a complex neural model like the Transformer from scratch is not an easy task and can present several challenges (e.g., the high cost of the data amount required and the computational resources required). With this attention, we proceeded to the Transfer Learning (TL) technique in order to build our Transformer-based neural seq2seq model for CDs removal. We fine-tuned the T5 (Text-to-Text Transfer Transformer) pre-trained model [16] in order to remove CDs from utterances in the TD.

4 Model Setup

Modern pre-trained language models based on transformers were trained on huge corpora. Recent literature is reporting considerable success on a variety of benchmark tasks with these models such as summarization, question-answer generation, translation, etc. We expect that disfluencies can also be captured by the pre-trained language models when fine-tuned to the task of disfluency processing in TD. Figure 1 presents the steps of the fine-tuned model adaptation.

4.1 Difluent-Fluent Dataset

We built a parallel corpus composed of utterance pairs (Disfluent, Fluent) where disfluent utterances are the source language and fluent utterances are the target

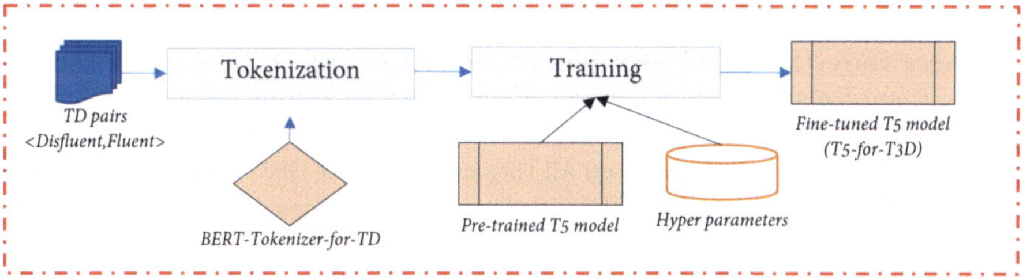

Fig. 1. Fine-tuning the T5 language model to TD

language. Disfluent utterances are extracted from the TD corpus DisCoTAT [3], and they contain only CDs phenomena (SDs are automatically removed using the approach proposed by [4]). Fluent utterances are the result of manually removing all types of disfluencies from the disfluent utterances set (Disfluency detection is guided by annotation rules presented in previous work). Table 3 resumes the dataset statistics.

Table 3. Disfluent-Fluent dataset statistics

Utterance pairs	2926
Complex repetitions	647
Insertions	226
Substitutions	396
Deletions	106

4.2 Tokenization

We built a subword-based tokenizer, called "BERT-tokenizer-for-TD", which consists of splitting utterance pairs into words and sub-words. Since the TD is an agglutinative language, the tokenizer relies on the principle that frequently used words should not be split into smaller subwords, but rare words should be decomposed into meaningful subwords. Subword tokenization allows the model to have a reasonable vocabulary size while being able to learn meaningful context-independent representations. In addition, subword tokenization enables the model to process words it has never seen before, by decomposing them into known subwords. The tokenizer is fine-tuned using the pre-trained BERT model [6], through the Hugging-Face platform[1], in order to improve the robustness of the MT model against TD vocabulary. Indeed, the initial vocabulary of BERT

[1] https://huggingface.co/. Hugging Face is an NLP-focused library with a large open-source community, around the Transformers library.

does not contain the TD lexicon and therefore, it will not allow dividing the utterances correctly. BERT is a bidirectional transformer pre-trained using a combination of masked language modelling objective and next sentence prediction on a large corpus. It makes use of a WordPiece algorithm that breaks a word into several subwords. We collected all the utterances of DisCTAT and other TD transcripts [1] for building the BERT-tokenizer-for-TD vocabulary.

4.3 Model Training

Inside our translation-based approach, our model is implemented using PyTorch on top of the HuggningFace transformers library and based on T5 architecture [16]. We use the pre-trained T5 language model for translating disfluent TD transcriptions to fluent sequences. T5 is an encoder-decoder model pre-trained on a multi-task mixture of unsupervised and supervised tasks and for which each task is converted into a text-to-text format.

As a common practice, we split the Disfluent-Fluent pairs into a training set (80%) and a validation set (20%). Training and validation datasets are tokenized using the BERT-tokenizer-for-TD language model with a max-length = 64 feature. It returns a dictionary containing the token ids and attention masks of the inputs, and the token ids of the labels. For fine-tuning, we used T5ForConditionalGeneration and we tried t5-small, t5-base and t5-large model implementations and found slightly better performance with the base model. The T5-base model consists of 12 encoder and decoder layers with 12 attention heads as well. We fine-tune all the models for 8 epochs with learning rate 2e-5 and batch size 16 for both training and validation sets and with the AdamW optimizer. The proposed model is called 'T5-for-T3D' (T5 for Tunisian Dialect Disfluency Detection).

5 Evaluation

We conduct two different types of experiments, examining: *a)* the fact that we treat the disfluency detection problem as a translation task, *b)* the comparison of the T5-for-T3D model with other models proposed for the disfluencies processing in TD. Evaluations are conducted using the evaluation dataset part of DisCoTAT. It consists of 732 disfluent utterances composed of 951 disfluency phenomena.

5.1 Translation-Based Disfluency Detection Model

We take advantage of the BLEU score to measure the performance of our translation model T5-for-T3D. First, we evaluated the utility of using the BERT-tokenizer-for-TD language model instead of the BERT baseline tokenizer for the parallel corpus tokenization. We recall that we chose the BERT tokenizer because it uses a word piece algorithm which gives better results for agglutinative languages such as TD. Second, we carried out a translation model version where disfluent utterances input contains CDs as well as SDs. Figure 2 shows that fine-tuning the t5-base language model with the BERT-tokenizer-for-TD tokenizer made for TD gives better results.

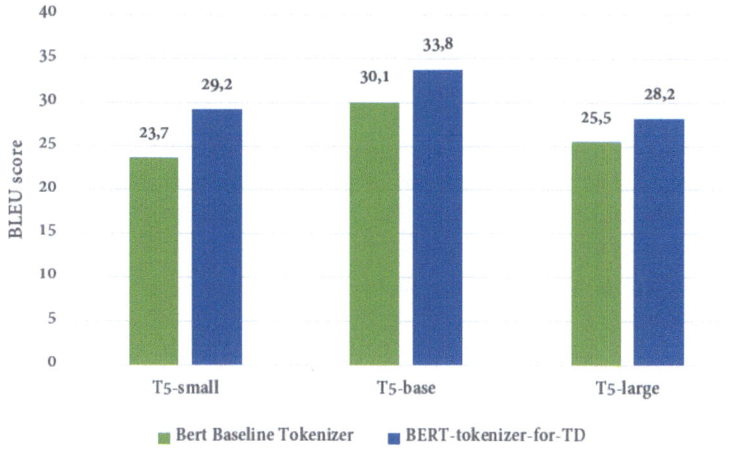

Fig. 2. Experiments results using the BERT-tokenizer-for-TD

Table 4 demonstrates that the fact of removing SDs in a previous step improves the CDs detection. We note that speech words are not removed with the other SD types as they help to detect CDs [4].

Table 4. The impact of removing SDs before detecting CDs

Model	BLEU score
With removing SDs	33.8
Without removing SDs	26.7

5.2 Comparison of Three Disfluency Detection Models for TD

This work presents an alternative model for detecting CDs in TD. Two other CDs detection models were proposed in previous works [4] and [5].

The first model is sequence-tagging-based. It classifies the utterance's tokens into six classes based on the reparandum-interregnum-repair structure and following the BIO encoding. Tokens can be labelled as B_RM (i.e., the beginning of the reparandum), I_RM (i.e., belongs to the reparandum part), B_RP (i.e., the beginning of the repair part), I_RP (i.e., belongs to the repair part), IP (i.e., Interregnum (i.e., belongs to speech words), or O: (i.e., fluent word). The sequence-tagging-based model is handled with the statistical CRF algorithm [2] and using morphosyntactic and other linguistic features.

The second model is transition-based. It detects and removes only the disfluent chunk of the utterance (i.e., reparandum and interregnum) with ignoring the rest of the utterance's words (i.e., fluent words, and repair) using a set of transition actions and without syntactic dependencies. The transition actions are selected using a binary classifier transitions model based on Support Vector Machine (SVM) algorithm [14] with the same features used in the sequence-tagging-based model.

We made a comparison study for the three CDs detection models using the F-Measure score (the harmonic mean of precision and recall metrics). Table 5 projects the evaluation rates of the CDs detection models in TD.

Table 5. Evaluation results of CDs processing in TD

	Sequence-tagging-based	Transition-based	Translation-based
Complex repetitions	89,10%	86,71%	81,50%
Insertions	76,13%	80,42%	71,95%
Substitutions	80,74%	80,92%	77,56%
Deletions	69,91%	71,21%	67,83%

While comparing different models, we found that the sequence-tagging-based and transition-based models give better results, they access the internal structure of a disfluency by identifying the reparandum, the repair and the interregnum boundaries. Indeed, the transition-based model has the best performance. The experiments have shown that the erroneous detection of the repair leads to degrading the performance of the delimitation task of the disfluent segment, insofar as detecting only the reparandum part is broadly adequate.

However, the main reason that explains the drop in the F-measure score for the deletion type disfluencies lies in the fact that these phenomena are quite frequent in the training corpus. Deletions present only 3.62% of the disfluencies types distribution.

Also, we experiment with the different models in which speech words are deleted as part of the SDs processing previous step in order to measure the impact of speech words in detecting CSs. F-measure scores for the CDs detection models were degraded by 17.55% (Sequence-tagging-based model), 16.6% (Transition-based model) and 21.3% (Translation-based model). Therefore, we prove the hypothesis that speech words influence the performance of CDs detection since they constitute the interregnum.

6 Conclusion and Future Work

This work investigated the disfluencies processing issue in TD. Disfluencies, especially CDs, lead to break the morphological and sysntacical flow of the spontaneous speech. We presented, in this paper, a transformer-based model by fine-tuning the pre-trained T5 language model. The model is inspired from the machine translation approach, it translates disfluent utterances to fluent sequences in TD transcripts. Also, we fine-tuned the BERT language model to build a TD tokenizer. The translation-based model is compared to a sequence-tagging-based model and a transition-based model, which are two other CDs processing models proposed for the TD. Evaluation show that the transition-based model gives better results.

Future work could extend the proposed method to other languages, where input data may be disfluent TD as well as MSA or other Arabic dialects transcriptions. We also aim to use the textual data augmentation techniques to expand our dataset.

References

1. Abdallah, N.B., Kchaou, S., Bougares, F.: Text and speech-based Tunisian Arabic sub-dialects identification. In: Proceedings of the 12th Language Resources and Evaluation Conference, pp. 6405–6411 (2020)
2. Alharbi, S., Hasan, M., Simons, A.J., Brumfitt, S., Green, P.: Sequence labeling to detect stuttering events in read speech. Comput. Speech Lang. **62**, 101052 (2020)
3. Boughariou, E., Bahou, Y., Hadrich Belguith, L.: Linguistic resources construction: towards disfluency processing in spontaneous Tunisian dialect speech. In: International Conference on Text, Speech, and Dialogue, pp. 316–328. Springer, Cham (2019)
4. Boughariou, E., Bahou, Y., Hadrich Belguith, L.: Classification based method for disfluencies detection in spontaneous spoken Tunisian dialect. In: Proceedings of SAI Intelligent Systems Conference, pp. 182–195. Springer, Cham (2020)
5. Boughariou, E., Bahou, Y., Hadrich Belguith, L.: Detecting speech disorders using a machine-learning guided method in spontaneous Tunisian dialect speech. SN Comput. Sci. **5**(5), 440 (2024)
6. Devlin, J., Chang, M.W., Lee, K., Toutanova, K.: BERT: pre-training of deep bidirectional transformers for language understanding. arXiv preprint arXiv:1810.04805 (2018)
7. Dong, Q., Wang, F., Yang, Z., Chen, W., Xu, S., Xu, B.: Adapting translation models for transcript disfluency detection. In: Proceedings of the AAAI Conference on Artificial Intelligence, vol. 33, pp. 6351–6358 (2019)
8. Georgila, K.: Using integer linear programming for detecting speech disfluencies. In: Proceedings of Human Language Technologies: The 2009 Annual Conference of the North American Chapter of the Association for Computational Linguistics, Companion Volume: Short Papers, pp. 109–112. Association for Computational Linguistics (2009)
9. Johnson, M., Charniak, E.: A TAG-based noisy-channel model of speech repairs. In: Proceedings of the 42nd Annual Meeting of the Association for Computational Linguistics (ACL 2004), Barcelona, Spain, pp. 33–39 (2004). https://www.aclweb.org/anthology/P04-1005
10. Liu, Y., Shriberg, E., Stolcke, A., Hillard, D., Ostendorf, M., Harper, M.: Enriching speech recognition with automatic detection of sentence boundaries and disfluencies. IEEE Trans. Audio Speech Lang. Process. **14**(5), 1526–1540 (2006)

11. Lou, P.J., Anderson, P., Johnson, M.: Disfluency detection using auto-correlational neural networks. arXiv preprint arXiv:1808.09092 (2018)
12. Lou, P.J., Johnson, M.: Disfluency detection using a noisy channel model and a deep neural language model. arXiv preprint arXiv:1808.09091 (2018)
13. Masmoudi, A., Bougares, F., Khmekhem, M.E., Estève, Y., Hadrich Belguith, L.: Automatic speech recognition system for Tunisian dialect. Lang. Resour. Eval. **52**(1), 249–267 (2018)
14. Mathur, A., Foody, G.: Multiclass and binary SVM classification: implications for training and classification users. IEEE Geosci. Remote Sens. Lett. **5**, 241–245 (2008). https://doi.org/10.1109/LGRS.2008.915597
15. Neifar, W., Bahou, Y., Graja, M., Jaoua, M.: Implementation of a symbolic method for the Tunisian dialect understanding. In: Proceedings of 5th International Conference on Arabic Language Processing, Oujda, Maroc (2014)
16. Raffel, C., et al.: Exploring the limits of transfer learning with a unified text-to-text transformer. J. Mach. Learn. Res. **21**(140), 1–67 (2020). http://jmlr.org/papers/v21/20-074.html
17. Saini, N., Khatri, J., Jyothi, P., Bhattacharyya, P.: Generating fluent translations from disfluent text without access to fluent references: IIT Bombay@ IWSLT2020. In: Proceedings of the 17th International Conference on Spoken Language Translation, pp. 178–186 (2020)
18. Vaswani, A., et al.: Attention is all you need. In: Advances in Neural Information Processing Systems, vol. 30 (2017)
19. Wang, S., Che, W., Zhang, Y., Zhang, M., Liu, T.: Transition-based disfluency detection using LSTMs. In: Proceedings of the 2017 Conference on Empirical Methods in Natural Language Processing, pp. 2785–2794 (2017)
20. Yoshikawa, M., Shindo, H., Matsumoto, Y.: Joint transition-based dependency parsing and disfluency detection for automatic speech recognition texts. In: Proceedings of the 2016 Conference on Empirical Methods in Natural Language Processing, pp. 1036–1041 (2016)
21. Zayats, V., Ostendorf, M., Hajishirzi, H.: Disfluency detection using a bidirectional LSTM. arXiv preprint arXiv:1604.03209 (2016)
22. Zribi, I., Ellouze, M., Hadrich Belguith, L., Blache, P.: Spoken Tunisian Arabic corpus "STAC": transcription and annotation. Res. Comput. Sci. **90**, 123–135 (2015)
23. Zribi, I., Graja, M., Khmekhem, M.E., Jaoua, M., Hadrich Belguith, L.: Orthographic transcription for spoken Tunisian Arabic. In: International Conference on Intelligent Text Processing and Computational Linguistics, pp. 153–163. Springer, Cham (2013)

BERT-Based Model for Sarcasm Detection in Arabic Texts

Amal Mezghani(✉), Rahma Boujelben, and Mariem Ellouze

ANLP Research Group, MIRACL Lab, FSEGS, University of Sfax, Sfax, Tunisia
mezghani.amal@gmail.com, mariem.ellouze@fsegs.usf.tn

Abstract. In the current digital era, social networks have become privileged spaces of expression for individuals, allowing them to share their opinions. However, the informal and often ambiguous nature of communications on these platforms can make it difficult to interpret the tone and intent of messages. In particular, the form of irony, sarcasm, can be particularly tricky to pin down. The automatic detection of sarcasm in Arabic texts is a recent area of research that has experienced significant growth in recent years. Indeed, understanding sarcasm is essential for an accurate analysis of feelings and opinions expressed on social networks. Our article is part of this dynamic by presenting our research work on the detection of sarcasm in tweets in dialectal Arabic. We developed a corpus of annotated data, composed of 19669 tweets labeled according to their sarcastic polarity. This corpus was then used to train and evaluate different classification models based on the BERT (Bidirectional Encoder Representations from Transformers) architecture. Our results indicate that the **Arabert-V2-twitter model**, finely tuned to our corpus of data, achieves an accuracy score of **81.24%** for sarcasm classification.

Keywords: Sarcasm detection · BERT · Arabic text

1 Introduction

Sarcasm is a form of irony where the opposite of what is meant is expressed, often to criticize or mock. Detecting sarcasm requires understanding context and linguistic clues. Recent research, such as by Ashwitha et al. [1], highlights the importance of context and emotion in sarcasm detection, demonstrating the capabilities of deep language processing models. Focusing on sarcasm detection in languages other than English, such as Arabic, presents unique challenges due to its complex grammar and prevalent use in online discussions. In Arab cultures, sarcasm is a common form of humorous and critical expression, especially in sensitive discussions about politics, religion, and society on social media. Accurate sarcasm detection can prevent misunderstandings and conflicts among users, emphasizing the need for a context-specific approach that considers linguistic diversity. Sarcasm significantly impacts sentiment analysis on social media, as it can mask true feelings and affect user interactions. Properly understood sarcasm can enhance community engagement and lively exchanges. Given the vast

L. H. Belguith and K. Shaalan (Eds.): LPKM 2024, LNNS 1303, pp. 91–100, 2025.
https://doi.org/10.1007/978-3-031-85067-7_9

amount of social media content, advanced text classification methods have been developed to improve sarcasm detection accuracy, which is crucial for precise sentiment analysis. The article begins with a literature review on sarcasm detection in Sect. 2, summarizing previous work, existing approaches, and relevant results, thereby contextualizing the current research within the field's advancements. Section 3 provides a detailed description of the dataset used in the study, highlighting its essential characteristics. In Sect. 4, the experimental details are discussed, covering the methodology employed to train the models, including data preprocessing techniques, selected model architectures, chosen hyperparameters, and adopted evaluation methods. Finally, Sect. 5 offers a conclusion that summarizes the results obtained from the study and proposes avenues for future research.

2 Related Work

Detecting sarcasm in social media, particularly in the Arabic language, has received increasing attention in recent years. In light of the results obtained by Abu Farha et al. [4], these models were tested on the AraSarcasmV2 dataset, and it was observed that the Marbert model stood out by obtaining the best F1-sarcastic score, reaching 0.584. Transformer-based models have demonstrated the ability to understand relationships between words even when they are separated by many other words, allowing them to capture the complex contexts of language. Thus, Abu Farha et al. [4] found that models based on language-specific transformers outperform multilingual models in sarcasm detection. Based on related works in this field, Hazem et al. [10] presented an innovative approach to improve sarcasm detection in the Arabic language using data augmentation techniques, contextual word integration, and the random forest model. They achieved a precision of 0.5189 in the task of sarcasm and sentiment detection in Arabic using the ArSarcasmV2 shared dataset [10]. Additionally, the work of Abuzayed et al. [6] has enriched this field by using BERT-based models, tuned fine-tuning, and enriched by data augmentation techniques to address data imbalance issues. Their MARBERT model showed a significant F-score improvement of 15% for sentiment analysis and sarcasm detection tasks on the AraSarcasmV2 dataset [10]. Per the findings of Rahma et al. [7] they conducted a study using the AraSarcasmV2 dataset [10]. In their study, a summary table of the evaluation metrics was presented as shown in Table 1. In this field, researchers like Shah et al. [15] are focusing on text classification in NLP. This task is often challenging due to the diversity of dialects, the presence of sarcasm, and the subjectivity of sentiments. To tackle these challenges, these researchers propose a modified transformer model called MST (Modified Switching Transformer). This model is evaluated on the ArSarcasm dataset and yields results that outperform other approaches such as Naive Bayes and Random Forest Classifier. These results reach up to an 83% F_measure for sarcasm detection. In their work Mansour et al. [13] evaluated several contextualized models for sarcasm detection using the ArSarcasmV2 and Saudi dialect Irony datasets, finding that the AraBERTv2 model achieved an F1 score of 72.15% on the ArSarcasmV2 dataset. This latest model was also tested on TunCorSS (Tunisian Corpus for Sentiment Analysis and Sarcasm Detection) and achieved the highest accuracy results for sarcasm detection with a score of 0.865 in the work of Asma et al. [14]. Despite the progress made, accurate detection of sarcasm in Arabic remains a complex challenge requiring a multidisciplinary approach and further research to overcome current obstacles and anticipate future developments in this constantly evolving field.

Table 1. Summary of Research Studies on Sarcasm Detection in Arabic (2021) Based on ArSarcasmv2 Dataset [7]

Description	Models	Performance
Build BERT-based models for sarcasm and sentiment detection	AraBERT and AraElectra	Precision: 72.64%, Recall: 71.47%, F1-score: 72.00%
Build BERT-based models for Arabic sarcasm and sentiment detection	MARBERT	Precision: 70.6%, Recall: 70.2%, F1-score: 70.4%
Build a BERT-based model with data augmentation for Arabic sarcasm and sentiment detection	MARBERT, ARBERT, AraBERT-v02, QARIB, ArabicBERT and mBERT	F1-score: 64.7%
Build a multi-task model using static and contextualized embeddings for Arabic sarcasm and sentiment detection	MAR-BERT + static word embeddings in CNN-LSTM network	F1-score: 62.3%
Build an ensemble-based system combining the static and contextualized representations for Arabic sarcasm detection	AraBERT + CNN-BILSTM	Precision: 70.31%, Recall: 74.47%, F1-score: 61.40%, Macro F1-score: 70.96%
Build a deep learning ensemble model for Arabic sarcasm sentiment analysis	XLM-R and AraBERT	F1-score: 61.27%, Macro F1-score: 73.10%
Build a deep learning model for Arabic sarcasm detection and sentiment analysis	MARBERT, attention layer+ CLS embedding	Precision: 0.7268, Recall: 0.7122, F1-score: 0.6000
Build a BERT-based model with ensemble technique for Arabic sarcasm detection	AraBERT-v01, AraBERT-v02 and XLM-R	Precision: 0.7268, Recall: 0.7235, F1-Score: 0.5989, Macro F1-score: 0.7251

3 Data Set

This work involves the selection of two large Arabic datasets specifically designed for text classification tasks. The first dataset, the AraSarcasm-V2 dataset [5], incorporates the original ArSarcasm dataset along with additional sarcastic tweets from the DAICT dataset [2]. Tweets were collected from various sources to enhance ArSarcasm-V2, ensuring a balanced representation of sarcasm. Annotators provided labels for each tweet, covering sarcasm, sentiment, and dialect. The final dataset consists of 15548 tweets, with 12548 tweets for training and 3000 tweets for testing. The second dataset, the Mawqif dataset [8], consists of 4121 annotated tweets covering three topics: "COVID-19 vaccine" with 1373 tweets, "digital transformation" with 1348 tweets, and "women empowerment" with 1400 tweets. It is a multi-label dataset where each tweet is annotated for stance, sentiment, and sarcasm. Out of these, 3502 tweets are allocated for training and 619 for testing.

3.1 DataSet Statistics

For this study, we created a new dataset by combining data from the Mawqif and ArSarcasmv2 datasets. Our dataset comprises **19669 tweets**, with 16050 allocated for training and 3619 for testing. Table 2 and Fig. 1 display the statistics of the new dataset, indicating that 16.2% of the data is sarcastic (3179 tweets).

Table 2. Data split statistics of sarcasm

	Non-Sarcastic	Sarcastic	Total
Train	13718	2332	16050
Test	2772	847	3619

We worked on a corpus of data including several Arabic dialects. In order to analyze the distribution of tweets according to their dialect, we used the pre-trained model CAMeLBERT-MSA DID NADI [11]. This model, specialized in dialect identification (DID), was developed by refining the CAMeLBERT model for Modern Standard Arabic (MSA). Figure 2 shows the static distribution of Arabic dialects in our corpus, where Modern Standard Arabic (MSA) is the most used dialect. Figure 3 below illustrates instances of sarcastic tweets sampled from our dataset.

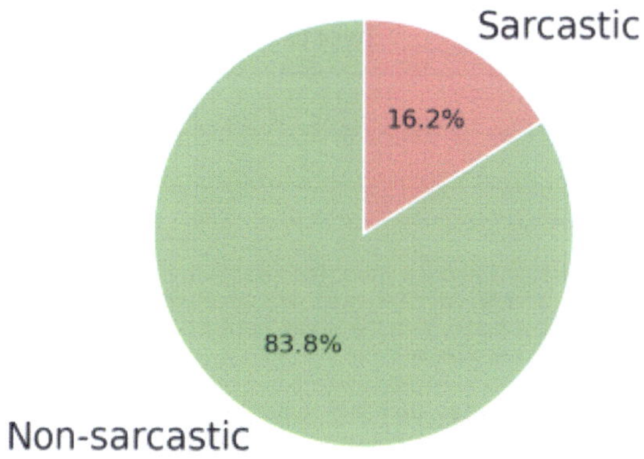

Fig. 1. Ratio of sarcasm

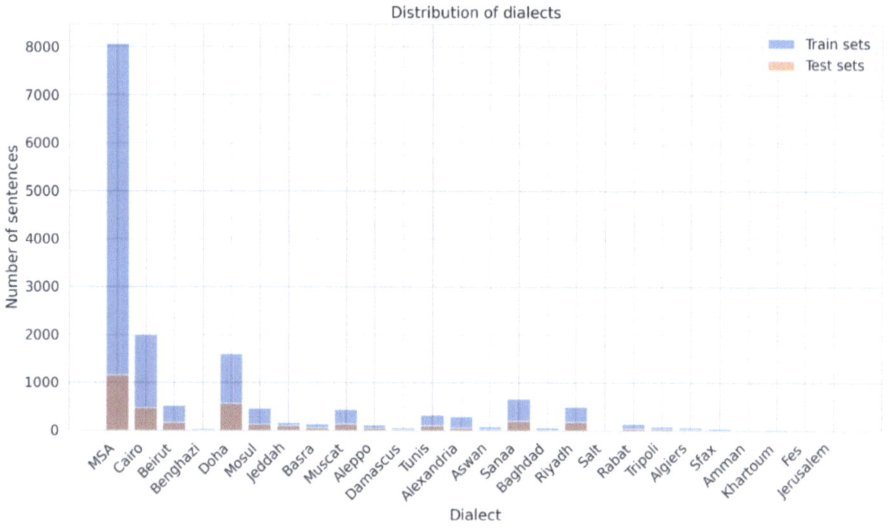

Fig. 2. Distribution of Arabic dialects in the dataset

4 Experimental Details

4.1 Data Preprocessing Techniques

During the preprocessing phase, we refine and optimize the text data by systematically eliminating elements that may introduce noise or ambiguity. We remove special characters such as commas, question marks, and exclamation points to simplify the text, along with hashtags commonly used in social media contexts. Additionally, unnecessary spaces and underlines are removed to improve readability. To ensure efficient and accurate preprocessing tailored to the specific needs of Arabic text analysis, we apply a specialized preprocessing function from AraBERT, which accounts for the nuances of Arabic language structure.

Tweets
ريم النجم: تطعيم كورونا راح يحولكم زومبي !   Reem Al-Najm: Corona vaccination will turn you into a zombie!
بالمناسبة اذا انت شب واسمك يبلش بحرف الميم لا تختصر ايميلك باول حرف من اسمك بعدين العيلة لانك راح تقلب مسخرة #افهموها   By the way, if you are a young man and your name begins with the letter M, do not shorten your email with the first letter of your name, then the family, because you will turn into a mockery. #Understand it
محاضرة حمزاوي فرصة ذهبية للمقتنعين أن ثورة مصر مخطط لها من الغرب منذ ١٠ سنوات، خصوصا وانه كان يعمل في مؤسسة كارنيغي يعني عميل خائن.   Hamzawy's lecture is a golden opportunity for those who are convinced that Egypt's revolution was planned by the West 10 years ago, especially since he was working at the Carnegie Foundation, meaning a traitorous agent.
الأمير: الوو مين؟ الكوتش:أنا سامي الجابر الأمير:وش تبي؟ الكوتش: نبي نفوز على التعاون. الأمير:طوط..طوط..طوط #التعاون_الهلال #النصر   Prince: Hello, who? Coach: I am Sami Al-Jaber. Prince: What do you want? Coach: We want to win through cooperation. Prince: Tot.. Tot.. Tot #Al-Taawon Al-Hilal #Victory

Fig. 3. Examples of some sarcastic tweets

Tokenization is then carried out using the AraBERT preprocessor, dividing each text into meaningful linguistic units called tokens, while considering the particularities of the Arabic language. Encoding is performed with a pre-trained tokenizer from the AraBERT preprocessor, which assigns a unique ID to each token and handles tasks such as adding special tokens, padding, and creating attention masks. This encoding process converts the tokens into a numeric form suitable for use as input by the natural language processing model.

4.2 Models Fine-Tuning

In this section, we introduce the models utilized for sarcasm detection. Four pre-trained models based on BERT have been explored: **CAMeLBERT-da** [12], developed as part of the CAMeL (Collaborative Arabic Modeling Lab) project, is pre-trained on 5.8 billion tokens in dialectal Arabic, **MARBERT** [3], trained on 15.6 billion tokens from 1 billion Arabic tweets, **AraBERT** [9], a model trained on 8.6 billion tokens from five datasets composed of Modern Standard Arabic (MSA) texts and **AraBERT-twitter** [9], an extended version of AraBERT (v0.2) trained on 60 million Arabic tweets. By using these models, we aimed to evaluate their relative effectiveness in detecting sarcasm, taking into account the linguistic and contextual particularities specific to each training corpus. This diversity of models allows for a more comprehensive and nuanced perspective on how sarcasm is expressed and detected in Arabic.

4.3 Selected Hyperparameters

We refine the hyperparameters by exploring different combinations in order to obtain the best results. Table 3 below presents the optimal configuration of the parameters used in our study. We selected this configuration after a series of experiments aimed at maximizing the performance of fine-tuning models. We found that this specific combination achieves the most satisfactory results in terms of accuracy and classification.

Table 3. Used hyperparameters

1	Epoch	10
2	Number of classes	2
3	Batch size	32
4	max_length	100
5	Optimizer	Adam
6	Learning Rate	2e−5

4.4 Evaluation Methods

In this part, we will examine the performance of the four fine-tuning models. We will evaluate their performance based on a set of metrics, including precision, recall, F1-score, and accuracy.

Results: The model performances are shown in Tables 4 and 5, for the development and test sets, respectively. The development set was generated by dividing the training set into five parts and then training the model using cross-validation. As shown in Tables 4 and 5, the AraBERT model demonstrated the best accuracy compared to other models. This performance is explained by the fact that this model was trained with the objectives of Masked Language Modeling (MLM) and Next Sentence Prediction (NSP), which allowed it to better understand long-term dependencies between sentences [8]. The AraBERT-V2-twitter model emerges as the best-fitting model, providing higher F1, recall, precision, and accuracy scores. This success is attributed to its training on dialectal and Twitter-specific Arabic data, combined with a bidirectional transformer architecture with a sophisticated attention mechanism, making it particularly effective for detecting sarcasm in Arabic tweets. In contrast, the MARBERT and CAMeLBERT-da models demonstrated similar, although relatively poor, performance for all evaluation metrics. It is important to note that the MARBERT model performs poorly in comparison, as it was exclusively trained on the Masked Language Modeling task, aiming to capture relationships between words [8]. Likewise, the poor results of CAMeLBERT-da can be explained by the fact that the amount of our data is limited compared to the amount of training data on which it was trained.

Table 4. Sarcasm detection results on the development set

Model	Precision	Recall	F1-score	Accuracy
Arabert-V2-twitter	0.7932	**0.6852**	**0.7207**	**0.8858**
MARBERT	0.4273	0.5000	0.4608	0.8546
CAMeLBERT-da	0.4273	0.5000	0.4608	0.8546
AraBERT	**0.7937**	0.6149	0.6485	0.8754

Table 5. Sarcasm detection results on the test set

Model	Precision	Recall	F1-score	Accuracy
Arabert-V2-twitter	0.7513	**0.6680**	**0.6910**	**0.8124**
MARBERT	0.3830	0.5000	0.4337	0.7660
CAMeLBERT-da	0.3830	0.5000	0.4337	0.7660
AraBERT	**0.7794**	0.6036	0.6179	0.8022

5 Conclusion and Future Directions

Our study examined different fine-tuning models for sarcasm detection to determine the best ones in terms of performance in detecting sarcasm in Arabic. Using two datasets, ArSarcasm-V2 and MAWQIF, we found that the best model for this classification, in terms of **recall**, **F1-Score** and **accuracy**, is **Arabert-V2-twitter**, with respective values of **66.80%**, **69.10%** and **81.24%**. However, **ArabBERT** achieved the highest **precision** with **77.94%**. Looking ahead, we plan to explore other fine-tuning models to see if they offer even better performance. Additionally, we plan to further enrich our corpus by including data from other Arabic dialects, in order to make our models more robust and generalizable.

References

1. Ashwitha, A.S.G.S.H.R., Shruthi, G., Shruthi, H.R., Upadhyaya, M., Ray, A.P., Manjunath, T.C.: Sarcasm detection in natural language processing. Mater. Today Proc. **37**, 3324–3331 (2021). https://doi.org/10.1016/j.matpr.2020.09.124. https://www.sciencedirect.com/science/article/pii/S2214785320368164. International Conference on Newer Trends and Innovation in Mechanical Engineering: Materials Science
2. Abbes, I., Zaghouani, W., El-Hardlo, O., Ashour, F.: DAICT: a dialectal Arabic irony corpus extracted from twitter. In: Calzolari, N., et al. (eds.) Proceedings of the Twelfth Language Resources and Evaluation Conference, pp. 6265–6271. European Language Resources Association, Marseille, France (2020). https://aclanthology.org/2020.lrec-1.768

3. Abdul-Mageed, M., Elmadany, A., Nagoudi, E.M.B.: ARBERT & MARBERT: deep bidirectional transformers for Arabic. In: Zong, C., Xia, F., Li, W., Navigli, R. (eds.) Proceedings of the 59th Annual Meeting of the Association for Computational Linguistics and the 11th International Joint Conference on Natural Language Processing (Volume 1: Long Papers), pp. 7088–7105. Association for Computational Linguistics, Online (2021). https://doi.org/10.18653/v1/2021.acl-long.551. https://aclanthology.org/2021.acl-long.551
4. Abu Farha, I., Magdy, W.: Analyse comparative des modèles de langage basés sur un transformateur pour la détection des sentiments arabes et du sarcasme. In: Habash, N., et al. (eds.) Actes du sixième atelier de traitement du langage naturel arabe, pp. 21–31. Association pour la linguistique computationnelle (2021). https://aclanthology.org/2021.wanlp-1.3
5. Abu Farha, I., Zaghouani, W., Magdy, W.: Overview of the WANLP 2021 shared task on sarcasm and sentiment detection in Arabic. In: Habash, N., et al. (eds.) Proceedings of the Sixth Arabic Natural Language Processing Workshop, pp. 296–305. Association for Computational Linguistics, Kyiv, Ukraine (Virtual) (2021). https://aclanthology.org/2021.wanlp-1.36
6. Abuzayed, A., Al-Khalifa, H.: Sarcasm and sentiment detection in Arabic tweets using BERT-based models and data augmentation. In: Habash, N., et al. (eds.) Proceedings of the Sixth Arabic Natural Language Processing Workshop, pp. 312–317. Association for Computational Linguistics, Kyiv, Ukraine (Virtual) (2021). https://aclanthology.org/2021.wanlp-1.38
7. Alaa, R., Azab, S., Mohammed, A.: A comprehensive survey on Arabic sarcasm detection: approaches, challenges and future trends. IEEE Access **11**, 18261–18280 (2023). https://doi.org/10.1109/ACCESS.2023.3247427
8. Alturayeif, N.S., Luqman, H.A., Ahmed, M.A.K.: Mawqif: a multi-label Arabic dataset for target-specific stance detection. In: Bouamor, H., et al. (eds.) Proceedings of the Seventh Arabic Natural Language Processing Workshop (WANLP), pp. 174–184. Association for Computational Linguistics, Abu Dhabi, United Arab Emirates (Hybrid) (2022). https://doi.org/10.18653/v1/2022.wanlp-1.16. https://aclanthology.org/2022.wanlp-1.16
9. Antoun, W., Baly, F., Hajj, H.: AraBERT: transformer-based model for Arabic language understanding. In: Al-Khalifa, H., Magdy, W., Darwish, K., Elsayed, T., Mubarak, H. (eds.) Proceedings of the 4th Workshop on Open-Source Arabic Corpora and Processing Tools, with a Shared Task on Offensive Language Detection, pp. 9–15. European Language Resource Association, Marseille, France (2020). https://aclanthology.org/2020.osact-1.2
10. Elgabry, H., Attia, S., Abdel-Rahman, A., Abdel-Ate, A., Girgis, S.: A contextual word embedding for Arabic sarcasm detection with random forests. In: Habash, N., et al. (eds.) Proceedings of the Sixth Arabic Natural Language Processing Workshop, pp. 340–344. Association for Computational Linguistics, Kyiv, Ukraine (Virtual) (2021). https://aclanthology.org/2021.wanlp-1.43
11. Inoue, G., Alhafni, B., Baimukan, N., Bouamor, H., Habash, N.: The interplay of variant, size, and task type in Arabic pre-trained language models. CoRR abs/2103.06678 (2021). https://arxiv.org/abs/2103.06678
12. Inoue, G., Alhafni, B., Baimukan, N., Bouamor, H., Habash, N.: The interplay of variant, size, and task type in Arabic pre-trained language models (2021)
13. Mansour, Y., Elnagar, A.: Sarcasm detection in Arabic text using contextualized models, pp. 278–283 (2023). https://doi.org/10.1109/IALP61005.2023.10337175

14. Mekki, A., Zribi, I., Ellouze, M., Belguith, L.: A Tunisian benchmark social media data set for Covid-19 sentiment analysis and sarcasm detection (2022). https://doi.org/10.21203/rs.3.rs-2321298/v1
15. Shah, S.M.A.H., Shah, S.F.H., Ullah, A., Rizwan, A., Atteia, G., Alabdulhafith, M.: Arabic sentiment analysis and sarcasm detection using probabilistic projections-based variational switch transformer. IEEE Access **11**, 67865–67881 (2023). https://doi.org/10.1109/ACCESS.2023.3289715

Typology of Event Data Imperfections

Nassira Achich(✉), Jihen Gargouri Krid, Fatma Ghorbel, and Bilel Gargouri

MIRACL Laboratory, University of Sfax, Sfax, Tunisia
nassira.achich@enetcom.usf.tn

Abstract. Handling data imperfection in events is a critical issue in many application domains ranging from journalism and social media to healthcare communication and financial forecasting. In fact, failure to handle these imperfections can have significant consequences and lead to incorrect analysis and decision-making. In the literature, several typologies of data imperfections are proposed. However, these typologies can not be applied to events due to their particular features which typically include a trigger, arguments, argument roles, and temporal/spatial information. This paper introduces a typology of data imperfection in events, in a two-step process: *(1)*, data imperfection types are identified based on an only single event mention. *(2)*, the arguments of the same event are compared across multiple event mentions to further investigate additional imperfection types. This typology distinguishes among imperfection types that can impact the event's entirety, types that specifically influence individual arguments within an event, and types that have the potential to affect both the complete occurrence of the event and its individual arguments. We finish by representing an illustrative example from an ontology- based memory prosthesis, handling imperfect events.

Keywords: Data Imperfection · Events · Typology · Ontology

1 Introduction

"Event data" is considered one of the most important data types used to conduct analysis processes [1]. In Fact, recognizing real-world events that unfold within spatial and temporal contexts, holds the potential to generate actionable knowledge. This knowledge acquisition process enables individuals to actively seek and obtain valuable advice, opinions, factual details, and timely updates related to the identified events [2]. However, event data is usually of poor quality [3,4]. Indeed, according to Luckham in [4], the inherent attributes and properties specific to event data result in a greater incidence of quality challenges compared to alternative data types. The quality of acquired information, in the context of events, have a direct impact on the effectiveness of analysis and decision-making. For instance, an erroneous decision to invest in a company, may be made based on extracted events indicating fake positive financial performance. Therefore, as

L. H. Belguith and K. Shaalan (Eds.): LPKM 2024, LNNS 1303, pp. 101–111, 2025.
https://doi.org/10.1007/978-3-031-85067-7_10

events are vulnerable to various kinds of imperfections that can affect the quality of the extracted information, it becomes a fundamental necessity to address these imperfections, which involves the recognition of elements that introduce doubts concerning the trustworthiness and precision of data. Consequently, Creating a typology to classify data imperfections is crucial to address the challenges it presents.

In the literature, various typologies of data imperfection are proposed. Some are generic and others are specific to a giving domain. However, to the best of our knowledge, existing works cannot be applied to events due to their particular features. Actually, an event can be affected by imperfections that may impact its entire occurrence or only some arguments within it such as time dimension [5]. For instance, let us have the following event: *"Myriam met her cousin last month somewhere in Tunis"*. In this example, the time dimension, which is "last month" as well as the spatial one, which is *"somewhere in Tunis"*, are both affected by "imprecision" as an imperfection type.

In this paper, we propose a typology of event data imperfection. It is based on three categories. Firstly, there are imperfection types that have a direct influence on the entire occurrence of an event. For example, rumors can exclusively affect the entirety of an event by introducing false or unverified information that be- comes associated with the event as a whole. It creates a distorted perception of the event in its entirety, potentially leading to widespread misinformation and misinterpretation. Secondly, there are imperfection types that specifically affect individual arguments within an event. For instance, imprecision in events typically affects individual arguments within the event rather than the event as a whole. This occurs when specific details within the event are not precisely described. Consequently, instead of characterizing the event itself as imprecise, we characterize it as incomplete. Lastly, there are imperfection types that possess the potential to affect both the complete occurrence of the event and its individual arguments. Taking the example of uncertainty, where the reliability of an information within the event is questionable, and thereby casting doubt in the event as a whole.

This paper is organized as follows. Section 2 represents an overview of an event definition and typologies of data imperfection. Section 3 introduces the proposed typology. Section 4 is dedicated to showcasing a representative example derived from an ontology-based memory prosthesis. Section 5 draws conclusions and future directions.

2 Related Work

The present work is related to: *(A)* event definition and *(B)* typologies of data imperfection.

2.1 Event Definition

The concept of an event can vary across different domains, disciplines, contexts and other considerations. Indeed, the understanding of events differs based on whether we are considering events within event-driven systems or real-world events. In event- driven systems, an event is typically defined as a significant occurrence or action that triggers a response or a series of actions within the system. These events can be generated by software components, sensors, user interactions, or external systems. Each event carries relevant information, such as the event type and timestamp, which is used to initiate specific processes or trigger event handlers [6]. In the context of real-world events, an event refers to a happening or occurrence of interest that takes place within a specific time-frame [7]. In the literature, an event typically consists of two main components: event arguments and argument roles. An event argument refers to an entity mention, temporal expression, or value that plays a role in an event. It represents a participant or attribute associated with the event. Event arguments can be various elements, such as entities (e.g., people, objects), locations, time expressions, or other relevant information that contribute to the understanding and description of the event. An argument role denotes the relationship between an event and its associated argument. It specifies the semantic or syntactic role that the argument plays in the event. Argument roles define the nature of the participation of an argument within an event and how it contributes to the overall meaning or structure of the event. We note that the terminology and approaches to event arguments and argument roles used in the field of event representation vary across different research works and frameworks. However, the general concept of identifying participants and their roles within an event remains consistent. In fact, "identifying events in free text is ideally represented by identifying who did what to whom, when, with what methods, where and possibly why" [8]. For instance, [9] employs the event and argument model, where events are represented as triggers and arguments, and their relationships are defined using argument roles. [10] involves defining events as anchored structures with event triggers and arguments. Arguments are associated with specific roles, such as Time, Place, Victim, and Perpetrator. In the financial domain, event representation typically involves the following components [11]: An event Trigger: refers to a verb or verb phrase that indicates the occurrence of a financial event. It serves as the main indicator or anchor point of the event. Arguments: can include entities such as company names, currencies, prices, or other financial indicators. Argument Roles: Each argument in the event representation is assigned a specific role that indicates its semantic relationship to the event. Temporal Information: such as event date or time, helps in capturing the time context of the financial event. We admit that the principal participants within an event encompass the event trigger, event arguments, event arguments' roles, as well as temporal and spatial information.

Let E be an event, defined as follows:

$$E = (Tr, Ai, Ri, T, S) \tag{1}$$

- **Tr** represents the event trigger, which is a specific word or phrase within the event mention that directly signifies or triggers the occurrence of the event. It serves as the anchor for the event.

- **{Ai, Ri}** represents a set of event arguments (Ai) and their associated roles (Ri). Each argument (Ai) is paired with its respective role (Ri). Event arguments are the entities or elements associated with the event that play a role or are affected by the event. While the event arguments' roles define the specific function or relationship that each argument plays within the event. Examples of argument roles include "agent," "patient," "destination," or "source." By explicitly pairing each event argument with its corresponding role, we ensure that the relationship between the arguments and their roles is maintained.
- **T** represents temporal information associated with the event, such as the start time, end time, duration, or temporal relations with other events. It provides a temporal context for understanding the event.
- **S** represents spatial information associated with the event, such as the location or spatial relations between entities involved. It may include details about the place, area, or geographical coordinates associated with the event.

In our research, we focus on data imperfections that may affect an event. The relations between events is not within the scope of this study.

2.2 Typologies of Data Imperfection

In the literature, several typologies have been proposed [12]. We distinguish two main categories: generic typologies and domain-dependent typologies.

[13] proposed a typology of data non-precision, which includes uncertainty, imprecision, ambiguity, and generality. In her first work, [14] distinguishes uncertainty and imprecision as types of imperfection, and in her second work [15], she adds incompleteness as a third type. [16] focus specifically on two types of uncertainty: fuzziness and ambiguity. Similarly, [17] classifies data imperfection into three types: imprecision, inconsistency, and uncertainty.

Several typologies are also proposed to identify data imperfection forms in a specific domain. [18] presents a typology of information imperfection for analysts and decision makers, including incomplete information, inconsistency, complicated information, uncertainty, corrupt information, and presentation quality. [19] focuses on uncertainty in geographic data, categorizing them as either well-defined (subject to uncertainty)or poorly defined (imprecision, ambiguity, or incompleteness). [20] builds upon Fisher's typology, applying it to ethnographic textual data and identifying imprecision, uncertainty, level of detail, and incompleteness as imperfection types. [21] adapt Fisher's typology [19] to archeological data, classifying imperfection as uncertainty, imprecision, ambiguity, and incompleteness. [22] addresses imperfect data in smart cities, identifying uncertainty, imprecision, vagueness, and missing as imperfection types. [12] propose a typology of temporal data imperfection. It is divided into nine imperfection types of both numeric and natural language-based temporal data which are uncertainty, typing error, imprecision, missing, circumlocution, uselessness, inconsistency, incompleteness and redundancy. [23] proposes a typology of the imperfection of data entries in the application for Alzheimer's patients. It distinguishes twelve types of imperfection. [24] propose a typology of imperfection of data inputs in the context of Collective memory model. It offers nine direct and four indirect imperfection types.

2.3 Discussion

It is recognized that there is no universally applicable typology of data imperfection, and different typologies have varying degrees of relevance. Domain-specific typologies are predominant, and they commonly address imprecision, uncertainty, and incompleteness. Furthermore, there is no definitive definitions for terms used to describe imperfect data [12]. The interdependence of imperfection types is observed, such as the imprecision leading to incompleteness according to [15]. Event data can be affected by more imperfection types compared to the ones that are proposed in the generic typologies due to the specificity that it holds. Indeed, an event contains a number of features, which typically include a trigger, arguments, argument roles, and temporal/spatial information. It can be affected by imperfections in its entire occurrence or in only some arguments within it such as the spatial dimension. Thus, existing typologies can not be applied to event data. In the following, we propose our typology of event data imperfection.

3 Proposed Typology of Event Data

In this section, we detail the proposed typology, based on a two-step process as shown in Fig. 1. In the first step, specific data imperfection types are identified for a single event mention. Then, in the second step, the arguments of the same event are compared across multiple event mentions to explore additional imperfection types. The final typology is then defined, consolidating the identified types.

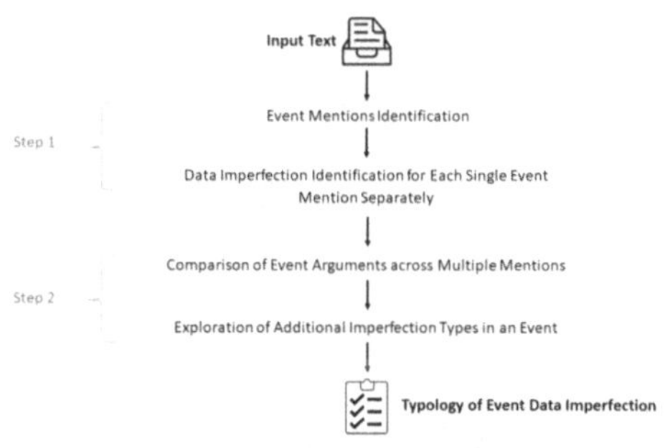

Fig. 1. Sequential actions for Event Imperfection Types Identification

3.1 Data Imperfection Types Considering a Single Event Mention

Considering events within a single mention involves examining the language used to describe the event, as well as looking for ambiguous or incomplete information that maybe needed to fully understand what happened. In this step, we delineate 8 imperfection types, each one will be explicated individually and exemplified accordingly.

Imprecision refers to the lack of specificity in the information provided about an event. This can include unclear descriptions of the timing, location, cause, or other arguments of the event. It is represented by the use of general words which leads to ambiguity. Imprecision may affect only individual arguments within an event. For example, the lack of precise specifications regarding the location of a meeting, described as *"somewhere in Europe"* introduces ambiguity in understanding the event.

Missing. We consider the same definition given in [23], which is simply the not found of the data. In our context, missing refers to the situation where an essential argument of an event is missing. It is true that this type of imperfection may consider only a single argument in an event, but it can significantly affect the understanding of this event. For instance, consider a news article that mentions a political rally without specifying the location or the time. In this case, readers may not be able to attend the event or understand its impact without this information.

Incomprehensibility refers to a lack of clarity or intelligibility in a manner that makes understanding difficult. Indeed, events can be difficult to understand when technical words or difficult jargon are used, particularly when the reader is not familiar with the relevant terminology. Incomprehensibility may affect only the whole event. In fact, it is a transitive imperfection type, where the non-understandability of an individual argument contributes to the overall non-understandability of the event. For example, the use of technical terms such as *"seismic tremor"*, *"magnitude"*, *"epicentral locality"*, *"temblors"* and *"aftershocks"* in an earthquake reporting, without providing clear explanations or simplifications can make the event incomprehensible for readers who are not familiar with the terminology used in seismology. Incomprehensibility can result in incompleteness by overlooking relevant aspects and hindering the understanding of the event's full scope and significance.

Incompleteness. According to [25], incompleteness refers to "the absence of information or knowledge, or the lack of a complete or accurate understanding of an event, resulting in ambiguity". Incompleteness considers the entire event. It arises from either the absence of information about individual arguments of an event or the inability to comprehend and perceive the complete meaning of it. So, incompleteness is an imperfection type which can be generated from imprecision, missing or incomprehensibility.

Uncertainty. According to [17], uncertainty is related to errors which make it challenging to determine the true nature or outcome of an event. An uncertain event involves the presence of multiple possible scenarios, and the inability to predict with certainty, which one will occur. Uncertainty can affect both the entire event and individual arguments within it. In some cases, uncertainty may be localized to a specific argument, while in other cases, it may affect the trigger and then the occurrence of the event. In addition to their arguments, events possess a number of properties related to features that implies if the event really took place. We distinguish two features which lead to uncertainty: tense and modality. Tense is determined with respect to the author. When the FUTURE is used, events have not yet occurred at the textual anchor time. For example, *"The Commission will mobilize additional support and respond to Turkey's latest request for more shelter capacity - in particular tents, blankets, heaters,..."*. Several modalities can lead to event uncertainty such as believed events, hypothetical events, desired events and promised events.

Sensationalism involves employing exaggerated language, or style to magnify the significance or influence of a specific event. It includes both Trivialization and Magnification. Trivialization is a minimization form in order to make some event seem less important [24], for example the downplaying of human rights violations. Sensationalism may consider only the entire Event. In contrast, the Magnification refers to the process of emphasizing the importance, or impact of a particular event, for example the exposure of discriminatory practices. Generally, sensationalism involves the use of hyperbole, metaphor or other rhetorical devices to create a sense of drama or urgency around a topic, as well as the use of statistical or other forms of data manipulation to inflate or distort the perceived significance of a particular issue. It may consider only the entire Event. For example, *"Miracle! 10 year old girl in Turkey rescued after 185 h trapped"*. While sensationalism can be an effective technique for capturing the attention of readers, it can also be controversial, as it can lead to a distortion of the truth.

Personal View can be defined as a personal belief regarding a particular occurrence or happening, which is influenced by attitudes, values, and emotions. It is a mental construct that reflects perception and understanding of an event and can vary among individuals based on their different backgrounds, knowledge, and perspectives. Views may also be shaped by external factors such as media coverage, social norms, and cultural values [26]. This type may consider the occurrence of the event or individual arguments within it or both. For example, the presence of subjective expressions, such as *"I think that,"* before the trigger or an argument of an event, reduces the credibility of the event. These expressions indicate a lack of confidence in the truthfulness of information.

Incoherence refers to a lack of logical or conceptual harmony within a set of data or information. In our context, incoherence is the absence of logical consistency among the arguments of an event. For instance, the marriage between a human and a doll results in an incoherent event.

3.2 Data Imperfection Types Considering Multiple Event Mentions

In the second step, we reconcile event arguments across multiple mentions of this event. More imperfections are identified: contradiction, rumor and inconsistency.

Contradiction. GRIM defines contradiction as the impossibility for something to simultaneously be and not be the same thing [27]. Contradiction arises when two or more event mentions pertaining to the same event present conflicting information or claims about an argument, showcasing opposing facts. Contradiction may affect only individual arguments within an event. For example, two mentions report conflicting information about the magnitude of the earthquake in Hatay Province, Turkey. One states it as a "6.3-magnitude quake", while the other mentions it as a "6.4- magnitude earthquake", creating a discrepancy in the reported magnitude.

Rumor refers to events disseminated with no discernible source generally on trending topics, in order to make the buzz. Rumor is a type of event that represents unverified or unconfirmed information that is spread among people [28]. Therefore, Rumor is a typical data imperfection of events that may affect only the whole occurrence of it. Rumors can spread rapidly through various means of communication, including social media, email, and text messages [29]. The reader will not be able to determine whether the event is subject to a rumor unless he investigates it through other event mentions. For example, the spreading of false information regarding the death of a famous actor.

Inconsistency. According to [17] and [23], inconsistency refers to conflicting data. In the context of events, inconsistency refers to the presence of falsehood or contradictions within the event, undermining its reliability or accuracy. Inconsistency can arise from uncertainty, sensationalism, personal view, incoherence, contradiction or rumor.

3.3 Definitive Typology of Data Imperfections in Textual Event Mentions

The definitive typology of data imperfections in textual event mentions is illustrated by Fig. 2. We distinguish 11 types of imperfections. Among these, 3 types exclusively affect individual arguments within an event: imprecision, missing, and contradiction. 6 types are specifically associated with the entirety of the event: incomprehensibility, sensationalism, incoherence, rumor, incompleteness, and inconsistency. Additionally, there are 2 imperfection types that have the potential to influence both individual arguments within an event and the overall occurrence of the event which are uncertainty and personal view. We assert that events can be susceptible to multiple types of imperfections simultaneously. For instance, an event may exhibit a missing time argument along with an uncertain spatial one. We acknowledge the interdependence of imperfection types, where the presence of one type can influence or be influenced by another. For example, the presence of imprecision in data can lead to incompleteness. We observe that the inconsistency of the data is associated with 7 imperfection types: uncertainty, sensationalism, personal view, incoherence, contradiction, rumor, and inconsistency. These types contribute to the erroneous nature of the information given by an event.

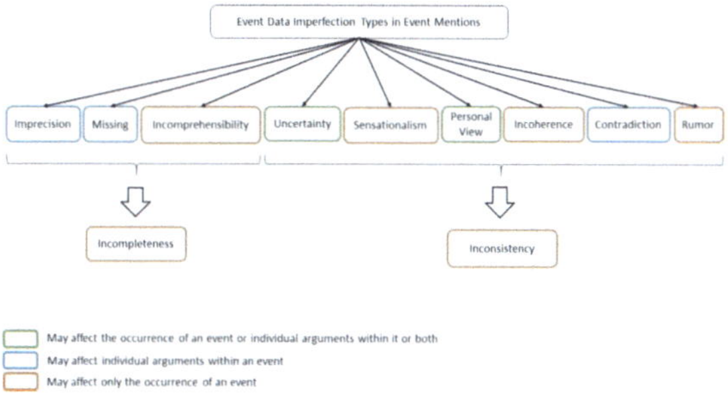

Fig. 2. The proposed typology of event data imperfection

4 Example of Imperfect Event Representation in Ontology

In this section, we present an example of an imperfect event within the context of Captain Memo [30], a memory prosthesis designed to assist the elderly with memory-related challenges. This prosthesis supplies a set of services. Among these services, one is devoted to "remember lived events". In Captain Memo, personal data are structured semantically using an ontology called, PersonLink [31].

Let us have the following example: *"Peter was married to Stephanie in Paris after 1960 and it lasts 15 years"*. According the definition of the event: The trigger event is "the marriage of Peter and Stephanie". The arguments roles are Peter and Stephanie representing the husband and the wife. "after 1960" represents the date of their marriage which is the temporal information associated to the event. "In Paris" represents the location of their marriage which the spatial information associated to the event. In this example, the imperfection affect an argument within the event and not the entire one, which is the temporal information. In fact, "after 1960" is imprecise. We represent the imprecise temporal dimension in events, based on a previous work [32]. This work proposes an ontology called CrispTimeOnto[1]. Based on this ontology, we propose a prototype to facilitate the interaction with this Ontology. It is integrated into Captain Memo, to manage the temporal dimension within the context of PersonLink. It involves merging the CrispTimeOnto and PersonLink ontologies. To represent the imprecise temporal data, the provided information of the wedding's start and end is converted into two intervals sequentially. Let $I = [I^-, I^+]$ be an imprecise time

[1] https://cedric.cnam.fr/isid/ontologies/files/CrispTimeOnto.html.

interval representing the duration of the marriage, where $I^- = \{1960...1963\}$ presents the interval of the beginning bound and $I^+ = \{1975...1978\}$ the interval of the ending bound.

5 Conclusion and Future Directions

In this paper, we propose a typology of data imperfections in event mentions, based on a two-step process. It involves the initial recognition of imperfections within individual event mentions, followed by analyzing multiple mentions of the same event to further explore additional imperfections. In our typology, we differentiate between imperfection types that exclusively impact the occurrence of an event, types that solely affect the arguments within an event, and types that have the potential to influence both the overall occurrence of the event and its associated arguments. Work in progress aims to adapt an event detection system that incorporates an ontological representation. It will be designed to handle various types of imperfections within the detected events. We also plan to explore techniques for reasoning and inference over the imperfect events.

References

1. Chen, W.: E-research event data quality. In: 2013 IEEE 29th International Conference on Data Engineering Workshops (ICDEW), pp. 314–317. IEEE (2013)
2. Atefeh, F., Khreich, W.: A survey of techniques for event detection in twitter. Comput. Intell. **31**(1), 132–164 (2015)
3. Friedman, T.: Findings from primary research study: data quality issues create significant cost, yet often go unmeasured. Gartner (2010). http://www.gartner.com/Display-Document
4. Luckham, D.: The Power of Events, vol. 204. Addison-Wesley Reading (2002)
5. Muller, H., Freytag, J.C.: Problems, methods, and challenges in comprehensive data cleansing. Professoren des Inst. FurInformatik (2005)
6. Laigner, R., et al.: From a monolithic big data system to a microservices event-driven architecture. In: 2020 46th Euromicro Conference on Software Engineering and Advanced Applications (SEAA), pp. 213–220. IEEE (2020)
7. Xiang, W., Wang, B.: A survey of event extraction from text. IEEE Access **7**, 173111–173137 (2019)
8. Tanev, H., Piskorski, J., Atkinson, M.: Real-time news event extraction for global crisis monitoring. In: International Conference on Application of Natural Language to Information Systems, pp. 207–218. Springer, Cham (2008)
9. Pyysalo, S., et al.: Overview of the ID, EPI and REL tasks of BioNLP shared task 2011. BMC Bioinform. **13**, 1–26 (2012)
10. Doddington, G.R., Mitchell, A., Przybocki, M.A., Ramshaw, L.A., Strassel, S.M., Weischedel, R.M.: The automatic content extraction (ace)program-tasks, data, and evaluation. In: LREC, Lisbon, vol. 2, pp. 837–840 (2004)
11. Ding, X., Zhang, Y., Liu, T., Duan, J.: Using structured events to predict stock price movement: an empirical investigation. In: Proceedings of the 2014 Conference on Empirical Methods in Natural Language Processing (EMNLP), pp. 1415–1425 (2014)

12. Achich, N., Ghorbel, F., Hamdi, F., Metais, E., Gargouri, F.: A typology of temporal data imperfection. In: 11th International Conference on Knowledge Engineering and Ontology Development (KEOD 2019), pp. 305–311. SciTePress-Science and Technology Publications (2019)
13. Niskanen, V.: Introduction to imprecise reasoning, uncertainty, decision making and knowledge engineering. Publications of the Society for Artificial Intelligence in Finland (1989)
14. Bouchon-Meunier, B.: La logique floue, Presses universitaires de France (1994)
15. Bouchon-Meunier, B.: La logique floue et ses applications, Editions Addison-Wesley France (1995)
16. George, K., Bo, J.Y.: Fuzzy sets and fuzzy logic, theory and applications (2008)
17. Smets, P.: Imperfect information: imprecision-uncertainty. In: Uncertainty Management in Information Systems: From Needs to Solutions, pp. 225–254 (1997)
18. Gershon, N.: Visualization of an imperfect world. IEEE Comput. Graph. Appl. **18**(4), 43–45 (1998)
19. Fisher, P., Comber, A., Wadsw, R.: Nature de l'incertitude pour les données spatiales (2005)
20. Olteanu, A., Mustière, S., Ruas, A., Desveaux, E.: L'apport de données spatiales pour une base de donnèes ethnographique. Actes de Confèerèence SAGEO (2006)
21. Desjardin, E., Nogent, O., De Runz, C.: Prise en compte de l'imperfectiondes connaissances depuis le saisie des données jusqu'à la restitution 3D. Archeologia e Calcatori, Supplemento **3**, 385–395 (2012)
22. Sta, H.B.: Quality and the efficiency of data in "smart-cities". Futur. Gener. Comput. Syst. **74**, 409–416 (2017)
23. Achich, N., Ghorbel, F., Hamdi, F., Métais, E., Gargouri, F.: TimeOntoImperfection': an OWL 2 ontology to representing temporal data imperfection using imperfection theories. J. Intell. Fuzzy Syst. (Preprint) 1–16 (2024)
24. Kharfia, H., Ghorbel, F., Gargouri, B.: Typology of data inputs imperfection in collective memory model. In: Abraham, A., Gandhi, N., Hanne, T., Hong, T.-P., Nogueira Rios, T., Ding, W. (eds.) ISDA 2021. LNNS, vol. 418, pp. 1198–1206. Springer, Cham (2022). https://doi.org/10.1007/978-3-030-96308-8_111
25. Li, X., Xu, J., Wang, Y., Zhang, Y., Liu, X.: A survey on incomplete information extraction. Knowl. Inf. Syst. **49**(2), 461–481 (2016)
26. Petty, R.E., Cacioppo, J.T.: Communication and Persuasion: Centraland Peripheral Routes to Attitude Change. Springer, Cham (2012)
27. Grim, P.: What is a contradiction. In: The Law of Non-contradiction: New Philosophical Essays, pp. 49–72 (2004)
28. Zhang, W., Guo, X., Chen, X.: The detection and evolution of rumors in social media. ACM Trans. Intell. Syst. Technol. **7**(4), 43 (2016)
29. Zeng, H., Cui, X.: SimCLRT: a simple framework for contrastive learning of rumor tracking. Eng. Appl. Artif. Intell. **110**, 104757 (2022)
30. Métais, E., et al.: Memory prosthesis. Non-pharmacol. Therapies Dementia **3**(2), 177 (2012)
31. Herradi, N., Hamdi, F., Métais, E., Ghorbel, F., Soukane, A.: PersonLink: an ontology representing family relationships for the CAPTAIN MEMO memory prosthesis. In: Jeusfeld, M.A., Karlapalem, K. (eds.) ER 2015. LNCS, vol. 9382, pp. 3–13. Springer, Cham (2015). https://doi.org/10.1007/978-3-319-25747-1_1
32. Ghorbel, F., Hamdi, F., Métais, E.: Dealing with precise and imprecise temporal data in crisp ontology. Int. J. Inf. Technol. Web Eng. (IJITWE) **15**(2), 30–49 (2020)

Towards Sentiment Analysis for Libyan Dialect

Hassan Ebrahem[1(✉)], Imen Touati[2], and Lamia Hadrich Belguith[2]

[1] Faculty of Information Technology, University of Tripoli, Tripoli, Libya
hasanalihasan@yahoo.com

[2] Faculty of Economic Sciences and Management, University of Sfax, Sfax, Tunisia

Abstract. In this paper, we are interested in developing a sentiment analysis (SA) system on the basis of machine learning (ML) techniques. We propose a Libyan dialect twitter dataset which we have built. It contains 6,000 comments and tweets cleaned, pre-processed, and annotated. To evaluate the performance of sentiment classification of the tweets and comments, four machine learning classifiers have been applied. These results show that all classifiers achieved good results. Furthermore, we have created a basic sentiment lexicon which includes words and phrases with sentiment polarity (positive, negative, and neutral). This sentiment lexicon is a valuable resource for the Libyan dialect.

Keywords: Libyan Dialect · Libyan Airlines · Corpus · Sentiment Analysis

1 Introduction

In this digital age, the majority of people have already signed into different social media platforms to share, updates, check on friends and family, keep track of global and local issues and engage in other online activities. In addition, people tend to use online social media to express their concern, frustration, satisfaction and other feelings over the services and products which companies offer. With such a significant impact social media has nowadays, it is of vital importance for airlines to be attentive to their customers' sentiments shared on social media platforms, such as Facebook, Twitter, Instagram and others, and adapt their services in response to the feedback posted on these channels. Thus, to help airlines companies to reach this objective, we propose a machine learning based sentiment analysis system which can identify neutral, negative, and positive sentiments for Libyan dialects.

SA studies performed for Arabic dialect concern different types of datasets. However, few studies have investigated customer sentiment towards Arabic airline companies, particularly in the context of the Libyan dialect of Arabic. By conducting a sentiment analysis in this context, the study aims to provide insights on customer opinions and satisfaction related to Libyan airline companies and demonstrate the potential of machine learning techniques for analyzing

L. H. Belguith and K. Shaalan (Eds.): LPKM 2024, LNNS 1303, pp. 112–121, 2025.
https://doi.org/10.1007/978-3-031-85067-7_11

expressed customer sentiment in the Libyan dialect. However, to the best of the researchers' knowledge, no studies have investigated sentiment classification of tweets and comments related to airline companies for Libyan dialect. In this work, we are interested in analyzing sentiment for Libyan dialect.

In the airline sector, millions of individuals utilize social networking sites such as Twitter, Facebook, and TripAdvisor to exchange information about aircraft services and express their feelings and opinions. Our dataset is a set of Libyan comments scraped from social networks (mainly Twitter and Facebook). These tweets and comment describes the quality of services provided by some Libyan airline companies (Afriqiyah_Airways, AlRahela, Buraq_Airline, Ghadames_Air, Libyan_Airlines and Libya_Wings).

In other words, we attempt to study the effect of services presented by Libyan airlines companies on the sentiment provided by social media users towards these companies. The major contributions of the study are: 1. Building an annotated dataset for Libyan dialect for sentiment analysis concerning airlines domain. 2. Building an effective models for Libyan datasets by applying machine learning algorithms. 3. Evaluating various machine learning algorithms for classifying the polarity of tweets.

The rest of the study is organized as follows. Section 2, describes Libyan dialect. Section 3, focuses on the most closely related studies on Arabic sentiment analysis for the Libyan dialect. Section 4, presents the proposed method and the undertaken preliminary experiments. Finally, the conclusion is presented in the last section.

2 Libyan Dialect

Libyan dialect is among the most famous Maghreb Arabic dialects (Darija دارجة). These dialects are the languages spoken in the different countries that make up the Maghreb which are Morocco, Mauritania, Algeria, Tunisia and Libya. Libyan Arabic is a generic term designating the different varieties of dialectal Arabic, mainly spoken in Libya and in south-eastern Algeria and Tunisia. In Libya, the mostly spoken language in everyday conversation is Libyan dialect. It's not written formally with preference to Modern Standard Arabic. However, it is used in folk poetry, media (TV dramas, comedies, songs, as well as in cartoons) and informal online communication (forums, emails, messaging). For example, the comment (شن مواعيد مزبوطة والا فيها تشحيطة الليل) written in Libyan dialect is equivalent to (هل المواعيد صحيحة أو الرحلة سوف تتأخر) in Modern Standard Arabic.

3 Related Work

One of the most active fields of study in Natural Language Processing (NLP) is sentiment analysis. Nowadays the use of sentiment analysis is not limited only for Modern standard Arabic but it have been carried out for many Arabic dialects.

As for other languages, techniques employed in sentiment analysis for Arabic dialects are divided in three approaches: machine learning, lexicon-based and hybrid approaches [3]. The system SAMAR, created by [1], is among pioneer Arabic sentiment analysis systems. It's a supervised system for Arabic social media. The system was built using an SVM classifier. Using a machine learning approach, [23] suggested to classify views or opinions at the expression level in Political Arabic press articles. The authors performed a semantic categorization of opinion expression, along with polarity and intensity classification [24]. The authors in [8] experimented the SVM model for Arabic sentiment classification. The dataset is a collection of tweets was collected from Twitter, which treats multiple social issues in Saudi Arabia. They reported the efficiency of SVM model in this task. [10] performed an empirical study to tackle the impact of machine translation impact on Arabic sentiment analysis. They found that machine translation can effectively preserve the sentiment prediction.

[7] showed the performance of deep learning approach in classifying the sentiment of Arabic tweets, by using Convolutional neural networks (CNN) and bi-directional long short memory (BiLSTM) as deep learning algorithms. The authors collected a dataset containing Saudi tweets considering the COVID-19 pandemic in Saudi Arabia. [9] proposed a polarity weighting technique based on a lexicon which they have constructed. This lexicon serves to assign weights to polarity words based on their associations with non-polarity words, then computes the sentiment score. This technique is assisted with some linguistics rules. [4] are interested in analysing sentiments in user reviews of healthcare services. They have carried out some experiments using several machine learning methods (Logistic Regression, Support Vector Machine (SVM), and Naive Bayes) as well as Convolutional Neural Network (CNN). Many other researchers were interested to develop sentiment analysis systems for different Maghreb dialects: Tunisian [19], Algerian [11], Moroccan [13], and Mauritanian [20].

An approach for creating a richly annotated corpus for Tunisian dialect (TD) was suggested by [14]. The authors compare transfer learning and machine learning models in order to develop fine grained sentiment analysis models. They accomplished the best performance with The BERT model Calibrated with TD information. [17] implemented AlgBERT which is an automated annotation platform for sentiment analysis in Algerian dialect. They used the obtained annotated corpus with BERT model as deep learning technique and the system gave excellent results.

As shown, several studies have been done for many Arabic dialects. However, few works have been carried out for Libyan dialect. [6] have employed lexicon-based technique to develop a sentiment analysis system using a Libyan dialect twitter corpus. [21] performed sentiment analysis of opinions expressed towards major Libyan telecommunication companies. Five machine learning models was used with an imbalance dataset. The authors showed the performance of Naïve Bayes model in the task.

[5] applied machine learning models (Naive Bayes, SVM and Decision Tree) for sentiment analysis for the Libyan dialect. A dataset of 2296 Libyan dialect

tweets, randomly retrieved over the period from the 20th of March to the 23rd of July 2023, was used to evaluate and compare models. [2] focuses on identifying emotion in poetry composed in the sub-dialect of Misurata Arabic, which is spoken in Misurata in Libya.

Many works in SA were relative to airlines domain. [12] categorize the sentiment of Twitter messages related to various airlines. The tweets are extracted, pre processed, and then classified into positive, neutral, or negative sentiments. This allows for the classification of recent tweets regarding different airlines based on sentiment. [25] using the TOPSIS-VIKORAISM methodology, they created a comprehensive model for assessing the level of service provided by airlines, they looked at the utility, Euclidean distance, and individual regret values of several airlines using the created framework, and came up with a ranking trade-off solution, and they also used the AISM approach to investigate a variety of indicators hierarchically to create directed topological hierarchy diagrams. [18] have collected consumer feedback on air transport management from multiple websites. They utilized the WEKA tool to apply machine learning approaches, and ratings were gathered for various aspects such as seat comfort, cabin service, food and beverage, and inflight entertainment, among others.

Verma, et al. [24] developed an approach to extract and analyze implicit characteristics and opinions from airline evaluations. Their method involved enhancing pre-trained word embedding for sequential modeling using stochastic gradient-descent optimized conditional random fields. Machine learning and ensemble learning techniques were then employed to extract and categorize inferred elements from opinionated texts. [24] presented an innovative research study focused on extracting and analyzing latent elements and opinions from English-language airline reviews. In this study, a novel two-way technique was developed. Firstly, pre-trained word embeddings were enhanced for sequential modeling using stochastic gradient descent optimized conditional random fields (CRF). Subsequently, machine learning and ensemble learning algorithms were employed to classify the implied aspects. The technique utilized an aspect-based annotated corpus specifically tailored to the airline domain. [15] conducted sentiment analysis on travelers' reviews of us airlines. They employed feature analysis to select the best features, resulting in improved overall model performance and reduced training time. Our work is the first interested in this domain for Arabic.

Understanding customer sentiment towards Libyan airlines is essential for airlines seeking to retain loyal customers and maintain a competitive edge. By identifying key factors that influence customer satisfaction, airlines can make targeted improvements to their services and enhance the overall customer experience.

4 The Proposed Method

This study has analyzed customer sentiment towards Libyan airline companies. The participants were customers of the Libyan airline companies, and the study did not involve any direct contact with the participants. The tweets and comments were publicly available on social media platforms.

5 Corpus Description

The dataset is composed by 6,000 tweets and comments posted by customers on social media platforms such as Facebook and Twitter from 2009 to 2023. To build the Corpus ASALDA, we have collected Libyan comments from Facebook concerning airline companies. Also, we have performed a process of scrapping Twitter to extract tweets reviewing services of Libyan airline companies. After that, since cleaning and pre-processing data is an essential step ([8,22]), we proceeded by operating several techniques as deleting superfluous elements like non-Arabic words, characters (like %, +, $), numbers, user mentions (@user), and punctuation (like ",",".",";") in order to reduce noise in the data. Additionally, data gone through a normalization step which involves the unification of some Arabic letters.

Tweets and comments were manually annotated for sentiment as positive, neutral, or negative on sentence level and also on aspect-based level. Our dataset is balanced with 2000 positive, 2000 negative, and 2000 neutral tweets and comments. Figure 1 shows the number of comments for each airline company.

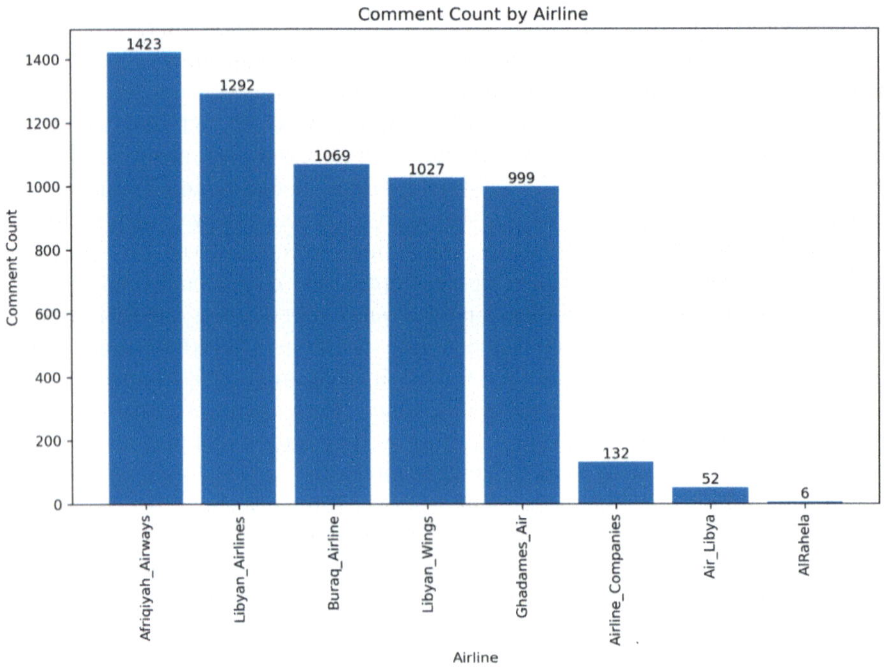

Fig. 1. Number of tweets and comments for airline companies.

In order to perform later aspect-based sentiment analysis, we annotate opinions regarding several aspects related to airline companies. We consider **15**

aspects (late flight, cancelled flight, lost luggage, customer service, flight booking, damaged luggage, food, flight, price, clean, general, flag, airport, plane, and crew). The ASALDA corpus is stored in a standard XML format as shown in Fig. 2.

```
<?xml version='1.0' encoding='utf-8'?>
<airlines> <Airlines name="Afriqiyah_Airways">
<Comment id="8"> <comment_text>انجاز عظيم انه في موعدها</comment_text>
<Polarity>positive</Polarity> <Late flight>positive</Late flight>
</Airlines>
</airlines>
```

Fig. 2. Corpus ASALDA's structure in XML file.

6 Preliminary Experiments and Results

In this work, we have conducted some preliminary experiments on ASALDA corpus using machine learning approach. Indeed, we have used some ML algorithms implemented in Waikato Environment for Knowledge Analysis (WEKA) data platform and in the python library sklearn for ML.WEKA is a software that was completely redesigned from scratch. It is well-liked in both academia and business, and it is used as a ground-breaking system for data mining and machine learning. The aim of the WEKA project is to provide a comprehensive set of machine learning algorithms and data preparation tools to both researchers and practitioners [16]. It's an open-source program that may be used for many different data mining tasks as well as for data preparation, feature selection, clustering, regression, and visualization. We started by evaluating the performance of the proposed method using Naïve Bayes, SVM, Logistic regression and Decision Tree J48 implemented in WEKA. Experiments were performed considering two settings. Firstly, we consider as feature, only the content or the text of the tweet or the comment. Secondly, we handle the same experiment considering the existence or not of words in the sentiment lexicon which we have previously mentioned.

We have carried the same previous experiments using sklearn library. The classification of the sentiment of the tweets/comments were tested using the same algorithms.

7 Discussion

In the current study, several machine learning algorithms were exploited using WEKA and sklearn library for performing sentiment analysis in the ASALDA dataset. Sentiment lexicon was also used in conjunction with the corpus in both experiments. As shown in Table 1, we noticed that when the sentiment lexicon is introduced the results are enhanced. Also, we found that the implementation of all classifiers in sklearn library produce much better results than WEKA. The Logistic Regression classifier using sklearn library shows increase of about 11% in result. So, we can conclude that for voluminous dataset sklearn will be more suitable. As the results shown in Table 1, considering the first setting, SVM classifier achieved the best result with an accuracy rate of 83.38%. In the second setting and with the addition of sentiment lexicon feature. The SVM and Logistic Regression achieved the best accuracy of rate 87.17%. We noticed that, after the introduction of the sentiment lexicon, results are improved for all classifiers.

Table 1. Accuracy rate with WEKA and ML

	Algorithm	Without lexicon feature	With lexicon feature
WEKA	Naïve Bayes	73.28%	78.20%
	SVM	75.25%	78.19%
	Logistic Regression	71.23%	75.09%
	J48	75.28%	78.20%
Sklearn library	Naïve Bayes	82.76%	87.01%
	SVM	**83.38%**	**87.17%**
	Logistic Regression	82.84%	**87.17%**
	J48	72.48%	81.35%

Assessing the performance of machine learning models by class is crucial to understand the strengths and weaknesses of each model. Based on experiments, we have noticed that all classifiers accomplished the best results with sentiment lexicon feature. Thus, we present in Table 2 the classification reports of used machine learning models with sentiment lexicon as feature. We consider three classes (positive, negative and neutral) in the sentiment analysis task for Libyan dialect. As shown in Table 2, all classifiers ensure accurate prediction for all classes regarding the f1-score values. Achieving the best accuracy, Logistic Regression classifier fulfilled the best results for positive and negative classes. For future research, it would be beneficial to explore more advanced sentiment analysis techniques that account for the nuances of natural language. This could involve the use of machine learning models and deep learning algorithms to classify sentiment more effectively. Furthermore, future studies could investigate additional factors that influence customer sentiment toward Libyan airlines. For instance, examining the impact of socio cultural factors on customer sentiment would provide valuable insights.

Table 2. Classification reports

Naive Bayes with feature lexicon

Accuracy: 0.8701700154559505
Classification Report:

	precision	recall	f1-score	support
negative	0.86	0.86	0.86	462
neutral	0.82	0.85	0.84	370
positive	0.92	0.90	0.91	462
accuracy			0.87	1294
macro avg	0.87	0.87	0.87	1294
weighted avg	0.87	0.87	0.87	1294

SVM with feature lexicon

Accuracy: 0.8717156105100463
Classification Report:

	precision	recall	f1-score	support
negative	0.77	0.98	0.86	462
neutral	0.95	0.81	0.87	370
positive	0.96	0.82	0.88	462
accuracy			0.87	1294
macro avg	0.89	0.87	0.87	1294
weighted avg	0.89	0.87	0.87	1294

Logistic Regression with feature lexicon

Accuracy: 0.8717156105100463
Classification Report:

	precision	recall	f1-score	support
negative	0.80	0.95	0.87	462
neutral	0.91	0.81	0.86	370
positive	0.93	0.84	0.89	462
accuracy			0.87	1294
macro avg	0.88	0.87	0.87	1294
weighted avg	0.88	0.87	0.87	1294

J48 with feature lexicon

Accuracy: 0.8137557959814529
Classification Report:

	precision	recall	f1-score	support
negative	0.74	0.83	0.78	462
neutral	0.87	0.81	0.84	370
positive	0.85	0.81	0.83	462
accuracy			0.81	1294
macro avg	0.82	0.81	0.82	1294
weighted avg	0.82	0.81	0.81	1294

8 Conclusion and Future Work

In conclusion, to achieve the best sentiment classification accuracy for Libyan dialect, several machine learning models were used with two types of setting. The experimental results showed good outcomes. This study aims to contribute to the creation of the ASALDA corpus and a sentiment lexicon that can be used to better understand customer preferences and opinions in Libyan airlines. This study provides a sentiment analysis of customer tweets and comments towards Libyan airline companies. The ASALDA created in this study can be used for future research in the field of customer sentiment analysis and can also be useful for the airlines themselves to improve their services and enhance customer satisfaction, and the sentiment lexicon can be used to classification in the Libyan dialect research. Future research could explore more sophisticated sentiment analysis techniques namely deep learning models and additional factors that influence customer sentiment towards Libyan airlines.

References

1. Abdul-Mageed, M., Kuebler, S., Diab, M.: Samar: a system for subjectivity and sentiment analysis of Arabic social media (2012)
2. Abugharsa, A.: Sentiment analysis in poems in misurata sub-dialect–a sentiment detection in an Arabic sub-dialect. arXiv preprint arXiv:2109.07203 (2021)
3. Al-Twairesh, N.: Sentiment analysis of Twitter: a study on the Saudi community. Ph.D. thesis, King Saud University Riyadh, Saudi Arabia (2016)

4. Alayba, A.M., Palade, V., England, M., Iqbal, R.: Arabic language sentiment analysis on health services. In: 2017 1st International Workshop on Arabic Script Analysis and Recognition (ASAR), pp. 114–118. IEEE (2017)
5. Alfared, R.A., Aljarm, H.M.: Sentiment analysis of Libyan tweet using machine learning algorithms. Malays. J. Ind. Technol. (MJIT) **8**(1) (2024)
6. Alhammi, H.A., Haddar, K.: Building a Libyan dialect lexicon-based sentiment analysis system using semantic orientation of adjective-adverb combinations. Int. J. Comput. Theory Eng. **12**(6), 145–150 (2020)
7. Alqarni, A., Rahman, A.: Arabic tweets-based sentiment analysis to investigate the impact of covid-19 in KSA: a deep learning approach. Big Data Cogn. Comput. **7**(1), 16 (2023)
8. Alyami, S.N., Olatunji, S.O.: Application of support vector machine for Arabic sentiment classification using twitter-based dataset. J. Inf. Knowl. Manag. **19**(01), 2040018 (2020)
9. Assiri, A., Emam, A., Al-Dossari, H.: Towards enhancement of a lexicon-based approach for Saudi dialect sentiment analysis. J. Inf. Sci. **44**(2), 184–202 (2018)
10. Barhoumi, A., Aloulou, C., Camelin, N., Esteve, Y., Belguith, L.: Arabic sentiment analysis: an empirical study of machine translation's impact. In: Language Processing and Knowledge Management International Conference (LPKM2018) (2018)
11. Benmounah, Z., Boulesnane, A., Fadheli, A., Khial, M.: Sentiment analysis on Algerian dialect with transformers. Appl. Sci. **13**(20), 11157 (2023)
12. Das, D.D., Sharma, S., Natani, S., Khare, N., Singh, B.: Sentimental analysis for airline twitter data. In: IOP Conference Series: Materials Science and Engineering, vol. 263, p. 042067. IOP Publishing (2017)
13. El Ouahabi, S., El Ouahabi, S., Dadi, E.W.: Contribution to the moroccan darija sentiment analysis in social networks. Soc. Netw. Anal. Min. **13**(1), 138 (2023)
14. Fsih, E., Boujelbane, R., Belguith, L.H.: A deep sentiment analysis of tunisian dialect comments on multi-domain posts in different social media platforms. In: Proceedings of the 5th International Conference on Natural Language and Speech Processing (ICNLSP 2022), pp. 226–233 (2022)
15. Hakh, H., Aljarah, I., Al-Shboul, B.: Online social media-based sentiment analysis for us airline companies. New Trends Inf. Technol. 176 (2017)
16. Hall, M., Frank, E., Holmes, G., Pfahringer, B., Reutemann, P., Witten, I.H.: The weka data mining software: an update. ACM SIGKDD Explor. Newsl. **11**(1), 10–18 (2009)
17. Hamadouche, K., Bousmaha, K.Z., Bekkoucha, M.A., Hadrich-Belguith, L.: Algbert: automatic construction of annotated corpus for sentiment analysis in algerian dialect. ACM Trans. Asian Low-Resour. Lang. Inf. Process. **22**(12), 1–17 (2023)
18. Jain, P.K., Pamula, R.: Sentiment analysis in airline data: customer rating based recommendation prediction using weka. Mach. Learn. Algorithms Ind. Appl. 53–65 (2021)
19. Masmoudi, A., Hamdi, J., Belguith, L.H.: Deep learning for sentiment analysis of tunisian dialect. Computación y Sistemas **25**(1), 129–148 (2021)
20. N'Diaye, A.C.M., Chrif, M.E.M.E.A., El Mahmoud, B.M., El Beqqali, O.: Apply sentiment analysis technology in social media as a tool to enhance the effectiveness of e-government: application on Arabic and mauritanian dialect 'hassaniya'. In: 2021 Fifth International Conference On Intelligent Computing in Data Sciences (ICDS), pp. 1–5. IEEE (2021)

21. Omar, A., Essgaer, M., Ahmed, K.M.: Using machine learning model to predict Libyan telecom company customer satisfaction. In: 2022 International Conference on Engineering & MIS (ICEMIS), pp. 1–6. IEEE (2022)
22. Refaee, E., Rieser, V.: An Arabic twitter corpus for subjectivity and sentiment analysis. In: 9th International Language Resources and Evaluation Conference, pp. 2268–2273. European Language Resources Association (2014)
23. Touati, I., Graja, M., Ellouze, M., Belguith, L.H.: Towards arabic semantic opinion mining: identifying opinion, polarity and intensity. In: Proceedings of the Mediterranean Conference on Pattern Recognition and Artificial Intelligence, pp. 131–136 (2016)
24. Verma, K., Davis, B.: Implicit aspect-based opinion mining and analysis of airline industry based on user-generated reviews. SN Comput. Sci. **2**(4), 286 (2021)
25. Xie, H., Li, Y., Pu, Y., Zhang, C., Huang, J.: Airline service quality ranking based on combined topsis-vikor-aism framework. arXiv preprint arXiv:2212.06332 (2022)

Text Categorization Can Enhance Domain-Agnostic Stopword Extraction

Houcemeddine Turki[1(✉)], Naome A. Etori[2], Mohamed Ali Hadj Taieb[1], Abdul-Hakeem Omotayo[3], Chris Chinenye Emezue[4], Mohamed Ben Aouicha[1], Ayodele Awokoya[5], Falalu Ibrahim Lawan[6], and Doreen Nixdorf[7]

[1] Data Engineering and Semantics Research Unit, University of Sfax, Sfax, Tunisia
turkiabdelwaheb@hotmail.fr
[2] University of Minnesota, Minneapolis, MN, USA
[3] University of California, Davis, CA, USA
[4] Technical University of Munich, Munich, Germany
[5] McPherson University, Lagos, Nigeria
[6] Kaduna State University, Kaduna, Nigeria
[7] Masakhane Research Community, Johannesburg, South Africa

Abstract. This paper investigates the role of text categorization in streamlining stopword extraction in natural language processing (NLP), specifically focusing on nine African languages alongside French. By leveraging the MasakhaNEWS, African Stopwords Project, and MasakhaPOS datasets, our findings emphasize that text categorization effectively filters out domain-specific keywords for most examined languages, allowing a simpler identification of stopwords. Nevertheless, linguistic variances result in lower detection rates for specific languages. Interestingly, we find that while over 40% of stopwords are common across news categories, less than 15% are unique to a single category. Uncommon stopwords add depth to the text, but their classification as stopwords depends on context. Therefore, combining statistical and linguistic approaches creates comprehensive stopword lists, highlighting the value of our hybrid method. This research enhances NLP for African languages and underscores the importance of text categorization in stopword extraction.

Keywords: Stopword Extraction · Text Categoization · African Languages · Domain-agnostic Stopwords

1 Introduction

Stopword extraction plays a key role in text analysis by removing commonly occurring but semantically insignificant words, such as "the," "is," and "of" [26]. Hence, it significantly enhances the NLP model's performance on various tasks such as sentiment analysis, topic modeling, and information retrieval by reducing

L. H. Belguith and K. Shaalan (Eds.): LPKM 2024, LNNS 1303, pp. 122–132, 2025.
https://doi.org/10.1007/978-3-031-85067-7_12

noise, improving text understanding, and ensuring consistency in analysis [26]. Removing stopwords shifts the focus to meaningful content words, streamlining computational processes and improving interpretability [7]. However, language-specific lists may be necessary to tailor the list of stopwords to specific contexts, especially for languages with complex morphological structures [7]. Research in this area shows numerous advanced approaches that have evolved to efficiently extract stopwords from language-specific text corpora. Commonly used but with minimal semantic value, stopwords can hinder in-depth analysis and strain computational resources [10,25].

A common approach to removing stopwords combines linguistic and statistical methods [10]. Linguistic methods rely on curated lists of stopwords, encompassing common articles, conjunctions, prepositions, and other low-value words [10]. By matching words in the text corpus against these lists, stopwords can be identified and eliminated to a more meaningful analysis [10]. Statistical approaches employ data-driven algorithms and machine learning (ML) models to automatically pinpoint stopwords based on word frequencies and patterns within the corpus [12,25]. Techniques like TF-IDF and probabilistic modeling help isolate and exclude stopwords statistically, resulting in more meaningful outcomes [25]. Recent NLP and deep learning (DL) advancements have introduced sophisticated stopword extraction techniques. Models like BERT and GPT-3, fine-tuned for context-aware stopword identification, adapt to language nuances and enhance precision [23]. These methods strive for accuracy while accommodating diverse text corpora and languages' unique traits [6].

This paper explores the potential of text categorization to simplify stopword extraction by filtering out domain-specific terms. Stopwords such as articles, conjunctions, and prepositions are universally present in the text, regardless of the topic or language in focus [12]. Their removal during text analysis ensures meaningful insights and avoids straining computational resources. We validate our hypothesis for nine African languages (Hausa, Somali, Igbo, Luganda, Kirundi, Shona, Swahili, Yoruba, and Nigerian Pidgin) and French by examining the presence of stopwords in a categorized corpus of African news articles. The rest of this paper is organized as follows: We begin with explaining the state-of-the-art of stopword extraction (Sect. 2) and outlining the language resources used in our study (Sect. 3). Subsequently, we introduce our approach (Sect. 4). Then, we present and discuss findings (Sect. 5). Finally, we conclude by summarizing our insights (Sect. 6).

2 Related Work

The state-of-the-art (SOTA) stopword identification for African languages is an emerging and ongoing area of research. Given Africa's linguistic diversity and scarce resources, identifying stopwords in these languages poses challenges [8]. Experts are enhancing NLP methodologies for these languages by refining tailored stopword lists [19], harnessing models trained on African text data [13], and deploying rule-based tactics sensitive to linguistic subtleties [28]. Collaboration among linguists, NLP professionals, and African language native speakers is pivotal for progress [8]. In this context, techniques range from curated stopword lists informed by native expertise to frequency-based tools [?] like word frequency [19], TF-IDF [18] and Information Entropy [3], as well as part-of-speech tagging [11] for grammatical insights. Additionally, machine learning models such as Naive Bayes and Recurrent Neural Networks (RNNs) are employed [13]. While rule-based strategies probe linguistic patterns, dictionary-centric methods tap into dedicated lexicons, and hybrid solutions merge various techniques, aiming for heightened stopword detection precision [17]. Recent advances in the field include the use of machine translation for stopword identification across multiple languages [24].

3 Resources

We leveraged projects and datasets from the *Masakhane* community [21], an organization advancing African NLP to evaluate our hypothesis. Drawing from *MasakhaNEWS* dataset [2], *African Stopwords Project* [8], a curated collection of stopwords, and *MasakhaPOS*, part-of-speech dataset [5] for African languages. Our choice to develop our experiments using the Masakhane resources is motivated by the availability of these open datasets under permissive licenses.

3.1 MasakhaNEWS

The MasakhaNEWS dataset [2] addresses the scarcity of African language datasets in NLP research by providing a news topic classification benchmark for 16 major African languages. This includes English and French as the widely-used African official languages and encompasses multiple language families, including Niger-Congo, Indo-European, English Creole, and Afro-Asiatic, representing various African regions such as East, West, etc. It involves classifying news articles such as "sports", "business", "entertainment", and "politics", acting as a performance benchmark for large language models. Traditionally, NLP emphasized high-resource languages, leaving African languages underrepresented. Despite their potential, multilingual language models were limited by the absence of appropriate evaluation datasets. The MasakhaNEWS dataset provides the necessary resource for evaluating multilingual models. Masakhane community annotators used a two-stage annotation blending manual and active learning for optimal quality. The dataset is freely accessible at https://github.com/masakhane-io/masakhane-news.

3.2 The African Stopwords Project

The project curates stopwords for low-resource African languages, which is essential for NLP tasks like information retrieval. Unlike high-resource languages, African languages lack standardized stopwords, limiting NLP advancement [8,9]. The project seeks to curate stopwords for African languages, with progress in 10 languages thus far. The project intends to use monolingual data to discern domain-specific stopwords for African languages, with aspirations to incorporate them into NLP tools or a dedicated Python package. The dataset is currently available at https://github.com/masakhane-io/masakhanePreprocessor/tree/main/african-stopwords.

3.3 MasakhaPOS

MasakhaPOS [5] is a key part-of-speech (POS) dataset supporting NLP research for 20 diverse African languages. It annotates POS tags for tokens, addressing a historical gap in resources for these languages. The dataset adheres to the Universal Dependencies (UD) guidelines, ensuring consistent and comparable POS annotations across the diverse languages in MasakhaPOS. [5].

MasakhaPOS is vital for NLP research and practice and crucial for machine translation, parsing, text chunking, spell and grammar checking, and more. It boosts NLP for often-overlooked African languages due to scarce annotated datasets. Native linguistic experts ensured quality annotations. The dataset comprises training, development, and test sets, ideal for POS model training and evaluation [5]. The dataset is accessible at https://github.com/masakhane-io/masakhane-pos.

4 Approach

We validated our hypothesis using a sample of nine African languages: two Afro-Asiatic (Hausa, Somali), six Niger-Congo (Igbo, Luganda, Kirundi, Shona, Swahili, Yoruba), and one English Creole (Nigerian Pidgin). We selected African languages based on stopword availability in MasakhaPOS and the African Stopwords Project (Table 1). We also included French, supported by MasakhaNEWS but not by MasakhaPOS or the African Stopwords Project, for robust results. French stopwords were sourced from the Stopwords-ISO dataset (https://github.com/stopwords-iso), a comprehensive collection of stopwords for multiple languages.

The African Stopwords Project and Stopwords-ISO feature crowdsourced stopwords, from publicly accessible lists, particularly those derived using probabilistic methods like TF-IDF [8]. We sourced the African language lists from the African Stopwords Project and used them in our study. Similarly, we obtained the French stopwords from Stopwords-ISO. As for MasakhaPOS, we extracted terms labeled with tags: *Universal Dependencies* tags: *Auxiliary Verbs* (AUX), *Pronouns* (PRON), *Coordinating Conjunctions* (CCONJ), *Subordinating Conjunctions* (SCONJ), *Determiners* (DET), and *Particles* (PART). Subsequently, we combined stopwords from MasakhaPOS, African Stopwords Project, and Stopwords-ISO into one unified list of stopwords for each studied language. Then, we standardized them by lowercasing and removing duplicates for a concise set of stopwords for evaluation.

We then analyzed the MasakhaNEWS development set to study the distribution of these stopwords across 'news' categories. Our analysis sought to

Table 1. African languages availability in MasakhaPOS and the African Stopwords Project.

Language	MasakhaPOS	African Stopwords
Hausa (hau)	✓	✓
Igbo (ibo)	✓	×
Luganda (lug)	✓	×
Nigerian Pidgin (pcm)	✓	✓
Kirundi (run)	×	✓
Shona (sna)	✓	×
Somali (som)	×	✓
Swahili (swa)	✓	✓
Yoruba (yor)	✓	✓

discern the distribution of stopwords among different news item categories. We first broke down the text into words, removed punctuation, and standardized all words to lowercase. This standardization facilitated consistent and accurate analysis throughout the text. After standardization, we identified unique words for each 'news' category in the MasakhaNEWS dataset. This allowed us to consistently identify unique words in each 'news' category. Finally, we calculated the number of MasakhaNEWS categories that include every stopword. This evaluation seeks to ascertain if text classification can bolster the efficiency of stopword collection.

5 Results and Discussion

MasakhaNEWS features a broad range of news articles with thousands of unique words across multiple languages, as shown in Table 2. These articles are sorted into distinct, non-overlapping categories, ranging from four to seven, including topics such as "business", "entertainment", "sports", and "technology" [2]. This categorization aids in testing our hypothesis on the role of text categorization in identifying domain-agnostic stopwords.

The combined use of crowdsourcing, human curation, and TF-IDF-generated word lists has proven highly effective in identifying stopwords for African languages (The African Stopwords Project) and French (Stopwords-ISO) as shown in Table 4, this collaborative approach has identified numerous stopwords for each language, underscoring the importance of both statistical techniques and human input in creating stopwords lists [8,25]. Additionally, using MasakhaPOS [5] to automatically filter POS tags and determine stopwords based on their grammatical functions has been equally successful. This method is based on the *Universal Dependencies* POS tags. This approach, recommended by [10] has yielded results comparable to statistical and crowdsourcing techniques. Merging

Table 2. Statistical information about the coverage of African languages in the development set of MasakhaNEWS.

Language	Categories	Unique words
French (fra)	5	18290
Hausa (hau)	7	10495
Igbo (ibo)	6	8441
Luganda (lug)	5	8186
Nigerian Pidgin (pcm)	5	8057
Kirundi (run)	6	15363
Shona (sna)	4	11551
Somali (som)	7	14389
Swahili (swa)	7	18532
Yoruba (yor)	5	8210

Table 3. Distribution of identified stopwords based on their presence across MasakhaNEWS categories. *N* corresponds to the number of categories where keywords exist.

Language	Found Stopwords	Available in *N* Categories						
		1	2	3	4	5	6	7
French (fra)	466 (67.5%)	69	43	61	64	229	N/A	N/A
Hausa (hau)	319 (97.0%)	16	20	19	28	38	52	146
Igbo (ibo)	49 (70.0%)	5	4	5	2	6	27	N/A
Luganda (lug)	90 (62.1%)	11	14	11	16	38	N/A	N/A
Nigerian Pidgin (pcm)	84 (86.6%)	2	7	11	11	53	N/A	N/A
Kirundi (run)	55 (93.2%)	3	0	0	1	6	45	N/A
Shona (sna)	175 (86.6%)	21	30	31	93	N/A	N/A	N/A
Somali (som)	25 (83.3%)	0	1	1	0	0	0	23
Swahili (swa)	151 (96.8%)	1	3	5	3	11	59	69
Yoruba (yor)	62 (38.8%)	3	3	3	4	49	N/A	N/A

the stopwords from these two methods shows that each can detect unique stopwords, underscoring the merit of combined approaches. This supports the idea that blending multiple NLP techniques enhances stopword identification over a single method [6].

In the MasakhaNEWS dataset, most languages had a favorable stopword detection rate of over 80%, as shown in Table 3. Yet, French had a rate of 67.5%, and three Niger-Congo languages-Igbo 70.0%, Luganda 62.1%, and Yoruba 38.8%-had lower rates. The agglutinative nature of languages like Yoruba [4], Igbo [20], and Luganda [15], which merges stopwords with subsequent terms, might have contributed to these variances. This issue complicates stopword iden-

Table 4. Statistical information about the identification of stopwords from MasakhaPOS and the African Stopwords Project (ASP, for African languages) or Stopwords-ISO (S-ISO, for French).

Language	MasakhaPOS	ASP or S-ISO	Stopwords
French (fra)	N/A	690	**690**
Hausa (hau)	90	321	**329**
Igbo (ibo)	70	N/A	**70**
Luganda (lug)	145	N/A	**145**
Nigerian Pidgin (pcm)	95	33	**97**
Kirundi (run)	N/A	59	**59**
Shona (sna)	202	N/A	**202**
Somali (som)	N/A	30	**30**
Swahili (swa)	97	103	**156**
Yoruba (yor)	122	60	**160**

tification, emphasizing the need of agglutinated stopwords in stopword lists. The apostrophe (') in French [22] might have also impacted its rate, especially since punctuation was removed during data pre-processing.

The analysis shows over 40% of considered stopwords appear across all Masakhane NEWS categories, with high rates in Somali 92.0%, Kirundi 81.8%, Yoruba 79.0%, and Nigerian Pidgin 63.1% (Table 3). Less than 15% are unique to one category, with low instances in Somali 0.0%, Swahili 0.6%, Nigerian Pidgin 2.4%, and Yoruba 4.8%.

Uncommon stopwords, highlighted in bold in Table 5, add depth to texts with nouns, verbs, and adverbs, often acting as verbal cues [27], especially in narrative-driven African languages [1]. An example is *dana* (son) in Hausa. However, not all are definitively stopwords. But context matters; terms like numerals, frequency-related adverbs, and adjectives, time, or color (italicized in Table 5) are not always stopwords. Some words carry meaning in specific contexts and structures [14,16]. Examples can be *dari* (one hundred) and *balle* (let alone) in Hausa. Thus, text categorization efficiently allows the elimination of odd terms (retrieved using other approaches such as *TF-IDF*) from domain-agnostic stopwords.

Table 5. List of stopwords featured in a unique MasakhaNEWS category for the nine considered African languages (with meaning).

Language	Uncommon stopwords
Hausa (hau)	**dana** (son), *dari* (one hundred), **lalle** (certainly), yayinda (while), shima (too), *milyan* (million), **guji** (avoid), *balle* (let alone), kanka (yourself), basa (they are not), namu (ours), sune (they are), **kwarai** (absolutely or extremely), wadancan (those), **daman** (right side), *daukacin* (all)
Igbo (ibo)	i (you), o (user), *ozo* (again), **ihi** (reason), *imirikiti* (many)
Luganda (lug)	teyali (it was not), **ebyali** (he was eating), ekyali (which was), alabika (appears), byali (were), **okwaliwo** (which occurred), **yalina** (it was raining), **kaali** (cabbage), *egisinga* (most of them), bakyali (they are still), kyayo (its own)
Nigerian Pidgin (pcm)	sey (as if), non (on point)
Kirundi (run)	kw (on, at, in, and from), nk (seems like),
Shona (sna)	ry (there), **vashaye** (miss them), haafanirwe (he should not), inove (it is), *uchiri* (you are still), racho (its), ndave (I have been), **vaifanirwa** (they deserved it), *dzinogara* (they last), **panga** (sword), *dzinenge* (almost), dzainge (were), raro (its), sezviri (as it is), vavari (who they are), chavari (what they are), mumwe (one), neku (and), **haro** (necessarily), hunogona (it can), vangadai (they would have), **dzimwe** (others)
Somali (som)	N/A
Swahili (swa)	yasio (not)
Yoruba (yor)	í (it), kì (do not), é (yes)

6 Conclusion

Text categorization's influence on stopword extraction on MasakhaNEWS dataset offers great insights. Linguistic methods, statistics, and Masakhane community experts have found that text categorization reliably identifies common stopwords across news categories. However, detection methods vary, especially in languages with intricate linguistic characteristics, hence, language-specific considerations are needed in stopword identification. We have identified unique stopwords that add depth and meaning to news categories, such as nouns, verbs, and adverbs-context in stopword extraction is crucial.

Future research will focus on expanding language coverage, using context-aware techniques for stopword extraction, multilingual models, and hybrid techniques, addressing challenges with agglutinative languages, standardized metrics, domain-specific stopwords, tailored tools, cross-language methods, and integrating ethical aspects.

Acknowledgments. We thank David Adelani (University College London, United Kingdom) and Akintunde Oladipo (University of Waterloo, Canada) for providing useful comments and discussion regarding this work. Certain portions of this work have undergone language proofreading and editing with the assistance of *ChatGPT*, a robust chatbot powered by OpenAI's advanced language model.

Data Availibility Statement. Source data links are provided throughout this manuscript. The meaning of the uncommon stopwords in Table 5 are collected from Google Translate (https://translate.google.ca), Hausa Dictionary (https://hausadictionary.com), Kirundi Study and Dictionary (https://www.matana.de/index1.php), Glosbe (https://glosbe.com/), Naija Lingo (http://www.naijalingo.com/), and the Masakhane Community.

Disclosure of Interests. This work is done within the framework of *Masakhane*, the African grassroots community for natural language processing.

Source Code. The source code is available at https://github.com/masakhane-io/masakhanePreprocessor/blob/main/african-stopwords/Masakhane_Stopword_Extraction_and_Validation.ipynb for reproducibility purposes.

References

1. Abdi, A.A.: Oral societies and colonial experiences: Sub-Saharan Africa and the de-facto power of the written word. In: Education, Decolonization and Development, pp. 39–56. BRILL (2009). https://doi.org/10.1163/9789087909260_004
2. Adelani, D.I., Masiak, M., Azime, I.A., et al.: MasakhaNEWS: News Topic Classification for African languages (2023). https://doi.org/10.48550/ARXIV.2304.09972
3. Asubiaro, T.V.: Entropy-based generic stopwords list for Yoruba texts. Int. J. Comput. Inf. Technol. **2**(5) (2013). https://www.academia.edu/9153147/Entropy_Based_Generic_Stopwords_List_for_Yoruba_Texts
4. Babarinde, O.: Linguistic analysis of the structure of yoruba numerals. Lang. Matters **45**(1), 127–147 (2014). https://doi.org/10.1080/10228195.2013.857362
5. Bamba Dione, C.M., Adelani, D.I., Nabende, P., et al.: MasakhaPOS: part-of-speech tagging for typologically diverse African languages. In: Proceedings of the 61st Annual Meeting of the Association for Computational Linguistics (Volume 1: Long Papers), pp. 10883–10900. Association for Computational Linguistics (2023). https://doi.org/10.18653/v1/2023.acl-long.609
6. Chekima, K., Alfred, R.: An automatic construction of malay stop words based on aggregation method. In: Berry, M.W., Hj. Mohamed, A., Yap, B.W. (eds.) SCDS 2016. CCIS, vol. 652, pp. 180–189. Springer, Singapore (2016). https://doi.org/10.1007/978-981-10-2777-2_16

7. Dolamic, L., Savoy, J.: When stopword lists make the difference. J. Am. Soc. Inf. Sci. Technol. **61**(1), 200–203 (2009). https://doi.org/10.1002/asi.21186
8. Emezue, C., Nigatu, H., Thinwa, C., et al.: The African Stopwords project: curating stopwords for African languages (2023). https://doi.org/10.48550/ARXIV.2304.12155
9. Emezue, C., et al.: The african stopwords project: curating stopwords for african languages. arXiv preprint arXiv:2304.12155 (2023)
10. Ferilli, S., Esposito, F., Grieco, D.: Automatic learning of linguistic resources for stopword removal and stemming from text. Procedia Comput. Sci. **38**, 116–123 (2014). https://doi.org/10.1016/j.procs.2014.10.019
11. Ganesh, B.R., Gupta, D., Sasikala, T.: Grammar error detection tool for medical transcription using stop words parts-of-speech tags ngram based model. In: Bhateja, V., Tavares, J.M.R.S., Rani, B.P., Prasad, V.K., Raju, K.S. (eds.) Proceedings of the Second International Conference on Computational Intelligence and Informatics. AISC, vol. 712, pp. 37–49. Springer, Singapore (2018). https://doi.org/10.1007/978-981-10-8228-3_5
12. Gerlach, M., Shi, H., Amaral, L.A.N.: A universal information theoretic approach to the identification of stopwords. Nat. Mach. Intell. **1**(12), 606–612 (2019). https://doi.org/10.1038/s42256-019-0112-6
13. Gorro, K.D., Ali, M.F., Lawas, L.A., Ilano, A.S.: Stop words detection using a long short term memory recurrent neural network. In: Proceedings of the 2021 9th International Conference on Information Technology: IoT and Smart City, pp. 199–202 (2021)
14. Jóhannsdóttir, K.M.: Temporal adverbs in icelandic: adverbs of quantification vs. frequency adverbs. Nordic J. Linguist. **30**(2), 157–183 (2007). https://doi.org/10.1017/s0332586507001734
15. Katamba, F.: A nonlinear analysis of vowel harmony in luganda. J. Linguist. **20**(2), 257–275 (1984). https://doi.org/10.1017/s002222670001389x
16. Keenan, E.L., Stavi, J.: A semantic characterization of natural language determiners. Linguist. Philos. **9**(3), 253–326 (1986). https://doi.org/10.1007/bf00630273
17. Ladani, D.J., Desai, N.P.: Stopword identification and removal techniques on TC and IR applications: a survey. In: 2020 6th International Conference on Advanced Computing and Communication Systems (ICACCS). IEEE (2020). https://doi.org/10.1109/icaccs48705.2020.9074166
18. Miretie, S.G., Khedkar, V.: Automatic generation of stopwords in the Amharic text. Int. J. Comput. Appl. **975**, 8887 (2018). https://www.ijcaonline.org/archives/volume180/number10/miretie-2018-ijca-916161.pdf
19. Niyongabo, R.A., Hong, Q., Kreutzer, J., Huang, L.: KINNEWS and KIRNEWS: benchmarking cross-lingual text classification for Kinyarwanda and Kirundi. In: Proceedings of the 28th International Conference on Computational Linguistics, pp. 5507–5521. International Committee on Computational Linguistics, Barcelona, Spain (Online) (2020). https://doi.org/10.18653/v1/2020.coling-main.480
20. Onyenwe, I.E., Hepple, M., Chinedu, U., Ezeani, I.: Toward an effective igbo part-of-speech tagger. ACM Trans. Asian Low-Resour. Lang. Inf. Process. **18**(4), 1–26 (2019). https://doi.org/10.1145/3314942
21. Orife, I., Kreutzer, J., Sibanda, B., et al.: Masakhane - Machine Translation For Africa. arXiv preprint arXiv: 2003.11529 (2020). https://arxiv.org/abs/2003.11529v1
22. Panckhurst, R.: Texting in three European languages : does the linguistic typology differ? In: i-Mean 2009 Issues in Meaning in Interaction, Bristol, United Kingdom, pp. 119–136 (2009). https://hal.science/hal-00443016

23. Qiao, Y., Xiong, C., Liu, Z., Liu, Z.: Understanding the Behaviors of BERT in Ranking (2019). https://doi.org/10.48550/ARXIV.1904.07531
24. Rajwal, S.: Lihisto: a comprehensive list of hindi stopwords. Multimedia Tools Appl. **83**(17), 50047–50059 (2023)
25. Rani, R., Lobiyal, D.: Automatic construction of generic stop words list for Hindi text. Procedia Comput. Sci. **132**, 362–370 (2018). https://doi.org/10.1016/j.procs.2018.05.196
26. Sarica, S., Luo, J.: Stopwords in technical language processing. PLoS ONE **16**(8), e0254937 (2021). https://doi.org/10.1371/journal.pone.0254937
27. Treisman, A.M.: Verbal cues, language, and meaning in selective attention. Am. J. Psychol. **77**(2), 206 (1964). https://doi.org/10.2307/1420127
28. Yeshambel, T., Mothe, J., Assabie, Y.: Amharic Adhoc information retrieval system based on morphological features. Appl. Sci. **12**(3), 1294 (2022). https://doi.org/10.3390/app12031294

Normalized Orthography for Tunisian Arabic

Houcemeddine Turki[1(✉)], Kawthar Ellouze[2], Hager Ben Ammar[2], Mohamed Ali Hadj Taieb[1], Imed Adel[3], Mohamed Ben Aouicha[1], Pier Luigi Farri[4], and Abderrezak Bennour[2]

[1] Data Engineering and Semantics Research Unit, University of Sfax, Sfax, Tunisia
turkiabdelwaheb@hotmail.fr
[2] Association Derja, Tunis, Tunisia
[3] University of Sousse, Sousse, Tunisia
[4] VBScuola, Rome, Italy

Abstract. Tunisian Arabic (ISO 693-3: aeb) is a distinct variety native to Tunisia, derived from Arabic and enriched by various historical influences. This research introduces the "Normalized Orthography for Tunisian Arabic" (NOTA), an adaptation of CODA* guidelines for transcribing Tunisian Arabic using Arabic script. The aim is to enhance language resource development by ensuring user-friendliness and consistency. The updated standard addresses challenges in accurately representing Tunisian phonology and morphology, correcting issues from transcriptions based on Modern Standard Arabic.

Keywords: Tunisian · Normalization · Orthography · Arabic Script

1 Introduction

Tunisian (ISO 693-3: aeb) is a linguistic variety native to Tunisia [12], emerging from Arabic with influences from Latin, Berber, Italian, French, Spanish, Turkish, Mediterranean Lingua Franca, and possibly Neo-Punic [35]. These influences shaped Tunisian into a distinct entity, separate from Classical Arabic [35]. Adoption of the Arabic script in the 19th century facilitated systematic recording and study of the language. This transition was triggered in the 1930s as part of Nationalist efforts to promote independence from France and improve literacy among Tunisians [28]. Literary production in Tunisian, led by intellectuals like Ali Douagi, gained momentum [28]. The Tunisian pro-independence party Neo-Destour collaborated with linguist William Marçais to describe dialects and create text corpora for Tunisian [19, pp. 216-250]. This initiative waned following post-independence Arabization policies but resurfaced after the Tunisian revolution [28].

These efforts, drawing inspiration from Modern Standard Arabic orthography to transcribe Tunisian [39], were formalized as guidelines by Maamouri et al. (2004) [23] and Makhoul et al. (2005) [5]. Habash et al. (2012) [15] expanded this

L. H. Belguith and K. Shaalan (Eds.): LPKM 2024, LNNS 1303, pp. 133–142, 2025.
https://doi.org/10.1007/978-3-031-85067-7_13

into CODA, with adaptations for specific Arabic languages, including Tunisian [43,45] and Maghrebi Arabic [39]. These adaptations later amalgamated into CODA*, facilitating comparative analyses [2], part-of-speech tagging [1], sentiment analysis [33], morphological modeling [42], and corpus creation [3]. Despite consensus among computational linguists, CODA* has not yet seen widespread adoption or evaluation by the general audience, motivating further exploration. CODA* Guidelines, designed for transcribing Arabic dialects, may pose limitations for speakers unfamiliar with Modern Standard Arabic, resulting in conflicting transcriptions and impeding consistent morphological patterns for Tunisian Arabic. Trials to solve these issues by developing statistical and computational approaches for normalizing the transcription of Tunisian using the CODA* Guidelines have been proposed by Besdouri et al. (2021) [8] and Mekki et al. (2022) [29]. However, limited efforts have been made to develop an orthography normalization for Tunisian using the principles of Applied Linguistics.

In this research paper, we introduce our adaptation of the CODA* guidelines for transcribing Tunisian Arabic using the diacritized or undiacritized Arabic script. This will allow us to benefit from the achievements of the CODA* Guidelines instead of developing a new standard from scratch. We present the resulting "Normalized Orthography for Tunisian Arabic" (NOTA), prioritizing user-friendliness and consistency. Our paper is structured as follows: We provide a linguistic overview of Tunisian (Sect. 2), specify the CODA* Guidelines for Arabic Languages (Sect. 3), explain our approach to developing NOTA (Sect. 4), describe our guidelines for transcribing Tunisian using the Arabic script (Sect. 5), explain possible applications of the NOTA Guidelines related to computational linguistics (Sect. 6), and conclude with future research directions (Sect. 7).

2 Linguistic Overview of Tunisian

Tunisian Arabic, a unique variant of Arabic, is the primary spoken language in Tunisia and is widely used in media, including music, television, and radio [12]. It shares features with other Maghreb languages [9]. Notable aspects include conservative consonant sounds, the urban use of "إنتِي [ɪnti]" (you), and no gender distinctions in the second person verb forms [12]. Unlike Modern Standard Arabic (MSA), Tunisian Arabic lacks an indicative prefix, blurring indicative and subjunctive moods [12]. It employs a progressive aspect using "قاعد [qɛːʕɪd]" (to remain doing something) and "فِي [fi]" (in) in transitive clauses [34]. The future tense uses prefixes like "باش [bɛːʃ]" or "بِش [bəʃ]" combined with verbs [12].

Tunisian Arabic follows a Subject-Verb-Object (SVO) order and generally acts as a Null-subject language [12]. It shows more agglutinative structures compared to Standard Arabic and other varieties, influenced by Turkish during the 17th century [13]. An example is "شنعمل [ʃnaʕmɪl]" (What should I do), a fusion of "آش [ɛːʃ]" (What) and "نعمل [naʕmɪl]" (I do). Dialects of Tunisian Arabic include Northwestern, Southwestern, Tunis, Sahel, Sfax, Judeo-Tunisian, and Southeastern. These mostly belong to the Hilalian group, except for the Sfax and Judeo-Tunisian variants [12]. While the voiceless uvular stop [q] is used in Tunisian, Southeastern, Northwestern, and Southwestern varieties use the voiced velar plosive [g] [12]. Some words consistently maintain the [g] sound, especially in agriculture, such as "بڨرة [bægra]" (cow), where /g/ derives from an originally Arabic [q], and "دڨلة [dɪglæ]" (dates) [4].

Phonologically, Tunisian Arabic is distinct from MSA. It predominantly retains interdental fricatives, except in the Sahel dialect [32], and merges /dˤ/ ض with /ðˤ/ ظ [20]. It includes a wide range of consonant phonemes, with emphatic consonants being relatively rare [20]. Non-Arabic words sometimes include /p/ and /v/, usually substituted with /b/ [20]. The glottal stop /ʔ/ is generally omitted but present in formal registers [20]. Shadda "gemination" frequently occurs, extending to final consonants in closed syllables [20]. Tunisian Arabic vowels include /a/, /ɛ/, /i/, and /u/, with vowel length being crucial [20]. Stress placement depends on syllable structure, favoring the ultimate syllable if doubly closed or the penultimate syllable otherwise [20]. Various phonological processes highlight its phonetic diversity and divergence from MSA [20].

The lexicon of Tunisian Arabic blends native, Latin, and Berber words with borrowed vocabulary from Italian, Spanish, French, and Turkish [38]. For example, "تريسيتي [triːsiːti]" (electricity) is borrowed from French, and "برتمان [bortmɛːn]" (flat) and "بياسة [bjɛːsæ]" (coin) have French roots. Turkish influence appears in words like "ڨاوري [gɛːwri]" (foreigner) and the occupation suffix "جي /ʒi/" [38]. Differences between Tunisian and Standard Arabic often result from shifts in the meanings of Arabic roots. For example, "خدم [xdɪm]" means to serve in Standard Arabic but to work in Tunisian, while "عمل [ʕmɪl]" means to work in Standard Arabic but has a broader sense of to do in Tunisian [6]. Word fusion is a significant mechanism in Tunisian Arabic, especially in question words, such as "اش [ɛːʃ]" or "ش [ʃ]" [12]. This feature is absent in MSA, Levantine, and Gulf Arabic [12].

3 CODA* Guidelines

CODA* is a conventional orthography for Dialectal Arabic (DA), designed primarily for computational modeling. Its mission is to provide a unified framework for writing various Arabic dialects, which differ significantly from Modern Standard Arabic (MSA) and lack standard orthography [16]. The goals and intentions of CODA* include maintaining internal consistency and coherence for writing DA, using the Arabic script, striking a balance between dialectal uniqueness and MSA-DA similarities, and ensuring ease of learning and readability [16]. The design principles of CODA* resemble English spelling in some ways, with phonological and historical considerations and some exceptions. It uses only Arabic script characters and can be written with or without diacritics [16]. Each DA word in CODA* has a unique orthographic form that represents its phonology, morphology, and meaning [16]. CODA* generally follows MSA-like orthographic decisions and preserves the phonological form of dialectal words, with some exceptions for highly variant root radical letters and pattern vowels [16]. It also maintains dialectal morphology and syntax, except for separating negation and indirect object pronouns. CODA* rules are dialect-independent and include specific rules for certain word classes to preserve morphological information, dialect integrity, and readability [16]. Overall, CODA* aims to make writing and learning dialectal Arabic more accessible and consistent. This is enhanced through the development of several tools to adjust the spelling of raw texts in Arabic languages to meet CODA* Guidelines such as the MADAR CODA Corpus [11]. Detailed instructions for the CODA* Guidelines are available at https://camel-guidelines.readthedocs.io/.

4 Proposed Approach

In our study, we aimed to improve the usability and clarity of the CODA* Guidelines by addressing user-reported issues. We engaged both native and non-native Tunisian speakers in structured interviews, live tests, and discussions on May 4 and May 24, 2017, considering the linguistic nuances of Tunisian Arabic. A text excerpt was dictated to three groups: A group provided with the CODA Guidelines as adapted for Tunisian by Zribi et al. (2014) [43], a group allowed to use the Tunisian Arabic Corpus (https://www.tunisiya.org/) by McNeil and Faiza [27] to see how to transcribe words, and a group with no support, spontaneously transcribing what was dictated. These tests were conducted by the Derja Association. Each group's output was assessed by another group to identify limitations. Structured interviews explored participants' experiences using common transcription rules for Tunisian. Choices to adjust the CODA* guidelines were made by a group of applied linguists based on the observations.

5 Differences from CODA* Guidelines

We will begin by establishing transcription conventions for consonants, then explore short and long-vowel transcription methods following NOTA Guidelines (Sect. 5.1). Finally, we will provide a comprehensive explanation of the NOTA orthography rules for Tunisian in the Arabic Script that do not apply to CODA* (Sect. 5.2). We provide the CODA* Guidelines that we kept for the NOTA Guidelines at https://doi.org/10.6084/m9.figshare.25892974.

5.1 Consonants and Vowels

We follow the CODA* Guidelines for our revision [16] and introduce three supplemental letters from the Perso-Arabic Script to represent /p/(پ), /v/ (ڥ), and /g/(ڨ) [39]. These sounds are not adequately represented in Arabic, causing potential reading confusion [39]. Our additional letters align with common transcription practices among Tunisian speakers [39]. For the full table of consonant representations in Tunisian, refer to the table of consonants at https://doi.org/10.6084/m9.figshare.25892974.

As for the vowels, unlike Modern Standard Arabic, Tunisian Arabic features four vowel qualities (/a/, /ɛ/, /i/, and /u/) [20]. CODA* suggested representing long /ɛ/ as ِا (U+0650, U+0627) to distinguish it from long /a/ [16]. However, the use of ِا (U+0650, U+0627) might not be practical as it could potentially lead to confusion among Tunisian learners because ـِ (U+0650) is mainly used to denote /i/. Furthermore, the rule for differentiating between /mˤ/, /nˤ/, /bˤ/, /zˤ/, and /rˤ/ and their non-emphatic minimal pairs also applies when the respective letter is adjacent to a short /a/. Therefore, a new Arabic diacritic for /ɛ/ needs to be defined. In this context, we propose adopting the Perso-Arabic diacritics used for Pashto, a language that also has four vowel qualities. We consequently use Fatha (U+064e), Zwarakai (U+0659), Damma (U+064f), and Kasra (U+0650) as diacritics. The details of vowel transcription for Tunisian are provided in the table of vowels at https://doi.org/10.6084/m9.figshare.25892974.

5.2 Revised Rules

Applying several CODA* guidelines to Tunisian revealed conflicts and user confusion. Thus, we adjusted several rules for a better transcription experience. First, the original CODA* Guidelines transcribe final /a/ and /ɛ/ as an Alif Maqsura ى,, reflecting MSA etymology. For instance, /jɛ/ (arrive) is transcribed as جا because it appears in MSA as جاء [16,43]. We propose a morphology-aware rule: verbs are written with Alif Maqsura if the present form in the singular masculine third person ends with long /i/ ي. This applies to nouns derived from such verbs. For example, جاء in MSA becomes "جى [ʒæ] " in Tunisian because it conjugates as "يجي [jʒi]" in the present form of the singular masculine third person [6]. Another example is "ما كلموش [mæ kællmuːʃ]", meaning both *he does not call him* and "they did not call". To disambiguate, [uː] (meaning "him") can be replaced with هو [huː]. Second, morphological adjustments address irregularities in verb conjugation. In Tunis dialect, the imperative form of "كلا /klɛ/" is "كول [kuːl]" [6]. The second person singular present form "تاكل" [tɛːkɪl]" suggests removing "ت" /t/ to form the imperative. Sfax dialect also uses "آكل" [ɛːkɪl]" as the imperative, indicating the removal of exceptions could standardize Tunisian morphology, without favoring the Tunis dialect [12,18].

Third, we propose to disambiguate homonyms when possible. For example, "باش [bɛːʃ] " or "بِش [bɪʃ]" combined with verbs means *will* (modal verb) [12]. We write it as "بِش [bəʃ]" to distinguish it from "باش [bɛːʃ]", meaning *with what.* Fourth, unlike the CODA* Guidelines, the definite article /il/ is always transcribed as ال (U+0627, U+0644) without diacritics to differentiate it from similar structures in diacritized texts. Fifth, we maintain a space between prepositions and the nouns they follow, despite Tunisian's agglutinative pattern. The same applies to the conjunction "و /w/" (U+0648), kept separate from the following noun. These choices were informed by examining the Tunisian Arabic Corpus [27] and user tests. Sixth, foreign vowels are transcribed according to their phonological similarity to Tunisian vowels, with [y] assimilated to /i/ and [ə] to /u/. Foreign vowels are written as long vowels unless in stressed syllables. Examples include loanwords like "فريكساي ***fricassé*** *fricassé*" (fried bread) and "أوردينتور ***ordinateur*** *ordinateur*" (computer). Revised prepositions, foreign vowel conversion rules, and an example of how an excerpt is transcribed using spontaneous writing as well as NOTA and CODA* Guidelines are available at https://doi.org/10.6084/m9.figshare.25892974.

6 NOTA-Based Applications

The NOTA guidelines have potential across various domains, including education [21], language preservation [10], mass media [10], and cultural heritage [37]. However, the primary impact of this CODA* revision lies in fostering the development of linguistic resources and tools for Tunisian [41]. Current efforts to create language resources for Tunisian are limited to raw or annotated undiacritized corpora and bilingual lexicons [41], primarily supporting applications in computer science, including natural language processing and information retrieval such as sentiment analysis [24], machine translation [36], part-of-speech tagging [17], Arabic script transliteration of Latin script texts [25], Arabic diacritization [26], and morphological disambiguation [44]. Despite these efforts, Tunisian remains an under-resourced language, with text corpora containing only a few texts and lexicons with limited vocabulary [41]. The most extensive resource is the Tunisian Arabic Corpus, an online corpus with 2,462 texts totaling 1,085,493 words [27]. The revision of CODA*, through the NOTA guidelines, aims to address this limitation by enabling the development of large-scale corpora, lexicons, and resources for modeling Tunisian morphology.

6.1 Corpus and Lexicon Creation

Creating corpora and lexicons for Tunisian requires integrating linguistic expertise, computational methods, and ontological frameworks. This process is crucial for tasks like Arabic Diacritization and Named Entity Recognition (NER) [30]. Corpus creation involves compiling a representative collection of texts, annotated with linguistic information like diacritics and named entities [30]. Orthography guidelines ensure consistency in representing Tunisian Arabic, specifying how words and phrases, including diacritics, are written [43], enhancing corpus quality and supporting subsequent NLP tasks [30]. Orthography guidelines, particularly NOTA, are also essential for constructing well-organized lexicons. OntoLex and the Lexical Markup Framework (LMF) provide structured representations of lexical information, facilitating integration into semantic applications [14]. Adapting WordNet for Tunisian Arabic involves creating synsets and defining semantic relationships to represent the language's lexical nuances [30]. To incorporate Tunisian Arabic into multilingual knowledge graphs like Wikidata, specific Tunisian labels can be added using statements such as *rdfs:label* and *skos:altLabel*, establishing links to corresponding concepts and enabling cross-language information retrieval and knowledge sharing [40].

6.2 Morphology Modeling

Morphology modeling, crucial in natural language processing, studies the internal structure of words [7]. The Grammar Matrix is a framework enabling the construction of broad-coverage, cross-linguistically applicable grammars [7]. It allows customization by combining linguistic features [7]. The Grammatical Framework (GF) is another notable tool for morphology modeling [31], focusing on abstract syntax for generating sentences across languages [31]. The Arabic Treebank is significant for Arabic morphology modeling [22], providing annotated data for Arabic syntax and morphology, aiding computational model development [22]. While these initiatives support various languages, including Arabic, they have limited support for Arabic dialects like Tunisian. Efforts like CODA* have improved this situation, enabling the creation of small-scale treebanks for Tunisian [30]. NOTA guidelines aim to resolve conflicts in CODA*-based transcription of morphology rules.

7 Conclusion

In this research paper, we introduced the "Normalized Orthography for Tunisian Arabic" (NOTA), an adapted version of the CODA* guidelines tailored for transcribing Tunisian Arabic using the Arabic script. NOTA aims to bridge the gap between computational linguistics and wider acceptance of standardized Arabic script transcription for Tunisian Arabic. Through collaboration with native and non-native Tunisian speakers and extensive testing, NOTA emerges as a valuable resource for transcription and linguistic analysis, with future research focusing on its further refinement and utility in various applications.

Acknowledgments. We thank Mohamed Maamouri (University of Pennsylvania, United States of America), Nizar Habash (New York University Abu Dhabi, United Arab Emirates), Salam Khalifa (Stony Brook University, United States of America), Dominique Caubet (Alliance Sorbonne-Paris-Cité, France), Ramzi Hachani (United Nations Office for Project Services, United States of America), Ramzi Cherif (Derja Association, Tunisia), Rafik Zribi (IÉSEG School of Management, France), Lameen Souag (Alliance Sorbonne-Paris-Cité, France), Tariq Daouda (Université de Montréal, Canada), and Nassim Regragui (Copenhagen Business School, Denmark) for useful comments and discussion. Certain portions of this work have been proofread using *ChatGPT*, a robust chatbot powered by OpenAI.

Disclosure of Interests. All the co-authors except Mohamed Ali Hadj Taieb and Mohamed Ben Aouicha are members of Derja Association, the non-profit organization promoting the widespread use of Tunisian.

References

1. AlKhwiter, W., Al-Twairesh, N.: Part-of-speech tagging for Arabic tweets using CRF and Bi-LSTM. Comput. Speech Lang. **65**, 101138 (2021)
2. Alsudais, A., Alotaibi, W., Alomary, F.: Similarities between Arabic dialects: investigating geographical proximity. Inf. Process. Manag. **59**(1), 102770 (2022)
3. Althobaiti, M.J.: Creation of annotated country-level dialectal Arabic resources: an unsupervised approach. Nat. Lang. Eng. **28**(5), 607–648 (2021)
4. Baccouche, T.: Le phonème /g/ dans les parlers arabes citadins de Tunisie. In: Proceedings of the seventh International Congress of Phonetic Sciences / Actes du Septième Congrès international des sciences phonétiques, pp. 652–655. De Gruyter (1972)
5. BBN Technologies, Makhoul, J., Zawaydeh, B., Choi, F., Stallard, D.: BBN/AUB DARPA Babylon Levantine Arabic Speech and Transcripts. Linguistic Data Consortium (2005)
6. Ben Abdelkader, R.: Peace Corps English-Tunisian Arabic Dictionary. ERIC Clearinghouse (1977)
7. Bender, E.M., Flickinger, D., Oepen, S.: The grammar matrix: an open-source starter-kit for the rapid development of cross-linguistically consistent broad-coverage precision grammars. In: COLING-02 on Grammar Engineering and Evaluation. Association for Computational Linguistics (2002)
8. Besdouri, F.Z., Mekki, A., Zribi, I., Ellouze, M.: Improvement of the cota-orthography system through language modeling. In: 2021 IEEE/ACS 18th International Conference on Computer Systems and Applications (AICCSA). IEEE (2021)
9. Čéplö, S., Bátora, J., Benkato, A., Milička, J., Pereira, C., Zemánek, P.: Mutual intelligibility of spoken Maltese, Libyan Arabic, and Tunisian Arabic functionally tested: a pilot study. Folia Linguistica **50**(2) (2016)
10. Daoud, M.: The language situation in Tunisia. Curr. Issues Lang. Plan. **2**(1), 1–52 (2001)
11. Eryani, F., Habash, N., Bouamor, H., Khalifa, S.: A spelling correction corpus for multiple arabic dialects. In: Proceedings of the Twelfth Language Resources and Evaluation Conference, pp. 4130–4138. European Language Resources Association, Marseille, France (2020)
12. Gibson, M.: Tunis Arabic. Encyclopedia Arabic Lang. Linguist. **4**, 563–571 (2009)
13. Gibson, M.: Dialect levelling in Tunisian Arabic: towards a new spoken standard. In: Language Contact and Language Conflict in Arabic, pp. 42–58. Routledge (2013)
14. Gugliotta, E., Mallia, M., Panascì, L.: Towards a unified digital resource for Tunisian Arabic. In: Carvalho, S., et al. (eds.) Proceedings of the 4th Conference on Language, Data and Knowledge, pp. 579–590. NOVA CLUNL, Portugal, Vienna, Austria (2023)
15. Habash, N., Diab, M., Rambow, O.: Conventional orthography for dialectal Arabic. In: Proceedings of the Eighth International Conference on Language Resources and Evaluation (LREC 2012), pp. 711–718. European Language Resources Association (ELRA), Istanbul, Turkey (2012)
16. Habash, N., Eryani, F., Khalifa, S., Rambow, O., et al.: Unified guidelines and resources for arabic dialect orthography. In: Proceedings of the Eleventh International Conference on Language Resources and Evaluation (LREC 2018). European Language Resources Association (ELRA), Miyazaki, Japan (2018)

17. Hamdi, A., Nasr, A., Habash, N., Gala, N.: POS-tagging of Tunisian dialect using standard Arabic resources and tools. In: Proceedings of the Second Workshop on Arabic Natural Language Processing. Association for Computational Linguistics (2015)
18. Kallel, M.A.: Choix langagiers sur la radio Mosaïque FM. Dispositifs d'invisibilité et de normalisation sociales. Langage et société **138**(4), 77–96 (2011)
19. Kassab, A., Ounaïes, A., Ben Achour Abdelkéfi, R., Louati, A., Mosbah, C., Sakli, M.: Histoire Générale de la Tunisie: Tome IV - L'Époque contemporaine. Sud Éditions (2010)
20. Maamouri, M.: The Phonology of Tunisian Arabic. Cornell University (1967)
21. Maamouri, M.: Language education and human development: Arabic diglossia and its impact on the quality of education in the Arab region. ERIC Clearinghouse (1998)
22. Maamouri, M., Bies, A., Buckwalter, T., Diab, M., et al.: Developing and using a pilot dialectal Arabic treebank. In: Proceedings of the Fifth International Conference on Language Resources and Evaluation (LREC 2006). European Language Resources Association (ELRA), Genoa, Italy (2006)
23. Maamouri, M., Graff, D., Jin, H., Cieri, C., Buckwalter, T.: Dialectal Arabic Orthography-based Transcription & CTS Levantine Arabic Collection. the Parallel STT-NA Tracks Session of the EARS RT-04 Workshop, Palisades IBM Executive Center, New York, Nov. 10, 2004 (2004)
24. Masmoudi, A., Hamdi, J., Hadrich Belguith, L.: Deep learning for sentiment analysis of Tunisian dialect. Computación y Sistemas **25**(1) (2021)
25. Masmoudi, A., Khmekhem, M.E., Khrouf, M., Belguith, L.H.: Transliteration of Arabizi into Arabic script for Tunisian dialect. ACM Trans. Asian Low-Resour. Lang. Inf. Process. **19**(2), 1–21 (2019)
26. Masmoudi, A., Mdhaffar, S., Sellami, R., Belguith, L.H.: Automatic diacritics restoration for Tunisian dialect. ACM Trans. Asian Low-Resour. Lang. Inf. Process. **18**(3), 1–18 (2019)
27. McNeil, K.: Tunisian Arabic Corpus: Creating a Written Corpus of an 'Unwritten' Language, pp. 30–55. Edinburgh University Press (2018)
28. McNeil, K.L.: Tunisian Arabic as a Written Language: Vernacularization and Identity. Georgetown University (2023)
29. Mekki, A., Zribi, I., Ellouze, M., Belguith, L.: Cota 2.0: an automatic corrector of Tunisian Arabic social media texts. Jordanian J. Comput. Inf. Technol. **8**(4), 370–387 (2022)
30. Mekki, A., Zribi, I., Ellouze, M., Hadrich Belguith, L.: Critical description of TA linguistic resources. Procedia Comput. Sci. **142**, 230–237 (2018)
31. Ranta, A.: Grammatical framework. J. Funct. Program. **14**(2), 145–189 (2004)
32. Rosa, C.L.: Mahdia dialect: an urban vernacular in the Tunisian Sahel context. Languages **6**(3), 145 (2021)
33. Rosso, P., Rangel, F., Farías, I.H., Cagnina, L., et al.: A survey on author profiling, deception, and irony detection for the Arabic language. Lang. Linguist. Compass **12**(4) (2018)
34. Saddour, I.: The expression of progressivity in Tunisian Arabic: a study of progressive markers in oral retellings of simultaneous situations. Revue de Sémantique et Pragmatique **25-26**(Espace temps, Interprétations spatiales / Interprétations temporelles?), 265–280 (2009)
35. Sayahi, L.: Diglossia and Language Contact: Language variation and change in North Africa. Cambridge University Press (2014)

36. Sghaier, M.A., Zrigui, M.: Rule-based machine translation from Tunisian dialect to modern standard Arabic. Procedia Comput. Sci. **176**, 310–319 (2020)
37. Shiri, S.: The homestay in intensive language study abroad: social networks, language socialization, and developing intercultural competence. Foreign Lang. Ann. **48**(1), 5–25 (2015)
38. Singer, H.R.: Grammatik der arabischen Mundart der Medina von Tunis. DE GRUYTER (1984)
39. Turki, H., Adel, I., Daouda, T., Regragui, N.: A conventional orthography for Maghrebi Arabic. In: Proceedings of the International Conference on Language Resources and Evaluation (LREC) (2016)
40. Turki, H., Vrandecic, D., Hamdi, H., Adel, I.: Using WikiData as a multi-lingual multi-dialectal dictionary for Arabic dialects. In: 2017 IEEE/ACS 14th International Conference on Computer Systems and Applications (AICCSA). IEEE (2017)
41. Younes, J., Souissi, E., Achour, H., Ferchichi, A.: Language resources for Maghrebi Arabic dialects' NLP: a survey. Lang. Resour. Eval. **54**(4), 1079–1142 (2020)
42. Zalmout, N., Habash, N.: Adversarial multitask learning for joint multi-feature and multi-dialect morphological modeling. In: Proceedings of the 57th Annual Meeting of the Association for Computational Linguistics. Association for Computational Linguistics (2019)
43. Zribi, I., Boujelbane, R., Masmoudi, A., Ellouze, M., et al.: A conventional orthography for Tunisian Arabic. In: Proceedings of the Ninth International Conference on Language Resources and Evaluation (LREC 2014), pp. 2355–2361. European Language Resources Association (ELRA), Reykjavik, Iceland (2014)
44. Zribi, I., Ellouze, M., Belguith, L.H., Blache, P.: Morphological disambiguation of Tunisian dialect. J. King Saud Univ. Comput. Inf. Sci. **29**(2), 147–155 (2017)
45. Zribi, I., Graja, M., Ellouze Khmekhem, M., Jaoua, M., Hadrich Belguith, L.: Orthographic Transcription for Spoken Tunisian Arabic, pp. 153–163. Springer, Heidelberg (2013)

Innovations in Language Modeling and Text Processing

Deep Learning Approach for Early Prediction of Depression on Social Network

Manel Ben Amira(✉), Nabil Khoufi, and Chafik Aloulou

University of Sfax, ANLP Research Group, Miracl Lab, Sfax, Tunisia
manelbenamira1997@gmail.com, {nabil.khoufi,chafik.aloulou}@fsegs.rnu.tn

Abstract. We used deep learning methods to create an innovative system for early detection of depression based on user comments on social networks. This revolutionary approach exploits large amounts of textual data available online to identify depression. The data set originates from the eRisk 2022 competition. Thanks to natural language and statistical modeling techniques, our system can analyze user comments and detect their depression. We applied deep learning methods, in particular the BiLSTM(Bidirectional Long Short-Term Memory) model. Using the approach, we obtained an F-score of 42.96%.

Keywords: Natural language processing · Prediction of depression · Deep learning · BiLSTM

1 Introduction

In the healthcare field, the analysis of social networks offers innovative perspectives thanks to advanced Automatic Natural Language Processing (NLP) technology. By exploiting the vast quantities of data generated on these platforms, researchers can extract valuable information about patients' personalities, such as depression, paranoia, schizophrenia, etc. [4].

Depression is an illness characterized by great sadness, a feeling of hopelessness (depressed mood), loss of motivation and decision-making ability, a diminished sense of pleasure, eating and sleeping disorders, morbid thoughts and a sense of worthlessness as an individual [10].

Typically, people suffering from depression have periods of depression that can last for weeks, months or even years. Symptoms of depression are classified as mild, moderate or major (severe). Depression can lead to suicide in the most severe cases [7,10].

Early prediction of depression could help address these threats, but most early research relies on questionnaire-based patient diagnosis, which is costly and time-consuming, and most countries have public health systems that lack the support for such diagnoses [1,5].

Our project involves developing system that uses social networks to early predict depression.

L. H. Belguith and K. Shaalan (Eds.): LPKM 2024, LNNS 1303, pp. 145–154, 2025.
https://doi.org/10.1007/978-3-031-85067-7_14

Five main sections make up this article. The related works are presented in Sect. 2. Section 3 presented the architecture and our proposed method. The evaluation and results of our system are presented in Sect. 4. Finally, the conclusion is presented in Sect. 5.

2 Related Works

A great deal of research has been carried out to identify early signs of predisposition to depression.

[5] applied the MixUp method using a TF-IDF approach and linguistic statistical functions of Reddit writings to identify early signs of depression. They used a sentence classification approach that augmented the data by interpolation. The authors obtained results using a voting classifier, applying hard voting to textual features and SVM to textual and numerical features. The authors used the E-risk 2022 data-sets, which include the previous E-risk 2017 and 2018 editions, which included a collection of user group contributions from the Reddit platform. They also collected data from publications and comments. The authors used two types of features. The first is TF-IDF based functionality. The second is the textual feature. After applying these features, the authors used a pre-trained BERT model on the mental health data-set and obtained similar performance to their first BERT. They also used a voting classifier for text features, an XGBoost model for metadata features, a linear support vector machine classifier for text + metadata features and BERT with the MixUp method. The authors obtained the F1 score for each model run, they obtained an F1 for BERT model = 60,4%, voting Classifier = 64,4%, XGBoost model = 39%, support Vector Machines (SVM) = 64,7% and BERT techniques MixUp model = 58,3% [5].

[13] propose to use a linear model such as logistic regression, which have been developed from various representations of the document input space. The authors used the E-risk 2022 data-sets, which include training and test data from the E-risk 2017 and E-risk 2018 editions. They employed machine learning techniques, particularly focusing on natural language processing methods, to identify signs of depression. They examined two different methods of document representation. One is based on latent semantic analysis, while the other is based on contextual sentence transformation. For the first method, the authors used the Singular Value Decomposition **(SVD)** technique to apply a Latent Semantic Analysis **(LSA)** based on reduced n-gram features. They considered using a model from the distillBERT transformer library for the second method. The following SBERT Model variants, distillBERT, RoBERTa and XLM-RoBERTa, were used for contextual functionality. The authors used scikit-learn's logistic regression model as the classification model. For the above features, they created logistic regression classifiers with penalties. They developed a classifier for specific expressions on the training set and evaluated it on the development set. The authors obtained an F1 score for each model. Their distillBERT score was 44,3%, RoBERTa was 32,52%, XLM-RoBERTa was 33,33% and LSA was 25,45% [13].

[2] propose to create a tool to identify symptoms of depression in Reddit users. They proposed two strategies: A feature-based classifier with features based on text data, TF-IDF terms, use of first-person pronouns, mood analysis and depression terms. A deep learning classifier with pre-trained integration's. The authors used a data-set that included the official shared Task 2 E-risk 2022 data-set and the UNED-MED Reddit depression data-set. They then used two separate feature strategies, depending on whether the classification algorithm was a traditional machine learning algorithm or a deep learning. They used the traditional feature and the TF-IDF feature for traditional learning. They used Embeddings for the deep learning algorithm. Finally, the authors used a deep learning model, a convolutional neural network (CNN), as well as two convolutional machine learning models, Random Forest and XGBoost [2].

[12] created E-risk a depression early detection system that classifies social media users based on their posts. They presented two simple strategies for improving and adapting common text classification models for early detection. The authors used the E-risk 2022 data-sets, which included the previous E-risk 2017 and 2018 editions. They also collected data on Reddit. This data-set contains a chronological collection of 1707 posts and comments from individual users on various sub-Reddit. The authors created two methods to achieve this, and templates for the joint execution of the tasks. They examined three different methods for preparing inputs. The first method is no-concat, the second is concat1 and the last is concat2. To correct the imbalance caused by the chain strategy, we then used a threshold design. Finally, two models were used. The first, logistic Regression, produced a vector representation of the input sequence that averaged the incorporations of the last four layers of the pre-trained BERT model over the entire input sequence. The second model is DistilBERT. It directly refines the binary classification task. After testing various threshold scheduling optimizations and chaining strategies, the authors proposed five executions with the following models: LauSAn#0: proposes logistic regression integrating BERT and contactless optimized exponential threshold scheduling for ERDE5, LauSAn#1: uses BERT integrations for logistic regression, concat1 (5 messages), optimized exponential threshold scheduling for ERDE50, LauSAn#2: uses BERT integrations for logistic regression, concat1 for 5 messages and optimized exponential threshold scheduling for ERDE50 (ignoring the first 3 messages), LauSAn#3: presents logistic regression using BERT integrations, concat1 for 5 messages and exponential threshold scheduling optimized for F1, which is latency-weighted (ignoring the first 3 messages), LauSAn#4: presents an exponential threshold scheduling optimized for ERDE5, and a refined DIstilBERT. It directly refines the binary classification task. They obtained the F-score for the five execution LauSAn#0 = 23,5%, LauSAn#1 = 27,9%, LauSAn#2 = 29%, LauSAn#3 = 50,8% and LauSAn#4 = 31,5% [12].

[6] proposed a Time-Aware Affective Memories (TAM) network for early detection of depression risk in users based on their Internet message flow. The authors used the E-risk 2022 data-sets, which include training and test data from the E-risk 2017 and E-risk 2018 editions. Firstly, the affective processing

module expects the user's last post at the step and the time interval between the last post and the user's last post as input. The authors concatenated the title and body of the message, replacing user-dependent and task independent information with special tokens. Secondly, the user's emotions in the publication are then mapped onto emotional states based on the pre-trained DistilBERTE2 sentiment classification model. Thirdly, the user's emotional state is preserved by a Time-Aware Network LSTM (T-LSTM). Finally, we use linear layers to map hidden T-LSTM states to emotional memory. To integrate enriched affective memory and semantic integration into TAM, the authors used a Transformer Decoder Network. The Transformer Decoder is a multi-headed cross-attention architecture where each head requests elements from the input sequence A and obtains new values from the reference input sequence A and obtains new values from the reference input sequence B based on the similarity between elements A and B [6].

3 Proposed Method

This section is dedicated to the presentation of our work for early detection of depression.

Our method is based on the use of Deep Learning Techniques. In the following, we present the proposed architecture and the data-sets.

3.1 Corpus

We used a data-set for early detection of depression, which was created by [3]. This data-set originates from the 2017 and 2018 editions of E-Risk and consists of multiple files. Each file contains identifiers, titles, dates, texts, information and a binary class (0 or 1), where 0 indicates that the user is not depressed and 1 indicates that they are [3].

Our corpus comprises a total of 1707 files, divided between positive and negative data from the 2017 and 2018 editions. Specifically, the 2017 data includes 752 negative cases and 135 positive cases, while the 2018 data includes 741 negative cases and 79 positive cases [3].

Each file was originally in XML format, but we converted them into a single CSV file, retaining only the attributes ID, Text and Class. This conversion resulted in a file containing 1076582 rows (Fig. 1).

In order to make this corpus balanced, we removed files containing short sentences.

```
ID,TEXT,CLASS
subject1010,They're gonna go stab it,0
subject1010,Unidan?,0
subject1010,"This was, character by character, going to be my exact comment.
subject1010,Maribou Storks,0
subject1010,None,0
subject1010,"Me too, can't gain weight",0
subject1010,They clean fish of their internal fluids,0
```

Fig. 1. Extract of our dataset after CSV conversion

3.2 Architecture

In this section, we present the general architecture of our proposed method.

Our approach is divided into four main stages. Preprocessing is the first step. Segmentation is the second step. The third step is the training. In the final step, we can test our model (Fig. 2).

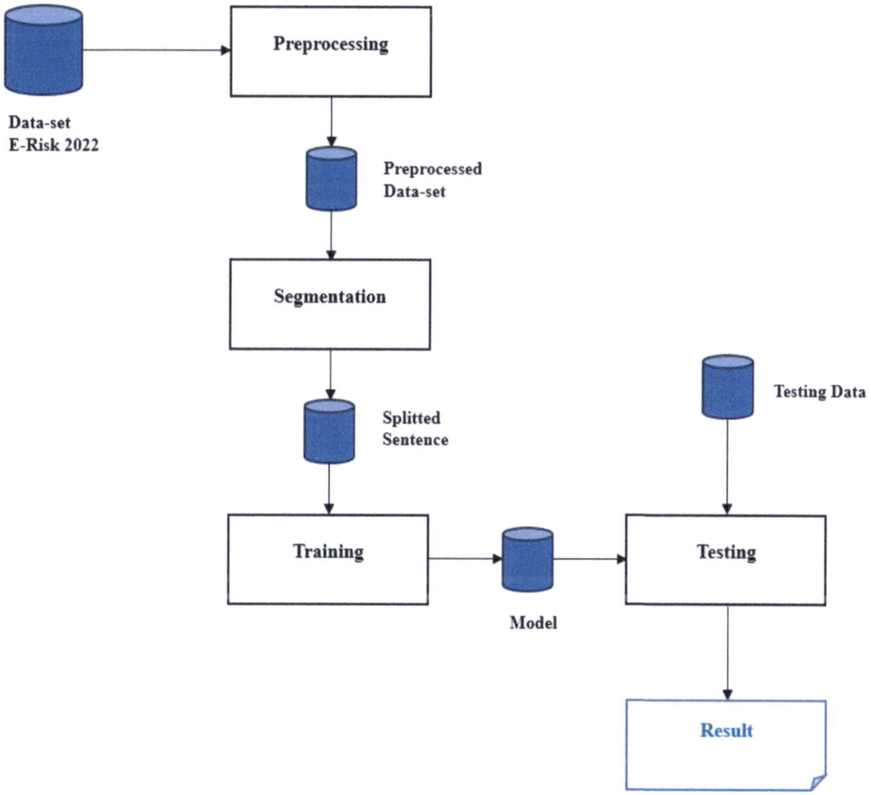

Fig. 2. Architecture for the proposed method

Preprocessing. The preprocessing stage involves preparing the data for the training stage.

We used the following preprocessing treatments:

- Punctuation removal: We excluded punctuation, including periods, commas, double quotes, apostrophes, hyphens, question marks, colons, exclamation points and semicolons because it is unlikely that punctuation provides a significant contribution, it is not necessary in our system.
- Convert Case: We converted the case of all the text to lowercase to reduce the vocabulary size, as the same word in lowercase and uppercase can have different meanings.
- Stop words remove: We excluded stop words like 'the', 'they', 'are', 'is', etc. because they have low discriminative power, there by reducing the vocabulary size.
- Sentence delete beginning with HTTP: We removed all links, such as those to Facebook, YouTube, websites, etc., as they were not necessary for our system.
- Empty lines delete: We removed all empty lines because an empty message does not constitute a valid text.

Segmentation. Most Machine learning applications(NLP) consider segmentation to be a crucial step. Segmentation enables a text to be broken down into sentences or words.

In this step, we used the tokenizer library to segment sentences into words. By adopting this approach, you will be able to break up sentences efficiently, creating distinct lexical units. This method will greatly simplify the text processing process, thereby improving the manipulation and analysis of linguistic.

After segmenting the text into words using the Tokenizer library, take the next step by performing vectorization using the Word2Vec algorithm. This approach involves converting words into vectors, which allows for a meaningful numerical representation of terms in the context of the model. The vectorization step with Word2Vec proves crucial to capture the semantic relationships between words, thus enriching the understanding and analysis of the text in a deeper and contextual way.

Word2vec is a set of widely used models in artificial intelligence and machine learning for the transformation of words into vectors, a process known as word embeddings. These models were developed by a team of researchers at Google, with [9] leading the way.

These models employ two-layer artificial neural networks to learn how to represent words in a manner that captures their linguistic context.

Word2Vec is a natural language processing technique. The Word2Vec algorithm gains an understanding of word associations by analyzing a large corpus of text using a neural network model. These Word2Vec models are used to generate lexical representations.

Training. After obtaining word representations using the Word2Vec method, we commence training using a deep learning technique.

There are several approaches in deep learning:

- Bidirectional Long Short-Term Memory **(biLSTM)** model: is a sequence processing model that consists of two LSTMs: one takes the input in the forward direction, and the other in the reverse direction. Bidirectional LSTMs effectively increase the amount of information available to the network by enhancing the context available to the algorithm, for instance, knowing which words immediately follow and precede a word in a sentence [11].

In this work, we focus on building our system using a BiLSTM model.
The following table shows the BiLSTM model parameters:

Table 1. BiLSTM model parameters

Epochs	2
batch-size	2
optimizer	adam
validation-split	0.2

Table 1 shows the parameters of the BiLSTM model. Detailed instructions for each parameter follow:

- Epochs: This parameter is set to 2. An epoch represents a complete training cycle on the training dataset.
- Batch-size: this parameter is set to 2. The batch size determines the number of data samples that will pass through the model before the internal parameters are updated.
- Optimizer: The optimizer used is "adam". Adam is a popular optimizer that combines the advantages of the moment gradient descent algorithm and learning rate adaptation to accelerate and improve model convergence.
- validation-split: this is set to 0.2, meaning that 20% of the training data will be used as the validation set.

4 Evaluation and Results

During this stage, we used a test dataset to assess the performance of our model. The test dataset we utilized was constructed following the methodology outlined by [3]. We applied the same preprocessing techniques to this test dataset that we used for the training dataset:

- Punctuation removal
- Convert Case

- Empty words remove
- Sentence delete beginning with HTTP
- Empty lines delete

To train and assess our models, we employed an approach based on deep learning techniques, with particular emphasis on the use of BiLSTM model. This strategic choice stems from BiLSTM's ability to capture long-term dependencies in text data sequences, an important aspect for accurate detection of depression. By taking advantage of this bidirectional recurrent neural architecture, our approach aims to model of user's linguistic interactions on social networks. Table 2 shows the results.

Table 2. Evaluation results

Precision	Recall	Accuracy	F-score
34.98%	55.64%	71.68%	42.96%

Our results are promising and offer encouraging prospects for detecting depression through the use of deep learning models. This conclusion is the result of a rigorous performance evaluation of our model, which demonstrated a satisfactory ability to identify the signs of depression in the comments of social network users.

After obtaining the results, we compared them with the results of related work. Thanks to this comparison, it is possible to assess the performance of our BiLSTM model in relation to existing methods in the literature (Table 3).

Our BiLSTM model obtained an F-score of 42.96%. Compared with the work of [5], our model performed worse than their BERT, Voting Classifier, XGBoost, SVM and BERT MixUp models, whose F-score ranged from 58.3% to 64.7%. Our F-score is 42.96%, which is better than the RoBERTa, XLM-RoBERTa and LSA models of [13], which have F-scores of 32.52%, 33.33% and 25.45%, respectively. However, it is slightly inferior to the DistilBERT model by [13] which has an F-score of 44.3%. Our model outperforms all models by [12] with F-scores ranging from 23.5% to 31.5%.

Although our BiLSTM model does not reach the performance of the best model by [5] it shows respectable performance, outperforming several models by [13] and all models by [12]. This shows that BiLSTM is a viable approach, even if further improvements are needed to reach the level of the best current models such as those based on BERT.

5 Conclusion

In this paper, we introduce innovative statistical methods for designing depression detection systems using advanced deep learning techniques [8]. Our approach relies on the implementation of complex models, the results of extensive

Table 3. Related works results

Author	Model	F-score
[5]	BERT	60.4%
[5]	voting Classifier	64.4%
[5]	XGBoost model	39%
[5]	Support Vector Machines	64.7%
[5]	BERT techniques MixUp model	58.3%
[13]	distillBERT	44.3%
[13]	RoBERTa	32.52%
[13]	XLM-RoBERTa	33.33%
[13]	LSA	25.45%
[12]	LauSAn#0	23.5%
[12]	LauSAn#1	27,9%
[12]	LauSAn#2	29%
[12]	LauSAn#3	50,8%
[12]	LauSAn#4	31,5%

research, to efficiently analyze signals of depression in user text data. Our approach uses several natural language processing and statistical modeling techniques, and aims to detect depression with greater accuracy. Our preliminary results demonstrate good performance, paving the way for future research and practical implementation of this system for the benefit of individuals. Our results are promising and represent the baseline for further future works.

References

1. Bsir, B., Khoufi, N., Zrigui, M.: Prediction of author's profile basing on fine-tuning bert model. Informatica **48**(1) (2024)
2. Campillo-Ageitos, E., Martinez-Romo, J., Araujo, L.: UNED-MED at erisk 2022: depression detection with TF-IDF, linguistic features and embeddings. Working Notes of CLEF, pp. 5–8 (2022)
3. Crestani, F., Losada, D.E., Parapar, J.: Early Detection of Mental Health Disorders by Social Media Monitoring: The First Five Years of the ERisk Project, vol. 1018. Springer (2022)
4. Ellouze, M.: Analyse des productions dans les médias sociaux : vers une évaluation de la personnalité des patients. (Analysis of productions on social media: towards an evaluation of patients' personality). Ph.D. thesis, University of Sfax, Tunisia (2023). https://tel.archives-ouvertes.fr/tel-04376647
5. Gînga, R.A., Manea, A.A., Dobre, B.M.: Sunday rockers at erisk 2022: early detection of depression. Working Notes of CLEF, pp. 5–8 (2022)
6. Kang, X., Dou, R., Yu, H.: Tua1 at erisk 2022: exploring affective memories for early detection of depression (2022)

7. Maalej, R., Khoufi, N., Aloulou, C.: Parsing arabic using deep learning technology. In: TACC, pp. 74–80 (2021)
8. Mechti, S., Faiz, R., Khoufi, N., Antit, S., Krichen, M.: Sentiment analysis: effect of combining bert as an embedding technique with CNN model for Tunisian dialect. In: International Conference on Information and Knowledge Systems, pp. 309–320. Springer (2023)
9. Mikolov, T., Chen, K., Corrado, G., Dean, J.: Efficient estimation of word representations in vector space. arXiv preprint arXiv:1301.3781 (2013)
10. Organisation mondial de la sante: Trouble dépressif (dépression) (2023). https://www.who.int/fr/news-room/fact-sheets/detail/depression
11. Papers with code: Bidirectional LSTM (2023). https://paperswithcode.com/method/bilstm
12. Säuberli, A., Cho, S., Stahlhut, L.: Lausan at erisk 2022: simply and effectively optimizing text classification for early detection. Working Notes of CLEF, pp. 5–8 (2022)
13. Tavchioski, I., Škrlj, B., Pollak, S., Koloski, B.: Early detection of depression with linear models using hand-crafted and contextual features. Working Notes of CLEF, pp. 5–8 (2022)

Adapting Large Language Models to Biomedical Domain: A Survey of Techniques and Approaches

Jaafer Klila[1(✉)], Sondes Bannour Souihi[2], Rahma Boujelben[1], Nasredine Semmar[2], and Lamia Hadrich Belguith[1]

[1] MIRACL Laboratory, University of Sfax, Sfax, Tunisia
klilajaafer@gmail.com, {rahma.boujelbane,lamia.belguith}@fsegs.usf.tn
[2] CEA, List, Palaiseau, Université Paris-Saclay, Gif-sur-Yvette, France
{sondes.souihi,nasredine.semmar}@cea.fr

Abstract. The adaptation of Large Language Models (LLMs) to specialized domains such as biomedicine holds significant promise for advancing Natural Language Processing (NLP) applications in healthcare and medical research. In this survey, we investigate various techniques and approaches for adapting LLMs to better understand and process biomedical text. We explore the challenges faced by LLMs in biomedical tasks, including the presence of specialized terminology and the scarcity of annotated datasets. The survey discusses adaptation techniques such as training LLMs from scratch, fine-tuning on biomedical data, and injecting domain-specific knowledge. Furthermore, we compare the efficacy of these techniques and highlight their advantages and challenges. Our examination reveals that while each approach offers distinct benefits, the choice depends on factors such as available resources, task requirements, and desired performance outcomes. Finally, we outline future research directions focusing on knowledge integration in the biomedical domain to enhance the capabilities of LLMs and their potential impact on healthcare and medical research.

Keywords: Natural Language Processing · Large Language Models · Biomedical Domain · Adaptation Techniques · Knowledge Integration

1 Introduction

Large language models (LLMs) such as BERT [6], ChatGPT[1], GPT-4 [1], and RoBERTA [18] undergo initial training on extensive datasets. They are then either directly used for downstream tasks or further fine-tuned on smaller datasets specific to natural language processing tasks. The impressive performance of these models across various NLP tasks has been widely proven [29]. Their ability to understand, produce, and interpret human language is crucial.

[1] https://openai.com/blog/chatgpt.

L. H. Belguith and K. Shaalan (Eds.): LPKM 2024, LNNS 1303, pp. 155–163, 2025.
https://doi.org/10.1007/978-3-031-85067-7_15

However, it is essential to clarify that when referring to these models 'understand' language, we are highlighting their capacity to process and produce text coherently and contextually relevant, rather than attributing them with human-like consciousness or comprehension. Furthermore, they play essential roles in language translation [3], content generation, and email filtering applications, demonstrating their versatility and usefulness in modern contexts. Despite their overall effectiveness, LLMs continue to face challenges when it comes to biomedical tasks. This difficulty arises from the presence of specialized terminology and a lack of sufficient training data.

2 Related Work

In this section, we initially present the notion of LLMs and then discuss the difficulties in Biomedical Tasks.

2.1 Large Language Models

Language models with a large number of parameters, known as LLMs, are developed based on the Transformer architecture and trained on hundreds of terabytes of textual data [26], such as BERT [6], GPT [22] (GPT: the initial pre-trained Transformer model, utilized for fine-tuning across diverse NLP tasks and achieving state-of-the-art results).

Unlike previous studies [3,29] that predominantly focus on fine-tuning, this paper makes several novel contributions to the field of biomedical NLP by addressing the broader aspects of knowledge integration and pretraining methods.

Large Language Models (LLMs) can be categorized based on their architectural components into three main types[2] (see Fig. 1):

(a) Encoder-only models: are particularly well-suited for tasks requiring robust comprehension of input data, such as sentence classification and named entity recognition.
(b) Decoder-only models: are particularly adept at generative tasks, such as text generation. These models excel in generating coherent and contextually relevant text based on provided prompts or conditions.
(c) Encoder-decoder models: also known as sequence-to-sequence models, are beneficial for tasks that involve generating output based on an input, such as translation or summarization.

As shown in Table 1, various types of language models have different architectural focuses.

[2] https://huggingface.co/learn/nlp-course/chapter1/5?fw=pt.

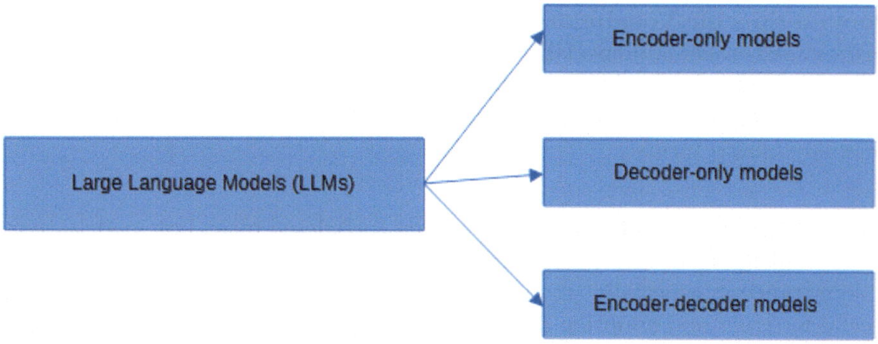

Fig. 1. Types of LLMs.

Table 1. Examples of different types of language models (Source: https://huggingface.co/learn/nlp-course/chapter1/5?fw=pt.)

Encoder-only models	Decoder-only models	Encoder-decoder models
BERT [6]	GPT [22]	T5 [24]
ALBERT [12]	Transformer XL [5]	BART [14]
DistilBERT [25]	GPT-2 [23]	Marian [11]
RoBERTa [18]	GPT-4 [1]	mBART [17]

2.2 Challenges of LLMs in Biomedical Tasks

LLMs have gained significant attention from both the community and the public with the emergence of large models like ChatGPT and GPT-4. Their demonstrated capabilities across various tasks have sparked great interest [8]. Despite containing substantial pre-training knowledge, ChatGPT still requires a knowledge injection method to address challenges in professional domains like healthcare [16,32]. Moreover, evaluating LLMs in the biomedical sector presents considerable difficulties due to the lack of standardized and inclusive datasets gathered from trustworthy and impartial sources [10]. Therefore in the field of biomedical studies the prevalence of medical terminology and abbreviations derived from Latin or Greek roots. Additionally, clinical notes follow a unique syntax and grammar distinct from that found in books or encyclopedias. These differences contribute to a gap in semantic understanding and domain-specific knowledge between general pre-trained language models and those designed for the biomedical field. As a result, numerous methods have been explored to adapt pre-trained language models from general domains to better suit the requirements of biomedical language processing [31].

Furthermore, using LLMs in the biomedical domain presents specific challenges, including:

Domain-Specific Vocabulary and Terminology. The biomedical field comprises an extensive and specialized lexicon, which encompasses medical jargon, drug nomenclature, and biological components. Language models may encounter difficulties in precisely comprehending and producing text that incorporates these specialized terms.

Limited Interpretability. LLMs are often considered black-box models, making it challenging to interpret their predictions and understand their decision-making process, especially in the biomedical domain where interpretability is crucial for trust and adoption in clinical settings

Domain Adaptation and Fine-Tuning. Pre-trained LLMs may need to be fine-tuned or adapted to perform well on specific biomedical tasks or datasets. However, acquiring labeled data for fine-tuning in the biomedical domain can be challenging due to data scarcity and the high cost of expert annotation.

3 Adapting LLMs to the Biomedical Domain

To address these challenges and effectively utilize large language models in the biomedical domain, we will now discuss various approaches for adapting LLMs, including training from scratch, fine-tuning, and knowledge injection techniques.

3.1 Training from Scratch

Training large language models from scratch for the biomedical field involves a thorough process focused on capturing the complexity and depth of biomedical text data. A key aspect of this effort is gathering a wide range of biomedical text sources, such as PubMed[3], clinical notes, and Electronic health record (EHR). These sources provide diverse linguistic patterns, domain-specific terminology, and contextual nuances necessary to effectively train LLMs in generating and understanding biomedical text. By incorporating various text sources, LLMs can grasp the intricacies of medical language - including specialized vocabulary, complex syntax, and nuanced semantics. Exposure to diverse biomedical texts also helps LLMs build strong linguistic representations and improve their ability to handle different tasks within the biomedical field.

PubMedBERT [9] was pretrained from scratch on 14 million PubMed abstracts, comprising 3.1 billion words and occupying 21GB. This search findings indicate that in domains with a large amount of text such as biomedicine, training language models specifically on the domain-specific corpus from the beginning yields better results than starting from a general-domain model. This approach leads to significant improvements in performance for biomedical NLP tasks. BioBERT [13] follows a continuous pre-processing process based on BERT,

[3] https://pubmed.ncbi.nlm.nih.gov/.

demonstrates a high performance across various biomedical text mining tasks such as Named Entity Recognition (NER), Relation Extraction (RE), and Question Answering (QA).

Table 2. Comparison of general and specialized domain language models on biomedical Named Entity Recognition task using F1 score: results from [9].

	BERT [6]	BioBERT [13]	PubMedBERT [9]	ClinicalBERT [2]
	General domain	biomedical domain	biomedical domain	biomedical domain
BC5-chem [15]	89.25	92.85	**93.33**	90.80
BC5-disease [15]	81.44	84.70	**85.62**	83.04
NCBI-disease [7]	85.67	**89.13**	87.82	86.32
BC2GM [27]	80.90	83.82	**84.52**	81.71
JNLPBA [4]	77.69	78.55	**79.10**	78.07

Table 2 presents a comparison of general and specialized domain language models on the biomedical Named Entity Recognition (NER) task using F1 scores. The results, obtained from [9], demonstrate varying levels of performance across different models and datasets.

In general, specialized domain language models, including BioBERT [13], PubMedBERT [9], and ClinicalBERT [2], consistently outperform the general domain model BERT [6] across all datasets. For example, on the BC5-chem dataset, BioBERT [13] achieves an F1 score of 92.85, while BERT [6] achieves 89.25. Similar trends are observed across other datasets such as BC5-disease, NCBI-disease, BC2GM, and JNLPBA.

3.2 Fine-Tuning

Fine-tuning LLMs for biomedical NLP [28] involves further training a pretrained model on a smaller, targeted dataset. The goal is to maintain the original capabilities of the pretrained model while adapting it for more specialized use cases.

LLMs may have extensive knowledge across various subjects, but with fine-tuning, it can develop specialized expertise in a particular field, like medical terminology.

The rationale behind fine-tuning lies in the need to adapt the general knowledge encoded in pre-trained LLMs to the nuances and intricacies of biomedical text.

Fine-tuning pre-trained LLMs such as BERT or BioBERT on annotated biomedical text has led to significant improvements in NER performance, enabling accurate identification and classification of biomedical entities such as genes, diseases, and proteins.

3.3 Knowledge Injection

Knowledge can stem from various sources, and its representation can vary depending on its origin and intended use. These sources may encompass scientific knowledge, world knowledge, and expert knowledge [30]. Furthermore, knowledge can be represented in diverse manners, such as knowledge graphs, logic rules, algebraic equations, and more [30]. The majority of knowledge integration techniques are applied during the training data preparation step.

In the biomedical field, the integration of domain specific knowledge plays a crucial role in improving the efficacy of models. A novel data-centric approach is proposed in [19] to enhance the language representations of biomedical transformer-encoder language models (LMs). This method entails extracting text sequences from the Unified Medical Language System (UMLS). In [20], a contextual embedding model was introduced that incorporates domain knowledge during the pre-training phase through a novel knowledge augmentation strategy. The work of [21] proposes an implementation that integrates the UMLS knowledgebase into a BERT-based language model with the aim of improving its performance in biomedical Named Entity Recognition.

Table 3 presents a comparison of the effect of knowledge integration in biomedical models, focusing on F1 score and accuracy for the MedNLI task. Two models, UmlsBERT and UMLS-KGI, incorporate knowledge integration, while the others do not. UmlsBERT and UMLS-KGI achieve higher scores in both F1 score and accuracy compared to the models without knowledge integration, suggesting that incorporating domain-specific knowledge enhances performance in biomedical tasks.

Table 3. Comparison of the effect of Knowledge integration in biomedical models on biomedical task using F1 score and accuracy for MedNLI task: results from [19,20]. KI refers to knowledge integration, "KS" refers to knowledge source, "Yes" mean model with knowledge integration otherwise "No".

Models	KI	KS	MedNLI	i2b2 2010	
Bert	No	-	77.9	85.2	
UmlsBERT [20]	Yes	UMLS	**83.0**	**88.6**	
			NCBI-Disease	BioRED-NER	JNLPBA04
PubMedBERT [9]	No	-	93.53	83.35	81.13
UMLS-KGI [19]	Yes	UMLS	**94.46**	**83.64**	**85.15**

4 Comparison of Adaptation Techniques

In comparing adaptation techniques for large language models (LLMs) in the biomedical domain, we can evaluate the effectiveness and suitability of different approaches. Here's a breakdown of the comparison: Training LLMs from scratch for the biomedical domain allows for a detailed representation of the complexity and depth of biomedical text data by incorporating diverse sources such as PubMed abstracts and clinical notes. This approach allows models to develop strong linguistic representations. However, it requires a large amount of data and computational resources, making it less feasible for organizations with limited resources. Additionally, the training process is time-consuming.

Fine-tuning pretrained LLMs on annotated biomedical datasets enables these models to specialize in biomedical terminology and nuances, leading to significant performance improvements for tasks like named entity recognition. However, this technique requires annotated biomedical datasets, which may be scarce or costly to obtain. The fine-tuning process can also be time-consuming, and the performance may not match models specifically designed for the biomedical domain.

Injecting domain-specific knowledge into LLMs enhances their efficacy in the biomedical domain by providing contextual understanding of terminology. This approach improves performance in biomedical tasks, but it requires careful integration of domain knowledge into the model architecture. Knowledge extraction and representation can be complex, necessitating domain expertise. Moreover, performance gains may not be as significant without proper integration of knowledge.

In summary, each adaptation technique has its advantages and challenges, and the choice depends on factors such as available resources, task requirements, and desired performance outcomes.

5 Conclusion and Future Work

The adaptation of large language models (LLMs) to the biomedical domain represents a significant advancement in natural language processing (NLP) with potential implications for healthcare and medical research. In this survey, we explored various techniques and approaches for adapting LLMs to better understand and process biomedical text. However, we acknowledge that our discussion has only scratched the surface in terms of detail.

Future research endeavors are likely to focus on knowledge integration in the biomedical domain, with an emphasis on methods and their efficacy. Specifically, there is a need to delve deeper into techniques for effectively integrating domain-specific knowledge into LLMs to improve their performance on biomedical tasks. This includes exploring novel approaches, refining existing methods, and evaluating their effectiveness in real-world applications.

References

1. Achiam, J., et al.: GPT-4 technical report. arXiv preprint arXiv:2303.08774 (2023)
2. Alsentzer, E., et al.: Publicly available clinical BERT embeddings. arXiv preprint arXiv:1904.03323 (2019)
3. Clusmann, J., et al.: The future landscape of large language models in medicine. Commun. Med. **3**(1), 141 (2023)
4. Collier, N., Ohta, T., Tsuruoka, Y., Tateisi, Y., Kim, J.D.: Introduction to the bio-entity recognition task at JNLPBA. In: Proceedings of the International Joint Workshop on Natural Language Processing in Biomedicine and its Applications (NLPBA/BioNLP), pp. 73–78 (2004)

5. Dai, Z., Yang, Z., Yang, Y., Carbonell, J., Le, Q.V., Salakhutdinov, R.: Transformer-XL: attentive language models beyond a fixed-length context (2019)
6. Devlin, J., Chang, M.W., Lee, K., Toutanova, K.: BERT: pre-training of deep bidirectional transformers for language understanding. arXiv preprint arXiv:1810.04805 (2018)
7. Doğan, R.I., Leaman, R., Lu, Z.: NCBI disease corpus: a resource for disease name recognition and concept normalization. J. Biomed. Inform. **47**, 1–10 (2014)
8. Fu, P., Zhang, Y., Wang, H., Qiu, W., Zhao, J.: Revisiting the knowledge injection frameworks. arXiv preprint arXiv:2311.01150 (2023)
9. Gu, Y., et al.: Domain-specific language model pretraining for biomedical natural language processing. ACM Trans. Comput. Healthcare (HEALTH) **3**(1), 1–23 (2021)
10. Hua, Y., et al.: Large language models in mental health care: a scoping review. arXiv preprint arXiv:2401.02984 (2024)
11. Junczys-Dowmunt, M., et al.: Marian: fast neural machine translation in C++. arXiv preprint arXiv:1804.00344 (2018)
12. Lan, Z., Chen, M., Goodman, S., Gimpel, K., Sharma, P., Soricut, R.: Albert: a lite BERT for self-supervised learning of language representations. arXiv preprint arXiv:1909.11942 (2019)
13. Lee, J., et al.: BioBERT: a pre-trained biomedical language representation model for biomedical text mining. Bioinformatics **36**(4), 1234–1240 (2020)
14. Lewis, M., et al.: Bart: denoising sequence-to-sequence pre-training for natural language generation, translation, and comprehension (2019)
15. Li, J., et al.: Biocreative V CDR task corpus: a resource for chemical disease relation extraction. Database **2016** (2016)
16. Liu, J., et al.: Benchmarking large language models on CMExam-a comprehensive Chinese medical exam dataset. In: Advances in Neural Information Processing Systems, vol. 36 (2024)
17. Liu, Y., et al.: Multilingual denoising pre-training for neural machine translation (2020)
18. Liu, Y., et al.: Roberta: a robustly optimized BERT pretraining approach. arXiv preprint arXiv:1907.11692 (2019)
19. Mannion, A., Chevalier, T., Schwab, D., Geouriot, L.: UMLS-KGI-BERT: data-centric knowledge integration in transformers for biomedical entity recognition. arXiv preprint arXiv:2307.11170 (2023)
20. Michalopoulos, G., Wang, Y., Kaka, H., Chen, H., Wong, A.: UmlsBERT: clinical domain knowledge augmentation of contextual embeddings using the unified medical language system metathesaurus. arXiv preprint arXiv:2010.10391 (2020)
21. Piat, G., Semmar, N., Allauzen, A., Essafi, H., Tourille, J.: Enriching contextualized representations with biomedical ontologies: extending knowbert to UMLs. In: Science and Information Conference, pp. 760–773. Springer, Cham (2022)
22. Radford, A., Narasimhan, K., Salimans, T., Sutskever, I., et al.: Improving language understanding by generative pre-training (2018)
23. Radford, A., Wu, J., Child, R., Luan, D., Amodei, D., Sutskever, I., et al.: Language models are unsupervised multitask learners. OpenAI Blog **1**(8), 9 (2019)
24. Raffel, C., et al.: Exploring the limits of transfer learning with a unified text-to-text transformer (2023)
25. Sanh, V., Debut, L., Chaumond, J., Wolf, T.: DistilBERT, a distilled version of BERT: smaller, faster, cheaper and lighter (2020)
26. Shanahan, M.: Talking about large language models. Commun. ACM **67**(2), 68–79 (2024)

27. Smith, L., et al.: Overview of biocreative II gene mention recognition. Genome Biol. **9**, 1–19 (2008)
28. Tinn, R., et al.: Fine-tuning large neural language models for biomedical natural language processing (2021)
29. Tinn, R., et al.: Fine-tuning large neural language models for biomedical natural language processing. Patterns **4**(4) (2023)
30. Von Rueden, L., et al.: Informed machine learning-a taxonomy and survey of integrating prior knowledge into learning systems. IEEE Trans. Knowl. Data Eng. **35**(1), 614–633 (2021)
31. Wang, B., et al.: Pre-trained language models in biomedical domain: a systematic survey. ACM Comput. Surv. **56**(3), 1–52 (2023)
32. Wu, C., Zhang, X., Zhang, Y., Wang, Y., Xie, W.: PMC-LLaMA: further finetuning llama on medical papers. arXiv preprint arXiv:2304.14454 (2023)

Tunisian Arabic Understanding: Resources Analysis and Evaluation

Fatma Zahra Besdouri[1(✉)], Inès Zribi[1,2], and Lamia Hadrich Belguith[1]

[1] ANLP Research group, MIRACL Labratory, University of Sfax, Sfax, Tunisia
fatma.basdouri2@gmail.com
[2] University of Monastir, Monastir, Tunisia

Abstract. Natural Language Understanding (NLU) is considered a core component in implementing Dialogue Systems (DS). This paper introduces an NLU module tailored to the Tunisian Dialect (TD) that can further be integrated with Automatic Speech Recognition (ASR), a Dialogue Manager (DM), a Natural Language Generator (NLG), and a Speech Synthesis (SS) module to build a fully working DS. Designed specifically for academic and research domain, our closed-domain TUN-NLU corpus undergoes preprocessing, including scraping, automatic normalization, data augmentation, and annotation. We employed the RASA NLU model (https://rasa.com/) to evaluate our corpus. The NLU model encompasses tokenization, featurization, intent detection, and entity recognition. Evaluation using standard metrics, including accuracy and F1 score, yielded promising results: 93% accuracy and 92% F1 score for intent detection, and 94% accuracy and 91% F1 score for entity recognition.

Keywords: Tunisian Dialect · Natural Language Understanding · Dialogue System · RASA

1 Introduction

The advent of artificial intelligence has led to the widespread adoption of intelligent dialogue systems (DS) like chatbots (customer service, health care, etc.), personal assistants, etc., which greatly simplify daily tasks for users. Dialogue systems (DS), aka conversational systems, are computer systems designed to converse with human users using natural language. The DS takes the input from the user, tries to extract the semantic information from the input, uses a policy to manage the status of the dialogue, and finally generates an output back to the user. These three steps are named respectively as Natural Language Understanding (NLU), Dialogue Management (DM), and Natural Language Generation (NLG). The output from one stage is used as input for the next stage. When the dialogue is vocal, however, the system adds two additional parts, Automatic Speech Recognition (ASR) and Speech Synthesis (SS), to translate between voice and text [9].

L. H. Belguith and K. Shaalan (Eds.): LPKM 2024, LNNS 1303, pp. 164–173, 2025.
https://doi.org/10.1007/978-3-031-85067-7_16

DS are increasingly intriguing in research and industry, but they often cater to a limited array of languages. The diversity of language dialects necessitates distinct linguistic models for each dialect, leading to some dialects being overlooked by DS. Although English, French, and Chinese are commonly supported, DS does not perform as well when it comes to Arabic. Some DS recognize Modern Standard Arabic (MSA) but neglect dialectal forms. Dialectal Arabic (DA) is prevalent in daily communication. Hence, there is a growing need to develop DS able to understand these Arabic variations [21]. These systems, with their sub-modules, are supervised, so they need annotated datasets to train [16].

This paper explores the feasibility of developing an NLU module as the central component of a DS for DA, specifically tailored to the Tunisian Dialect (TD). Our primary motivation is to automate the task of responding to inquiries from researchers and PhD students. This automation allows the doctoral school to dedicate more time and resources to addressing more complex and demanding responsibilities. Additionally, this work aims to tackle the unique linguistic challenges and specific needs of the Tunisian research community by providing a valuable tool that facilitates academic and educational interactions in the local dialect. This focus on TD is justified, since the majority of researchers in our faculty are Tunisian. Our study employs a straightforward method for data preparation and evaluation. It includes the creation of a TUN-NLU corpus of approximately 1,050 sentences within the educational and academic domains, tailored for academic purposes. These sentences are sourced from a variety of channels, including social media and direct input from Tunisian researchers. Following data collection, automatic normalization and annotation techniques are employed to enhance the quality and diversity of the data set. The proposed evaluation approach involves utilizing the RASA NLU[1] module to analyze the entries, classify them based on intent, and extract relevant entities.

The rest of the paper is organized as follows: Sect. 2 begins by presenting the main NLU challenges for the Arabic language (MSA and DA). Section 3 describes TUN-NLU corpus used to build the NLU model for TD. Section 4 presents the evaluation results of using our corpus in the RASA NLU module. The paper concludes in the last section and provides some future outlines.

2 Arabic NLU Challenges

The challenges of adapting NLU models to Arabic are multifaceted. In this section, we present the most relevant challenges encountered in NLU for the Arabic language in general and TD in particular.

2.1 Arabic Script and Absence of Normalization

The Arabic script utilizes a distinct writing direction, from right to left, and does not employ the Latin alphabet but rather the Abjad writing system. Also, Arabic text includes diacritics to denote certain short vowels, but their usage is optional,

[1] https://rasa.com/.

adding complexity to text processing. Moreover, in the dynamic world of digital communication, Arabizi stands out as a fascinating linguistic phenomenon where DA was transcribed into Latin characters, for example, "ta9s s5oun" instead of طقس سخون (Tqs sxwn) 'the weather is hot'. Besides, the complexity of DA is increased due to the absence of standardized writing conventions. Notwithstanding the existence of some works dealing with the normalization of DA (CODA [12,22], etc.), most of the resources available are non-normalized.

Addressing these challenges is crucial for the effective development and deployment of natural language processing tools and systems in Arabic-speaking contexts [1].

2.2 Morphological Ambiguity and Code-Switching

One of the key challenges in Arabic is its inherent ambiguity, which arises from the rich morphological structure of the language. The complex interaction of roots, stems, and affixes means that words can have multiple interpretations depending on their morphology. This ambiguity can result in misunderstandings, incorrect interpretations, and errors in natural language processing (NLP) applications.

In addition, Code Switching (CS) presents another challenge for DA, involving language changes within conversations. CS occurs at various levels, intersentential when switching languages from one sentence to another, for example: "c'est possible نبدل اختصاصي (nbdl AxtSASy) ?" 'it's possible to change my specialty'), or extra-sentential, when a loan word is borrowed from a secondary language, such as "l'inscription غدوة تبدا (gdwp tbdA)." 'The inscription will begin tomorrow.'. CS also encompasses code-mixing, where multiple languages are used within the same sentence. For example, "حبيت نختار (Hbyt nxtAr) Spécialité de thèse en deep learning." 'I wanted to choose a thesis specialty in deep learning.', as well as bilingual verbs that involve switching at the morpheme level, such as ".(mtAEk) متاعك [publication (Al) ال]] [partagi (n) ن]] (bA$) باش" ('I will share your post.').

In the Middle East, the integration of English and French is influenced by colonization, business, and education, leading to prevalent CS in Arab nations like Tunisia, Algeria, Morocco, and various others, with examples such as Arabic-French and Arabic-English. CS occurs at all linguistic levels, especially in morphologically rich languages like Arabic [13].

In NLU, efforts to address this issue involve techniques like transliteration or romanization to the Latin script [18].

3 Arabic NLU Resources

Data resources present a very important and challenging task for the Arabic NLU; we will see efforts made in this direction for MSA and DA.

3.1 MSA

Bashir et al. [5] present an open-domain corpus for building a "Home Automation" system. To collect data for this application, they used an online survey as there were no available sources in Arabic oriented to Home Automation. After filtering and labeling the collected data, their corpus was composed of 768 entries.

Mozannar et al. [17] introduced the open-domain Arabic Reading Comprehension Dataset (ARCD), which is composed of 1,395 questions posed by crowd workers in Wikipedia articles.

ARLUE [3] also targets Arabic NLU tasks and is composed of 42 datasets arranged into four task clusters: sentence classification, text classification, structured prediction, and QA. However, many of these datasets are not publicly available, which presents a barrier to widespread adoption.

ALUE [20] incorporates a total of 8 tasks such as the Emotion Classification, the Offensive and Hate Speech subtasks, etc. It covers a wide spectrum of Arabic dialects and NLP/NLU problems.

To create their data, Alruily et al. [4] used various datasets, including the one suggested by [5]. These datasets, along with scraped data from Arabic websites and newspapers, were used to generate the data for their proposed framework. In total, they created 2540 datasets containing seven entities for the entity extraction experiment.

Arabic-TOD [10] is the first human-annotated Arabic dataset for task-oriented DS, produced by translating 27% of the English BiToD dataset [15] dialogues into Arabic. The Arabic-TOD is the first dialogue dataset for the Arabic task-oriented DS that supports CS. The main limitation of this work is the small size of the Arabic-TOD dataset. A related limitation is using non-Arabic entities with the dataset CS due to entities in the original BiToD dataset. Essentially, they have turned this challenge into an opportunity by incorporating it into their model's training process, potentially making the model more adept at handling real-world scenarios where CS occurs.

ORCA [8] is a challenging and diverse benchmark for Arabic NLU. It involves 60 datasets that were arranged into 29 tasks and seven task clusters. To assess the current advancements in Arabic NLU, ORCA was employed to conduct a thorough comparison of 18 multilingual and Arabic language models. ORCA comes with an interactive publicly available leaderboard. Although ORCA currently covers both DA and MSA, some Arabic varieties, such as Iraqi and Yemeni, are not covered.

3.2 DA

Generally, DA resources are in-domain, contrary to MSA ones.

TUDICOI, short for TD Corpus Interlocutor [11], comprises 12,182 utterances in TD, tailored for addressing railway information requests in partnership with the National Company of Railway in Tunisia (SNCFT)[2]. The objective

[2] http://www.sncft.com.tn/.

was to devise a discriminative algorithm that utilizes conditional random fields (CRF) to semantically annotate spoken TD utterances that were not segmented into discrete units.

JANA corpus [7] comprises 4,725 utterances in Egyptian Dialect collected from multi-genre Egyptian banks and Egypt Air Company call centers. The aim was to develop a statistical dialogue analysis model for recognizing utterance dialogue acts.

Abdelhafez et al. [2] present another Egyptian dialect corpus proposed for the educational domain. This dataset consists of 45 standard queries, each with 25 variations. The authors used two methods to collect data: scraping student queries from popular social media sites and directly collecting data from students via forms. In the latter method, questions were asked in English, and students were requested to express in Arabic, providing as many variations as possible. This method yielded spontaneous forms of collected data, including aspects of CS used in the academic domain by Egyptian speakers.

For the Levantine dialect, [14] presented the first corpus in the food domain. It contains 873 manually tagged sentences. These sentences were collected from various sources, including crowdsourcing, Levantine tweets, and a dataset from a previous food order chatbot project.

Creating DS or NLU for specific domains requires significant time and effort, as the data collection process demands carefully annotated domain-specific content.

4 TUN-NLU Corpus

Many advancements in Arabic NLU mainly concentrate on open-domain ones specifically designed for MSA. However, open-domain ones exhibit limited effectiveness in industrial and defined-domain applications. Therefore, our work aims to develop closed-domain NLUs to fill this gap for TD. The proposed TUN-NLU corpus is focused on the academic field and designed for Tunisian PhD students and researchers.

In this work, we address several other challenging aspects of TD accordingly. Notably, we cite the absence of standardized writing and the CS. Considering that our NLU operates in the research domain, the integration of French is unavoidable, especially in technological fields where French usage is more widespread than Arabic. The Arabic-French CS manifests either through foreign words or foreign terms adapted to their origins, which are foreign words augmented with Arabic enclitics or proclitics [19].

Our TUN-NLU corpus preparation process is presented in Fig. 1.

4.1 Data Collection

Our data is collected from two sources: gathering data directly from students and scraping questions from Facebook groups.

In this method, we gather data from students directly. Seven PhD students are given a questionnaire that asks them to identify five topics related to their PhD journey that they find unclear and to formulate questions about these topics in their own words. The questionnaire comprises standard questions in French to prevent influencing student responses. Participants were requested to ask the French questions in TD and provide as many variations as possible. Scraping student queries from social media platforms, particularly Facebook, has become increasingly prevalent. Facebook has emerged as the primary communication platform for Tunisian researchers and students, with numerous groups dedicated to discussing topics such as supervisors, registration, scholarships, and more. This method of collection yields spontaneous text that closely represents the authentic TD in conversation. API scraping from those groups of most recent posts results in a corpus of 796 sentences, varying between long and short sentences.

4.2 Automatic Normalization

Now, we proceed to the preprocessing stage, where we prepare the data for testing its quality through exploration in RASA NLU. The resulting data is divided into two parts: one written in Arabic letters and the other in Latin characters (Arabizi). Given the non-standardized orthography of TD, the first part undergoes automatic normalization utilizing COTA-Orthography [6], employing the CODA-TUN convention [22]. This step aims to achieve greater uniformity in Tunisian writing and enhance overall consistency in the data.

The second part, constituting over 80% of our data, is first manually converted into Arabic characters. Subsequently, we normalize the text using the COTA-Orthography tool. Notably, since the normalization tool does not normalize the code-switching (CS) script, we made sure to retain this script without any intervention from the COTA-Orthography process.

4.3 Data Annotation

The next step involves data annotation. During this process, we manually enhanced the data by rephrasing some sentences with system replacements or altering their structure. This resulted in a dataset of approximately 1,050 sentences. Manually paraphrasing is justified by presenting this corpus as a carefully crafted golden dataset. For annotation, the data is manually labeled to identify intents and entities, facilitating effective processing and understanding within the RASA NLU framework. Rasa uses YAML as a flexible and comprehensive format for managing all data, including NLU data, which consists of sample user statements organized by intent and may contain entities-structured bits of information extractable from user messages. We begin with intent labeling or text classification, categorizing each standard query into classes such as Joint PhD-related or PhD registration-related. Following this, we proceed with entity labeling, identifying crucial information/entities relevant to each intent within the sentences. This process ensures that important details are accurately recognized and extracted for further analysis and processing. We identified 21 intents and 14 entities. The resulting dataset, TUN-NLU is divided into 80% for training and 20% for testing purposes.

5 Corpus Evaluation

Our proposal introduces a framework for testing data quality, leveraging the RASA NLU framework, a well-known open-source platform for developing dialogue systems. We selected Rasa for several reasons: it is free, user-friendly, and can be run locally, enhancing its convenience. Additionally, Rasa leads to creating and generating models efficiently doing simple and complicated tasks, even with minimal initial training data [21]. Another key factor in choosing RASA is its limited application in Arabic data science development thus far, making it particularly suitable for our work. The proposed method comprises four main stages: tokenization, feature extraction, intent classification, and relevant entity extraction. These steps form the basis of our closed-domain NLU. In these steps, only trainable modules are evaluated by RASA tests. We used the Whitespace-Tokenizer to perform whitespace tokenization, where words are separated by spaces. As an output, it creates a token for every whitespace-separated character sequence. Next, the LexicalSyntacticFeaturizer was employed to generate lexical and syntactic features for user messages to support entity extraction. It operates by moving with a sliding window over each token in the user message and creates features based on RASA's default configuration. For intent classification and entity recognition, we utilized the recommended DIETClassifier. Dual Intent Entity Transformer (DIET) is a multi-task architecture designed for both intent classification and entity recognition due to its effectiveness in training custom entities. It utilizes a shared transformer model for both tasks. For entity recognition, it predicts a sequence of entity labels using a Conditional Random Field (CRF) tagging layer placed on top of the transformer's output sequence, which corresponds to the input token sequence. For intent classification, the transformer's output for the entire utterance is combined with the intent labels, embedding them into a single semantic vector space.

We evaluated the NLU module using standard metrics, including accuracy and F1 score. Our framework achieves promising results, with 93% accuracy and 92% F1 score for intent classification and 94% accuracy and 91% F1 score for entity extraction.

There are errors in intent classification regarding greetings and goodbyes. Upon error analysis, we notice that the classifier struggles with distinguishing between closely related phrases such as عالسلامة (EAlslAmp) for 'hello' and بالسلامة (bAlslAmp) for 'goodbye'. Regarding the errors in entity recognition, the DIET classifier is unable to detect all components of an entity. For example, an entity like "dual-supervisor": الانكادرون الثاني (AlAnkAdrwn AlvAny) 'second supervisor' is often detected incorrectly as الانكادرون (AlAnkAdrwn) 'supervisor', which transforms into another entity, "thesis-supervisor". This misclassification results in incorrect intent detection, such as changing "dual-thesis-related" to "thesis-related". Another issue arises from entities that differ only in their prefixes or suffixes. For instance, the word رينسكربسيون (rynskrbsywn) "re-registration" is incorrectly considered synonymous with لنسكربسيون (lnskrb-sywn) "registration."

In our future work, to avoid some errors and enhance the intent and entity classifier, we propose refining the intent and entity annotation and augmenting the data size for a better evaluation and accurate entity recognition and intent detection.

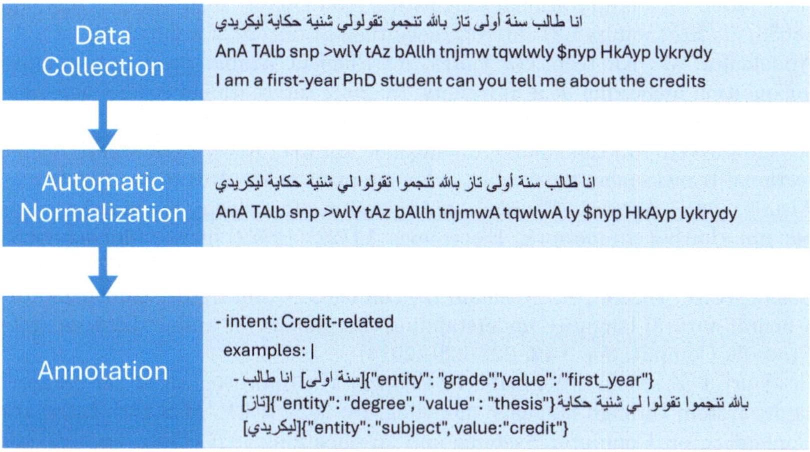

Fig. 1. TUN-NLU Data Preparation Process

6 Conclusion and Future Works

This paper presents a Tunisian NLU dataset called TUN-NLU. It consists of around 1,050 sentences within a closed domain designed for the academic field. These sentences were gathered from various sources, including social media and direct contributions from Tunisian researchers. After collecting the queries, the data underwent automatic normalization and manual augmentation and annotation to enhance its quality. To evaluate the dataset's quality, we utilized a simple Rasa NLU module. The results showed promising accuracy rates: 93% for intent classification and 94% for entity extraction tasks. Additionally, the NLU model achieved F1 scores of 92% for intent classification and 91% for entity extraction.

In future work, we plan to improve the labeling of intents and entities and increase the amount of data to achieve better evaluation and more accurate recognition and classification. We will use neural networks that have shown good performance in NLU for intent detection and entity recognition. Additionally, We aim to include ASR, DM, and NLG components followed by an SS module, using state-of-the-art techniques from existing literature. This comprehensive approach aims to establish a complete DS tailored to the Tunisian Dialect, providing Tunisian researchers with clarity and assistance in their academic pursuits.

References

1. Abboud, K., Golovneva, O., DiPersio, C.: Cross-lingual transfer for low-resource Arabic language understanding. In: Bouamor, H., et al. (eds.) Proceedings of the Seventh Arabic Natural Language Processing Workshop (WANLP), pp. 225–237. Association for Computational Linguistics, Abu Dhabi (2022). https://doi.org/10.18653/v1/2022.wanlp-1.21. https://aclanthology.org/2022.wanlp-1.21
2. Abdelhafez, M., Khateeb, G., Yahya, A.: Efficient Arabic query auto-completion for question answering at a university. In: 2022 International Arab Conference on Information Technology (ACIT), pp. 1–9. IEEE (2022)
3. Abdul-Mageed, M., Elmadany, A., Nagoudi, E.M.B.: Arbert & marbert: deep bidirectional transformers for Arabic. arXiv preprint arXiv:2101.01785 (2020)
4. Alruily, M.: Arrasa: Channel optimization for deep learning-based Arabic nlu chatbot framework. Electronics **11**(22) (2022).https://doi.org/10.3390/electronics11223745. https://www.mdpi.com/2079-9292/11/22/3745
5. Bashir, A.M., Hassan, A., Rosman, B., Duma, D., Ahmed, M.: Implementation of a neural natural language understanding component for arabic dialogue systems. Procedia Comput. Sci. **142**, 222–229 (2018)
6. Besdouri, F.Z., Mekki, A., Zribi, I., Ellouze, M.: Improvement of the cotaorthography system through language modeling. In: 2021 IEEE/ACS 18th International Conference on Computer Systems and Applications (AICCSA), pp. 1–7 (2021). https://doi.org/10.1109/AICCSA53542.2021.9686898
7. Elmadany, A., Abdou, S., Gheith, M.: Improving dialogue act classification for spontaneous arabic speech and instant messages at utterance level. arXiv preprint arXiv:1806.00522 (2018)
8. Elmadany, A., Nagoudi, E.M.B., Abdul-Mageed, M.: Orca: a challenging benchmark for arabic language understanding. arXiv preprint arXiv:2212.10758 (2022)
9. Feng, Y.: The latest progress in human-computer dialogue system. Highlights Sci. Eng. Technol. **85**, 9–19 (2024). https://doi.org/10.54097/gw3z2h19
10. Fuad, A., Al-Yahya, M.: Araconv: developing an Arabic task-oriented dialogue system using multi-lingual transformer model mt5. Appl. Sci. **12**(4), 1881 (2022)
11. Graja, M., Jaoua, M., Belguith, L.H.: Discriminative framework for spoken Tunisian dialect understanding. In: Dediu, A.-H., Martín-Vide, C., Mitkov, R., Truthe, B. (eds.) SLSP 2013. LNCS (LNAI), vol. 7978, pp. 102–110. Springer, Heidelberg (2013). https://doi.org/10.1007/978-3-642-39593-2_9
12. Habash, N., Diab, M., Rambow, O.: Conventional orthography for dialectal Arabic. In: Calzolari, N., et al. (eds.) Proceedings of the Eighth International Conference on Language Resources and Evaluation (LREC'12), pp. 711–718. European Language Resources Association (ELRA), Istanbul (2012)
13. Hamed, I., Denisov, P., Li, C.Y., Elmahdy, M., Abdennadher, S., Vu, N.T.: Investigations on speech recognition systems for low-resource dialectal Arabic-English code-switching speech. Comput. Speech Lang. **72**, 101278 (2022)

14. Joukhadar, A., Saghergy, H., Kweider, L., Ghneim, N.: Arabic dialogue act recognition for textual chatbot systems. In: Proceedings of The First International Workshop on NLP Solutions for Under Resourced Languages (NSURL 2019) co-located with ICNLSP 2019-Short Papers, pp. 43–49 (2019)
15. Lin, Z., et al.: Bitod: a bilingual multi-domain dataset for task-oriented dialogue modeling. CoRR arxiv:2106.02787 (2021)
16. Mahmoudi, K., Faili, H.: Pershop – a Persian dataset for shopping dialogue systems modeling (2024)
17. Mozannar, H., Hajal, K.E., Maamary, E., Hajj, H.: Neural Arabic question answering. arXiv preprint arXiv:1906.05394 (2019)
18. Muller, B., Anastasopoulos, A., Sagot, B., Seddah, D.: When being unseen from mBERT is just the beginning: handling new languages with multilingual language models. In: Toutanova, K., et al. (eds.) Proceedings of the 2021 Conference of the North American Chapter of the Association for Computational Linguistics: Human Language Technologies, pp. 448–462. Association for Computational Linguistics, Online (2021). https://doi.org/10.18653/v1/2021.naacl-main.38. https://aclanthology.org/2021.naacl-main.38
19. Nasr, L.I., Masmoudi, A., Belguith, L.H.: Natural Tunisian speech preprocessing for features extraction. In: Liu, Y., et al. (eds.) 23rd IEEE/ACIS International Conference on Computer and Information Science, ICIS 2023, Wuxi, China, 23–25 June 2023, pp. 73–78. IEEE (2023). https://doi.org/10.1109/ICIS57766.2023.10210270
20. Seelawi, H., et al.: Alue: Arabic language understanding evaluation. In: Proceedings of the Sixth Arabic Natural Language Processing Workshop, pp. 173–184 (2021)
21. Zribi, I., Belguith, L.H.: Toward developing an intelligent personal assistant for Tunisian Arabic. Jordanian J. Comput. Inf. Technol. **8**(4) (2022)
22. Zribi, I., Boujelbane, R., Masmoudi, A., Ellouze, M., Belguith, L., Habash, N.: A conventional orthography for Tunisian Arabic. In: Calzolari, N., et al. (eds.) Proceedings of the Ninth International Conference on Language Resources and Evaluation (LREC'14), pp. 2355–2361. European Language Resources Association (ELRA), Reykjavik (2014)

From Data to Decisions: An Ontology-Driven Method for Opinion Mining

Mehdi Belguith[1](✉), Wafa Khabou[1], Chafik Aloulou[1], and Bilel Gargouri[2]

[1] ANLP-RG MIRACL Laboratory FSEGS, University of Sfax, Sfax, Tunisia
belguith.mehdi2017@gmail.com
[2] MIRACL Laboratory FSEGS, University of Sfax, Sfax, Tunisia

Abstract. In this paper, we propose an ontology-driven method for opinion mining in social media. We first construct our Tunisian datasets from YouTube and social media content, mainly Twitter and Facebook. Then, based on natural language techniques and machine learning, we built four ontologies related to the Mobile phone, Election, Restaurant, and TV channel domains. We also propose an opinion mining method based on our domain ontologies.

Keywords: Domain ontologies · Opinion Mining · Aspects · Tunisian dialect · Machine learning · NLP

1 Introduction

The massive use of social media has reached significant proportions with 5.35 billion internet users worldwide in January 2024, representing 66.2% of the global population. Of this total, 5.04 billion are social media users, accounting for 62.3% of the global population. Each individual spends an average of 2 h and 31 min per day on social networks, thus facilitating the rapid and global dissemination of information. Given this abundance of information, Sentiment Analysis (SA) of online comments becomes tedious if not impossible for humans. Automating sentiment analysis becomes essential and unavoidable.

In this paper we focus on creating domain-specific ontologies specifically adapted for sentiment analysis in the Tunisian Dialect (TD). This paper is organized in seven sections. Section 2 presents the concepts and main categories of ontologies. Section 3 gives a brief overview of related works. Section 4 presents our proposed method. Section 5 describes the ontology building. Section 6 presents the evaluation results and finally we present in Sect. 7 the conclusion and some perspectives.

2 Ontology: Main Concepts and Categories

An ontology is a data model that mainly consists of concepts and relationships. The main concepts of an ontology are as follows:

L. H. Belguith and K. Shaalan (Eds.): LPKM 2024, LNNS 1303, pp. 174–182, 2025.
https://doi.org/10.1007/978-3-031-85067-7_17

- **Individuals**: They are the fundamental elements of an ontology such as specific entities or concrete instances (e.g., a cat named "Milou").
- **Classes**: They group individuals sharing similar properties (e.g., the class "Cat" in "Animal" domain ontology).
- **Properties**: They describe the relationships between individuals and classes (e.g., the property "Fur Color" associated with the class "Cat").
- **Relations**: They connect individuals and classes (e.g., the relation "Owns" between the classes Person and Cat).
- **Events**: They capture dynamic changes in an ontology (e.g., the change of the attribute "Age" of a cat over the years).
- **Metaclasses**: They represent sets of classes sharing common characteristics (e.g., grouping the classes "Cat" and "Dog" under the more general category "Pets").

Ontologies can be categorized based on their level of abstraction, domain specificity, and focus on particular tasks, with the choice depending on the specific objectives and needs of knowledge modeling. High-level ontologies offer a broad conceptual framework using a general vocabulary applicable to multiple domains. Domain ontologies provide detailed and specialized knowledge representation for specific fields. Task ontologies focus on modeling specific tasks with limited vocabularies, often inheriting from higher-level ontologies. Application ontologies are the most specific, combining domain knowledge with task-specific modeling to offer comprehensive and detailed representations for particular applications.

3 Related Work

The ontology-based opinion mining approach focuses on key methods for detecting opinions using ontologies [1–3]. Recent research focuses on aspect opinion mining rather than sentiment analysis [11,12,15]. Schouten et al. [13] developed a sentiment analysis method based on aspects enhanced by ontology using restaurant reviews from SemEval. Their dataset contains 254 reviews, each consisting of one or more sentences (i.e., a total of 1315 sentences). Each sentence is annotated with positive, negative, or neutral sentiment.

They used the linear Support Vector Machine (SVM) to classify sentiments. Similarly, Thakor et al. [16] implemented a sentiment analysis model based on ontology to detect negative sentiments expressed by customers regarding any postal service from their tweets. They used the Gate software to annotate verbs and nouns in tweets, then cleaned the data using a Python script and an Excel file. Nouns were considered as objects and verbs as object properties. Protégé software is used to build the ontology model and relationships between ontology concepts were manually defined. In the same context, Ceci et al. [5] developed a sentiment analysis model based on domain ontology for Spanish. Their method is case-based reasoning. It compares new cases with previous problems stored in a database with an SVD technique to determine adjective polarities. Sharma et al. [14] created OMASA, an Ontology-Driven Multi-Aspect Sentiment Analysis for the Hotel industry using three classifiers (SVM, NB, and MaxEnt). They demonstrated that the ontology helps to detect latent aspects. Carneiro et al. [4] focused on the importance of ontology in sentiment analysis for Portuguese without translation. They showed that results vary based on linguistic structures. Dera et al. [6] developed a semi-automatic method to build an ontology using synsets, demonstrating a reduction in execution time despite slightly lower performance than the manual method. Karoui et al. [7] proposed an ontology-based method to understand Tunisian dialect using two types of ontologies for annotation and semantic interpretation.

4 Proposed Method for Ontology Building from Raw Comments

4.1 Ontology Domains

In this research work, we focus on four domains and we propose to build the following domain ontologies:

- **RestOnto**: This ontology focuses on the restaurant domain, offering a deep semantic representation of the concepts and relationships associated with this field. It is based on a dataset of 2719 instances.
- **ElectOnto**: This ontology is dedicated to the election sector, providing a formal and organized structure of relevant elements extracted from a dataset containing 3076 instances that we collected in the context of this work.
- **PhoneOnto**: This ontology is related to the telephone domain. It is based on the dataset of Belguith et al. [1], which includes a total of 5813 instances.
- **ChannelOnto**: It is dedicated to television channels and is constructed based on the database of Mdhaffar et al. [9], which contains 13667 instances.

4.2 Ontology Structure

The structure of our ontologies is based on three main classes and a single object property called "hasAsentiment" that establishes the links between these three classes:

- **Sentiment**: It has three values (positive, negative, or neutral).
- **SentimentExpression**: Represents the textual expression of the sentiments and could be composed of subclasses.
- **Target**: This class groups the aspects. Note that our objective is to determine the sentiment (positive, negative, neutral) for each aspect in the comment.

This classification allows us to minimize the cost in terms of time and memory, as the target class will be used in machine learning to determine the domain of a comment and will also be used to identify the aspects to be extracted from the comment.

4.3 Main Steps of Our Proposed Method

Our method for building ontologies is based on 8 main steps as shown in Fig. 1.

- **Tokenization:** is the process of splitting the comments into units called "tokens" (words in our case) based on spaces and punctuation signs.
- **Stop Word Removal:** These words often include articles, prepositions, conjunctions, and other commonly used terms that generally do not provide much information for content analysis.
- **Part-of-Speech Tagging:** Part-of-speech (POS) tagging, is the process of assigning grammatical categories (parts of speech) to each word3. These categories include nouns, verbs, adjectives, adverbs, prepositions, conjunctions, etc.
- **Lemmatization:** It is a lexical processing step that involves extracting the canonical form of a lexeme (or a word), regardless of its variations such as gender, number, etc. This step allows grouping multiple forms of the same word into one. For example, the two words الطلاب and الطالبة will be grouped into a single word طالب .
- **Domain detection:** To determine the domain of a comment, we experimented machine learning models, deep learning models and a statistical method.
- **Machine Learning (ML):** We experimented models that are widely used in the literature such as Linear Regression, Support Vector Machine, Random Forest, K-Means, and MultinomialNB. Best results are reached by the MultinomialNB, a supervised classifier based on a probabilistic approach using the multinomial distribution.
- **Deep Learning (DL):** We experimented two models, the Sequential model4 and the Multilayer Perceptron (MLP), a neural network composed of multiple layers, including an input layer, one or more hidden layers, and an output layer. Each neuron in one layer is connected to all neurons in the next layer. Testing showed stable results starting from the tenth epoch for all algorithms. The trigram, with n-grams equal to 3, demonstrated the highest performance.
- **Statistical Method:** This method involves extracting the lemmas of the nouns in the comment. These lemmas are then compared with those in each ontology. The ontology with the most common lemmas is selected.
- **Sentiment Detection:** Once the domain of the comment is identified and the corresponding ontology is selected, we extract the sentiments based on the list of retained words, along with their POS. The aim is to determine the polarities (positive, negative, neutral) for aspects, comments and the whole dataset.
  - **Determination of Aspect Polarity:** We assign the polarity of each aspect based on the corresponding domain ontology. There are two sentiment categories. The first category represents sentiments that are directly associated with specific aspects whose ontology allows immediate identification without requiring further processing, such as سعر (Price) and طعم (Taste). The sentiment غالي (Expensive) expresses a negative opinion about the "Price" aspect, while بنين translates a positive sentiment

towards the "Taste" aspect. The second category represents ambiguous sentiments, such as احسن هاتف (better phone), for which the ontology alone does not allow determining the corresponding aspects (examples, battery, screen, design, etc.) the user is referring to. In this case, we assume that the aspect in question is the word following the sentiment, and proceed to determine the aspect from the comment.

- **Determination of comment polarity:** Once the polarity of each aspect is determined, we compute the overall polarity of a comment, which is the difference between the sum of positive polarities and the sum of negative polarities. Thus, in the comment خامج و بنين (disgusting and delicious), the aspect of cleanliness has a negative polarity while the aspect of taste has a positive polarity. So, the overall polarity of this comment will be neutral.
- **Determination of multiple comment polarity:** Based on the same technique, we determine the overall sentiment for a list of comments by computing the total number of positive comments and negative ones.

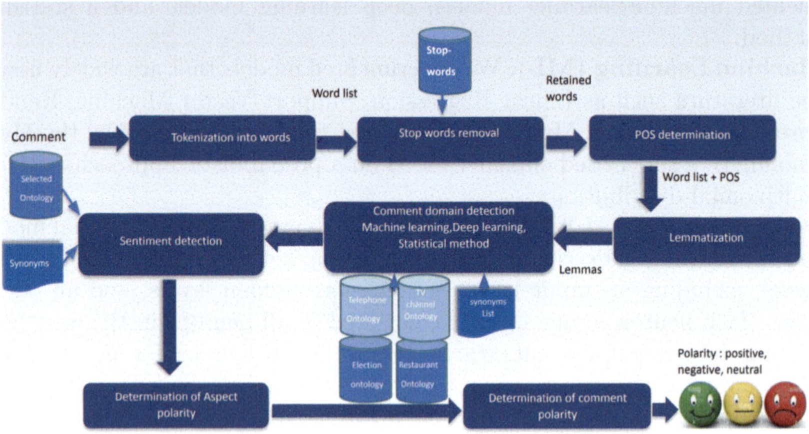

Fig. 1. Main steps of our proposed method

5 Ontology Building

Based on the proposed method, we build our four ontologies using the "Protégé" tool. We present in the following some extracts from our ontologies.

5.1 "Election" Ontology

Our "Election" ontology called ElectOnto contains 16 classes, 584 individuals, 1693 axioms, and 3 main classes (Sentiment, SentimentExpression, and Target). Figure 2 presents an example from ElectOnto Ontology.

The SentimentExpression class: corresponds to the sentiments expressed by people. It is composed of 5 subclasses according to the nature of the sentiments:

- Simple: حزين (sad), سعيد (happy), غاضب (angry) مندهش (surprised).
- Compound: are sentiments expressed by a phrase like: أشعر بالحزن لأن الجو بارد (I feel sad because the weather is cold). أنا سعيد بلقاء أصدقائي (I am happy to meet my friends).
- Proverb: sentiments expressed by known proverbs: كما تزرع تحصد , (You reap what you sow), الجار قبل الدار (The neighbor comes before the house).
- Character: sentiments expressed through comparison with famous persons as for examples : مثل حاتم الطائي (Like Hatim al-Tai), meaning someone is very generous, as Hatim al-Tai was known for his generosity. كأنه عنترة ابن شداد (Like Antarah ibn Shaddad), meaning someone is very brave, as Antarah ibn Shaddad was known for his bravery.
- Target: it contains 5 subclasses which correspond to the five aspects we determined for the "Election" domain according to our Datasets which refers to Tunisan election in 2019:
- President: represents the candidates for the election.
- People: refers to Tunisian people. (e.g. الشعب التونسي العظيم (Long live the Tunisian people), يحيا الشعب التونسي (The great Tunisian people)).
- Election: refers to the election process (e.g. انتخابات نزيهة (Fair elections), مهزلة الانتخابات (The farce of the elections)).
- Debate: refers to the debates between candidates (النقاش مفتوح (The discussion is open), نرحب بالنقاش (We welcome the discussion)).
- Presenter: refers to the person who is animated the debates (المذيع غير محترف (The presenter is not professional), هذه الصحفية غير كفؤة (This journalist is not competent), المذيع بارع في تقديم الأخبار (The presenter is excellent in delivering the news)).

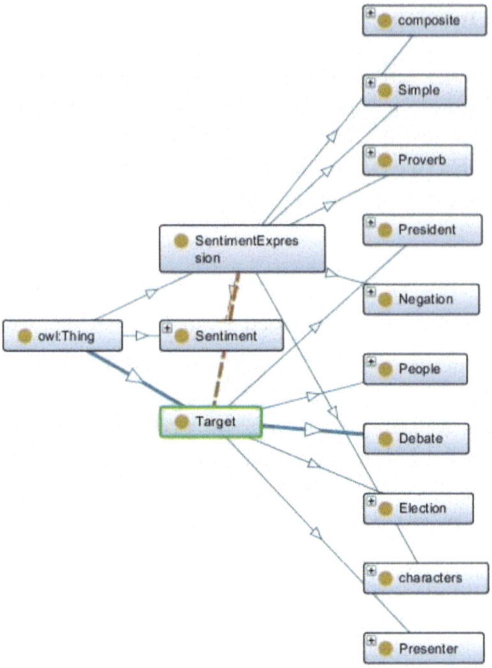

Fig. 2. The classes of ElectOnto ontology

5.2 "Restaurant" Ontology

This ontology is composed of three main classes : the Sentiment (positive, negative and neutral), the SentimentExpression and the Target. The Target Class has 7 subclasses that correspond to the aspects detected in the domain of "Restaurant" based on our dataset described below.

- Chef: refers to the person who is in charge of the food شاف معز أبدع فوق الابداع (Chief Moez, excelling beyond excellence).
- Restaurant: أحببت الأطباق اللذيذة بشكل كبير، الخدمة كانت بطيئة في بعض الأحيان (I loved the delicious dishes immensely, but the service was slow at times).
- Price: السعر مرتفع للغاية (The price is extremely high).
- Taste: refers to consumer satisfaction regarding taste: لذيذ جداً (very wonderful), رائع جداً ,(very bad), سيء جداً (very delicious).
- Cleanness: Refers to property. Note that our aim is to detect the client pinion about cleanness نظيف جداً (very messy) فوضوي جداً (very clean).
- Drink: Drink subclass consists of 4 subclasses: عصير (Juice), عصير حليب (Milk juice), كوكتال cocktail, and other drinks مشروبات مختلفة .
- Food: It is composed of 68 subclasses which correspond to food names.

6 Evaluation Results

Our first evaluations are conducted for our methods of detecting the domain of the comment. As shown in Table 1, the best results are achieved by the sequential model with an accuracy of 94%. Note that errors are due to the specificity of the Tunisian language. For example: عملة كالبصل (Act like an onion) was classified as a restaurant domain comment instead of an election domain comment.

Table 1. Evaluation results of our domain detection method

	Machine Learning	Deep Learning	Statistical Method
Accuracy (%)	MultinomialNB (84%)	Sequential (94%) Multilayer Perceptron (73%)	Sentiment words occurrences (70%)

7 Conclusion and Future Works

In this paper, we proposed a method for social media opinion mining based on ontologies. First, we proposed a method to build ontologies and we particularly focused on four domains which are restaurants, elections, television channels, and mobile phones. Then we proposed a method for aspect sentiment analysis based on the domain ontologies. Our proposed method for opinion mining relies on five steps. The first step involves data collection from YouTube and social media. The second step involves preprocessing which consists in data cleaning by eliminating symbols, empty words, and other superfluous units. The third step involves automatically detecting the domain of the comment. The fourth step involves determining the polarity (positive, negative, or neutral) of each aspect inside the comment. Then we determine in the fifth step the overall polarity of the comment. As future work, we plan to increase the coverage of the domains addressed, and we will perform an extrinsic evaluation by studying the effect of using our ontologies in enhancing the sentiment analysis process. We also intend to experiment with a Transfer learning model for the Tunisian dialect [8, 10] and compare it with deep learning models.

References

1. Belguith, M., Aloulou, C., Gargouri, B.: Aspect level sentiment analysis based on deep learning and ontologies. SN Comput. Sci. **5**(1), 58 (2023)
2. Belguith, M., Aloulou, C., Gargouri, B.: Building domain ontologies for Tunisian dialect: towards aspect sentiment analysis from social media. In: International Conference on Intelligent Systems and Pattern Recognition, pp. 252–267. Springer, Heidelberg (2023). https://doi.org/10.1007/978-3-031-46335-8_20

3. Belguith, M., Azaiez, N., Aloulou, C., Gargouri, B.: Social media sentiment classification for Tunisian dialect: a deep learning approach. In: International Conference on Intelligent Systems and Pattern Recognition, pp. 377–393. Springer, Heidelberg (2022). https://doi.org/10.1007/978-3-031-08277-1_31
4. Carneiro, E., Guedes, G.P., Belloze, K.T.: A review of ontology-based approaches for sentiment analysis: possible improvements on the Brazilian affective computing scenario. In: ONTOBRAS, pp. 127–136 (2022)
5. Ceci, F., Goncalves, A.L., Weber, R.: A model for sentiment analysis based on ontology and cases. IEEE Lat. Am. Trans. **14**(11), 4560–4566 (2016)
6. Dera, E., Frasincar, F., Schouten, K., Zhuang, L.: SASOBUS: semi-automatic sentiment domain ontology building using synsets. In: Harth, A., et al. (eds.) ESWC 2020. LNCS, vol. 12123, pp. 105–120. Springer, Cham (2020). https://doi.org/10.1007/978-3-030-49461-2_7
7. Karoui, J., Graja, M., Boudabous, M.M., Belguith, L.H.: Semi-automatic domain ontology construction from spoken corpus in Tunisian dialect: railway request information. Int. J. Recent Contribut. Eng. Sci. IT (iJES) **1**(1) (2013)
8. Masmoudi, A., Hamdi, J., Belguith, L.H.: Deep learning for sentiment analysis of Tunisian dialect. Computación y Sistemas **25**(1), 129–148 (2021)
9. Mdhaffar, S., Bougares, F., Esteve, Y., Hadrich-Belguith, L.: Sentiment analysis of Tunisian dialects: linguistic ressources and experiments. In: Third Arabic Natural Language Processing Workshop (WANLP), pp. 55–61 (2017)
10. Messaoudi, A., et al.: Tunbert: pretrained contextualized text representation for tunisian dialect. In: International Conference on Intelligent Systems and Pattern Recognition, pp. 278–290. Springer, Heidelberg (2022). https://doi.org/10.1007/978-3-031-08277-1_23
11. Mowlaei, M.E., Abadeh, M.S., Keshavarz, H.: Aspect-based sentiment analysis using adaptive aspect-based lexicons. Expert Syst. Appl. **148**, 113234 (2020)
12. Pour, A.A.M., Jalili, S.: Aspects extraction for aspect level opinion analysis based on deep cnn. In: 2021 26th International Computer Conference, Computer Society of Iran (CSICC), pp. 1–6. IEEE (2021)
13. Schouten, K., Frasincar, F., de Jong, F.: Ontology-enhanced aspect-based sentiment analysis. In: Cabot, J., De Virgilio, R., Torlone, R. (eds.) ICWE 2017. LNCS, vol. 10360, pp. 302–320. Springer, Cham (2017). https://doi.org/10.1007/978-3-319-60131-1_17
14. Sharma, S., Saraswat, M., Dubey, A.K.: Multi-aspect sentiment analysis using domain ontologies. In: Iberoamerican Knowledge Graphs and Semantic Web Conference, pp. 263–276. Springer, Heidelberg (2022). https://doi.org/10.1007/978-3-031-21422-6_19
15. Tamrakar, S., Bal, B.K., Thapa, R.B.: Aspect based sentiment analysis of nepali text using support vector machine and naive bayes. Tech. J. **2**(1), 22–29 (2020)
16. Thakor, P., Sasi, S.: Ontology-based sentiment analysis process for social media content. Procedia Comput. Sci. **53**, 199–207 (2015)

LLMs for Cyberbullying Detection in Political Social Media

Sahar Ben Bechir(✉), Asma Mekki, Mariem Ellouze, and Lamia Hadrich Belguith

ANLP Research Group, MIRACL Laboratory, University of Sfax, Sfax, Tunisia
saharbenbechir2018@gmail.com,
{mariem.ellouze,lamia.belguith}@fsegs.usf.tn

Abstract. As smartphones and social media usage grow among young people, particularly in Arab communities, the risk of encountering cyberbullying and harmful content increases. However, existing cyberbullying detection solutions are primarily tailored for English, leaving Arabic users underserved. This study addresses this gap by developing improved detection models for Arabic content. Through training various classifiers with annotated Arabic datasets, including traditional machine learning and deep learning techniques, this research aims to enhance the effectiveness of cyberbullying detection in Arabic online spaces, collected from three different platforms (Facebook, Twitter, and YouTube). Additionally, we conducted domain-specific data extraction from our existing datasets, focusing solely on political discourse. This was followed by testing the extracted data with deep learning algorithms. The results indicate that the proposed model outperforms other classifiers examined in the study. The overall enhancement achieved by the proposed model reaches an accuracy of 89% compared to the best-performing classifier.

Keywords: Cyber-Bullying · Arabic text · Social media content · Deep Learning

1 Introduction

The widespread prevalence of social networks exerts a significant impact on both individuals and communities, fostering increased connectivity and engagement within expansive networks comprising relatives, friends, followers, and acquaintances. However, it is unfortunate that some individuals exploit these digital platforms to propagate offensive content, intending to embarrass or harm others, thus perpetuating the troubling phenomenon known as cyberbullying. Its consequences are severe, leading to depression, emotional and physical stress, loss of self-esteem, and social isolation [11]. Although social media is commonly associated with cyberbullying, it is not limited to these platforms. Cyberbullying can occur across various digital spaces, including online games, mobile services, websites, and sharing platforms [13]. It can manifest in different forms, such

L. H. Belguith and K. Shaalan (Eds.): LPKM 2024, LNNS 1303, pp. 183–193, 2025.
https://doi.org/10.1007/978-3-031-85067-7_18

as text, images, videos, and audio. Despite extensive efforts in recent years to automate cyberbullying detection, instances of cyberbullying persist and even escalate. Various tools and methodologies have been developed to analyze textual content and identify cyberbullying behaviors, with Natural Language Processing (NLP) techniques being a common choice to process unstructured text data and extract relevant patterns. Arabic encompasses multiple forms, including Modern Standard Arabic (MSA), Classical Arabic, and various dialects [6]. This linguistic diversity complicates cyberbullying detection in Arabic. Solutions designed for other languages often prove ineffective when applied to Arabic due to its unique linguistic features and usage patterns. Therefore, developing effective cyberbullying detection systems for Arabic requires tailored approaches that account for these linguistic complexities. This study seeks to improve the effectiveness of current cyberbullying detection models for Arabic content by creating and refining a model ensemble. To attain this objective, a methodology comprising four phases is employed. The initial phase entails extracting comments relevant to our domain from multiple datasets containing instances of cyberbullying. The second phase is dedicated to data preprocessing, which involves tasks such as normalizing orthography, cleaning noise, tokenizing, removing stop words, URLs, and punctuation, and creating a corpus using NLP techniques alongside annotation. During the third phase, our attention is directed towards feature extraction and representation. We transform unstructured text features into structured and numerical word representations. This process is CountVectorizer. Next, deep learning algorithms are trained using the dataset. The top-performing classifiers from these trained models are then chosen to construct the proposed ensemble model. The F1-score, which is a harmonic mean of recall and precision, is employed to rank the classifier outputs. These scored outputs are aggregated to make decisions based on consensus, with each classifier contributing to the final decision. Ultimately, the model makes a final determination regarding the presence or absence of cyberbullying.

The findings suggest that the proposed model surpasses other trained classifiers in effectiveness. Section 2 reviews related literature, Sect. 3 details the methodology adopted in this study, Sect. 4 presents the experimental results and accompanying discussion, and Sect. 5 concludes the paper while suggesting directions for future research.

2 Related Work

Many studies have examined cyberbullying detection through sentiment analysis, focusing on various perspectives for assessing sentiment scores. Some research emphasizes textual sentiment analysis, while others consider the sentiment conveyed through emojis. A considerable body of literature has developed in this area, particularly regarding cyberbullying detection and analysis in Arabic dialects.

A notable contribution to this field is the work by BB. Sahar et al., [7] who introduced a dataset for cyberbullying detection in the Tunisian dialect. This dataset comprises 13,694 annotated comments. The researchers applied preprocessing techniques and numerical conversion before employing various machine learning algorithms, including Recurrent Neural Networks (RNN), Long Short-Term Memory (LSTM), Support Vector Machines (SVM), Transfer Learning (TL), and Deep Neural Networks (DNN). Among these, the Transfer Learning model demonstrated the highest performance, achieving an accuracy of 84%

Al.Dhayani et al., [2] developed an innovative hybrid model for Arabic cyberbullying detection, combining CNN, Bi-LSTM, and GRU with inverse document frequency. This approach eliminated the need for post-processing and effectively captured the complex nature of Arabic text, including spatial features and contextual information. Their model outperformed existing state-of-the-art ensemble models in the official deep learning challenge. The three-layer inverse document frequency LSTM classifier achieved the highest accuracy of 92.75%. This hybrid approach demonstrates the potential of combining multiple deep learning techniques to improve cyberbullying detection in Arabic text, capturing both local and long-range dependencies while weighting the importance of different words.

Mazari et al., [10] conducted a pioneering study on toxic text detection in the Algerian dialect, creating a dataset of 14,150 annotated comments from Facebook, YouTube, and Twitter. They utilized advanced natural language processing tools and various machine learning and deep learning models for analysis. The Bidirectional Gated Recurrent Unit (Bi-GRU) model proved to be the most effective, achieving a precision of 73.6% and an F1-score of 75.8%. This research significantly enhances the understanding of toxic communication in the Algerian context and contributes to the development of techniques for identifying harmful online interactions.

In their study, AM. Shiheb et al., [1] gathered data from Jazeera.net, sourcing content from various platforms such as YouTube, Facebook, Instagram, and Twitter. They employed the Term Frequency-Inverse Document Frequency (TF-IDF) method for feature extraction and utilized a combination of deep learning (DL) and machine learning (ML) algorithms for analysis. This approach resulted in an accuracy rate of 72%. Their research contributes to the understanding of toxic content detection in Arabic text across multiple social media channels, highlighting the effectiveness of integrating different algorithmic techniques in this domain.

F. Shannag et al., [14] developed the ArCybC, a multi-dialect Arabic Cyberbullying Corpus, by collecting 4,505 tweets from various domains such as gaming, sports, news, and celebrity platforms. They employed five annotators to manually classify the tweets into Offensive/non-Offensive and Cyberbullying/non-Cyberbullying categories, utilizing a harassment lexicon with profane words across four categories: sexual, racial, physical appearance, and intelligence. Their experiments compared machine learning techniques using bag-of-words (BoW) and word embedding models, revealing that the Support Vector Machine (SVM) with word embedding achieved an accuracy of 86.3% and an F1-score of 85%. The study also highlighted challenges in Arabic cyberbullying research, including the scarcity of available texts and annotation difficulties. By creating this dataset, the researchers aimed to provide a valuable resource for future Natural Language Processing tasks and enhance the understanding of cyberbullying in Arabic digital spaces.

In their study, Almutairi et al., [4] addressed the pressing issue of cyberbullying on social media platforms, particularly focusing on Twitter and the challenges associated with dialectal Arabic. They emphasized the scarcity of resources available for detecting and classifying cyberbullying in this linguistic context. To tackle this problem, the researchers employed Pointwise Mutual Information (PMI) for lexicon generation and utilized Support Vector Machine (SVM) algorithms for analyzing tweets. Their findings revealed a significant improvement in detection accuracy, with performance rising from 50% using PMI to 82% with SVM on resampled data. This notable increase underscores the effectiveness of their approach in addressing cyberbullying within the Saudi Arabian dialect on Twitter, contributing valuable insights to the growing field of Arabic cyberbullying detection.

In their comprehensive research, Khairi et al., [9] conducted a detailed analysis by dividing their collected data into three distinct sets. They employed the Term Frequency-Inverse Document Frequency (Tf-idf) method and systematically tested various machine learning algorithms, including K-Nearest Neighbors (KNN), Logistic Regression (LR), and Support Vector Machine (SVM), within two ensemble frameworks: SML and EML. The SML ensemble revealed SVM achieving an impressive 98% accuracy, following performances by LR and KNN. In the EML ensemble, which utilized Bagging, Boosting, and Voting techniques, the Voting method demonstrated superior performance with a 98.5% accuracy rate. Overall, both ensemble approaches showed significant improvements, with SML reaching 98% accuracy and EML achieving 98.8% accuracy, highlighting the effectiveness of ensemble machine learning techniques in classification tasks.

Kamel et al., [3] conducted a comprehensive study on cyberbullying detection in multiple Arabic dialects, addressing a significant research gap. Their approach involved collecting data from popular social media platforms, Twitter and YouTube, to capture a diverse range of Arabic linguistic variations. The researchers employed various machine learning algorithms to analyze this data, aiming to develop effective cyberbullying detection methods. Their evaluation revealed promising results across different classifiers. Notably, the Naive Bayes (NB) algorithm demonstrated superior performance, achieving an impressive 89% accuracy in detecting cyberbullying content. Support Vector Machine (SVM) and Logistic Regression (LR) also showed strong capabilities, both attaining 88% accuracy. These findings highlight the potential of machine learning techniques in identifying and combating cyberbullying in Arabic social media contexts, while also underscoring the effectiveness of NB in particular for this task.

Recent research has highlighted the importance of clustering algorithms and classification techniques in detecting cyberbullying and cyber-harassment in Arabic social media content. Kanan et al., [8] conducted a comprehensive study exploring various classification algorithms and leveraging Arabic Natural Language Processing (NLP) tools, including P-Stemming, to analyze their dataset. Their findings revealed the effectiveness of different algorithms across various contexts. The Random Forest (RF) algorithm emerged as the top performer, achieving an impressive F1-Measure of 94.9%. Notably, the Support Vector Machine (SVM) demonstrated superior performance when applied to specific subsets of data, such as Facebook posts and Twitter tweets. This research underscores the potential of machine learning approaches in addressing the complex challenge of identifying harmful content in Arabic social media, while also highlighting the importance of tailoring algorithms to specific data sources for optimal results.

Saadi et al., [12] conducted a significant study on cyberbullying detection in Arabic social media content, addressing a critical issue affecting young people online. Their research focused on developing an automated system to identify potentially harmful Arabic comments indicative of cyberbullying. The researchers employed a Support Vector Machine (SVM) algorithm, a powerful machine learning technique, for this task. They utilized term frequency-inverse document frequency (TF-IDF) vectorizer and count vectorizer methods for feature extraction from the text data. To further enhance the system's performance, they applied the cuckoo search algorithm to optimize the SVM's hyperparameters. This optimization process led to a notable improvement in the system's accuracy, increasing from 85.8% to 87.1%. This advancement in cyberbullying detection technology represents a significant step towards creating safer online environments, particularly for vulnerable youth populations in Arabic-speaking digital spaces.

In [5], the study focused on the automatic detection of cyberbullying in tweets using the RoBERTa model, enhanced with GloVe word vectors. This method is benchmarked against several machine learning techniques, deep learning models, and other transformer-based models such as FastText. The results demonstrate that the proposed model outperforms the alternatives, achieving a 95% accuracy rate in detecting cyberbullying tweets. The robustness and superiority of this method are further validated through k-fold cross-validation.

3 Proposed Method

In this study, we present a method composed of four distinct phases: dataset collection, preprocessing, feature representation, and model construction. These phases are carefully designed to tackle the particular challenges and nuances associated with Arabic text, aiming to enhance the accuracy of cyberbullying detection in this language. Figure 1 illustrates the proposed classification model leveraging a deep learning algorithm. In this study, we propose a method that consists of four phases: data collection, data preprocessing, feature extraction and representation, and model construction. These phases are designed to address the specific challenges and characteristics of Arabic text, ultimately enhancing the effectiveness of cyberbullying detection in this language.

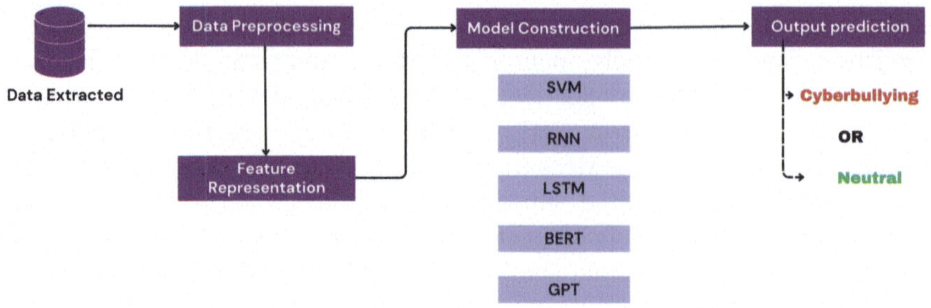

Fig. 1. Main steps of our proposed method

3.1 Dataset

To ensure the relevance of our analysis of political cyberbullying in Tunisia, we meticulously extracted a specific subset of political comments from an initial dataset of 13,694 comments collected from various social media platforms [7]. This process resulted in 3,906 political comments. The proportion of comments classified as cyberbullying in this subset is significantly higher than in the original dataset, highlighting the severity of the phenomenon within this context and the need for effective countermeasures. The extracted comments provided insights into the linguistic nuances of cyberbullying in the Tunisian dialect, featuring informal, non-standard vocabulary, repeated letters, unconventional abbreviations, onomatopoeic expressions, emoticons, and comments in both Arabic script and Arabizi (Latin script). This linguistic diversity introduced complexity to the analysis and processing of the data.

3.2 PreProcessing

During preprocessing, the text undergoes transformations to optimize it for machine learning algorithms. Key steps include word tokenization, orthographic normalization, and stemming. Tokenization breaks down the text into smaller units called tokens. Preprocessing techniques are applied to remove irrelevant components like numbers, hyperlinks, special characters, Latin words, symbols, dates, punctuation marks, stop words, sparse terms, and white spaces, which can interfere with classification accuracy. Stemming converts words to their base form. Orthographic normalization standardizes text to reduce variation and complexity. Our method normalizes text by replacing words or phrases with their standardized versions in Arabic comments, then returning the normalized text.

3.3 Feature Representation

During this phase, textual features are transformed into numerical representations using CountVectorizer. CountVectorizer, a class in Python's scikit-learn library, converts textual documents into a matrix of word occurrence counts. It creates a vocabulary of all unique words in the document collection, assigning each a unique integer identifier. Each document is represented by a vector with dimensions corresponding to the vocabulary words, indicating their occurrence counts. This numerical representation enhances the application of machine learning algorithms for tasks like text classification, clustering, and sentiment analysis. Additionally, SMOTE (Synthetic Minority Over-sampling Technique) addresses class imbalance by generating synthetic samples of the minority class, improving the model's performance and reliability in detecting political cyberbullying in the Tunisian context [12].

3.4 Construction Model

The dataset is partitioned into distinct categories, with 80% allocated for the training set and the remainder for the testing set. In this study, various machine learning algorithms, including SVM, RNN, LSTM, TL with bert-base-uncased model and Large Language Models (LLM) with gpt2 model, are investigated and compared. These algorithms are widely acknowledged and utilized in machine learning for their effectiveness in classification tasks.

4 Performance Analysis

4.1 Experimental Configurations

To tackle the class imbalance between normal and cyberbullying samples, oversampling techniques were applied specifically to balance the data by increasing the number of cyberbullying samples. The experimental setup utilized the Python programming language along with libraries like NLTK for natural language processing, scikit-learn for traditional machine learning classifiers, Imbalanced-learn for resampling imbalanced data, and Keras for deep learning, in addition to other pertinent libraries. These tools were applied across various stages, including dataset preprocessing, oversampling, feature extraction, representation, model training, and testing. For comparison with our proposed model, we employed five machine learning classifiers: SVM, TL, LSTM, RNN, and LLM. Among these classifiers, the five best-performing ones were chosen to construct the proposed ensemble model. We initiated the process by implementing SVM using the scikit-learn library. The dataset was divided into training and test sets using the "train-test-split" function. Subsequently, the SVM model was built using the "SVC" (Support Vector Classifier) class from scikit-learn. Initially, we utilized the linear kernel, which is known for its simplicity and is commonly used when the data is linearly separable. We assessed the performance of each kernel architecture by analyzing metrics like accuracy, precision, recall, and F1-score. This evaluation was conducted using the "classification-report" function from the scikit-learn library. Furthermore, we constructed the LSTM model using the Sequential class from Keras. The model consisted of three dense layers. The initial layer comprised 32 neurons with a ReLU activation function. To mitigate overfitting, we introduced a Dropout layer with a rate of 0.2. Subsequently, a second layer with 32 neurons and a ReLU activation function was added, followed by another Dropout layer with the same rate of 0.2. The final layer featured a single neuron with a sigmoid activation function. For compiling the model, we utilized the "binary-crossentropy" loss function and the "adam" optimizer. To optimize the model's performance and achieve the best results, we experimented with varying numbers of epochs and batch sizes. This iterative process allowed us to fine-tune the model's parameters and enhance its effectiveness in cyberbullying detection. Moreover, in our experimentation with the RNN algorithm, conducted using Python and TensorFlow's Keras framework, we commenced by partitioning the dataset into training and testing sets. Subsequently, we engineered an RNN model utilizing Keras' Sequential class, comprising three dense layers. The initial layer housed 64 neurons with ReLU activation, supplemented by a Dropout layer with a 0.2 rate to mitigate overfitting. The subsequent layer contained 64 neurons with sigmoid activation, followed by another Dropout layer. Lastly, a single neuron with a sigmoid activation function facilitated binary classification. Post-architecture definition, we compiled the model, specifying "binary-crossentropy" as the loss function and utilizing the "adam" optimizer. Evaluation of the model's performance was facilitated by the "accuracy" metric. During the training phase, we executed the model across 20 epochs, employing a batch size of 64. Validation data was leveraged to monitor the model's performance during training. Expanding in our article S.Ben Bechir et al., [7], our study wants to explore more about how the method they discussed might affect things. We also applied the GPT-2 LLM on the already processed data. We loaded the pre-trained model and compiled it with training arguments from the output directory, setting the number of epochs to 20, and defining a batch size of 32. In terms of frozen and fine-tuned layers, we adopted an approach where some layers of the pre-trained GPT-2 model remain frozen while others are fine-tuned. The initial layers of the GPT-2 model, which capture general language features, were kept unchanged, while the upper layers, including the added classification layer, were fine-tuned to adapt to the specifics of our comment classification task. The rationale behind this choice lies in the fact that the initial layers of the pre-trained GPT-2 model were trained on a large dataset to capture general language structure. By freezing these layers, we preserve these previously acquired knowledge and mitigate the risk of overfitting to our specific dataset. The upper layers of the model, including the classification layer, were fine-tuned to better adapt to our comment classification task. This process adjusted the model weights for improved discrimination between different comment types. We trained our model using GPT-2 and BERT, leveraging their capabilities despite our relatively small dataset. These large language models excel in various NLP tasks due to their understanding of linguistic subtleties and semantic nuances. Fine-tuning them allowed us to harness their strengths for our specific task. We evaluated the model's performance on a test set to ensure it generalized well to unseen data. GPT-2's contextual understanding and text generation abilities, combined with BERT's bidirectional attention mechanism, made them ideal for capturing the rich context and nuances of comments, enhancing our classification performance.

4.2 Performance Measures

The proposed cyberbullying detection model was evaluated using four widely recognized performance metrics commonly utilized in the literature. Firstly, classification accuracy (Acc) [1] was employed, which measures the proportion of correctly classified samples out of the total dataset. Secondly, precision (pre) [2] was utilized to assess the ratio of correctly classified normal samples to the total number of samples classified as normal. Thirdly, recall (Rec) [3] was considered, indicating the ratio of correctly classified normal samples to the total number of actual normal samples. Additionally, the F1-score, [4] a composite metric, was calculated. It is computed as the harmonic mean between precision and recall. Precision represents the ratio of true positive predictions to the total number of predicted positives, while recall represents the ratio of true positive predictions to the total number of actual positives. The F1-score offers a balanced evaluation, considering both precision and recall simultaneously, thereby providing a robust measure of the model's performance. These performance measures collectively offer a comprehensive assessment of the proposed model, considering various factors such as true negatives, balancing precision and recall, analyzing performance across different thresholds, and assessing average precision across multiple query instances.

$$Acc = \frac{Number\ of\ True\ Classified\ Samples}{Total\ Number\ of\ Samples} \tag{1}$$

$$Pre = \frac{Number\ of\ True\ Classified\ Normal\ Samples}{Total\ Number\ of\ Samples\ Classified\ Normal} \tag{2}$$

$$Rec = \frac{Number\ of\ True\ Classified\ Normal\ Samples}{Total\ Number\ of\ Normal\ Samples} \tag{3}$$

$$F1 - score = \frac{2 \times Pre \times Rec}{Pre + Rec} \tag{4}$$

5 Discussion and Results

Table 1 presents a comparison of our model's performance against various machine learning and deep learning algorithms, including SVM, RNN, LSTM, GPT-2, and BERT. Metrics such as accuracy, precision, recall, and F1 score were evaluated for each algorithm. BERT stands out for its superior performance across these metrics, thanks to its ability to grasp intricate contextual nuances in text, effectively discerning and classifying complex patterns in the data. While SVM, RNN, and LSTM show competitive performance, they often struggle to capture the nuanced linguistic features that BERT excels at. GPT-2, although a powerful language model, tends to lag behind in performance metrics. This could be due to task-specific requirements, the scale and quality of training data, and fine-tuning nuances. Unlike BERT, which is pre-trained with a masked language model objective followed by fine-tuning for classification tasks, GPT-2 is solely trained on a language modeling objective. This difference in pre-training likely contributes to GPT-2's comparatively lower performance in tasks requiring sequence classification, such as cyberbullying detection.

Table 1. Deep learning algorithms results

Classifier	Accuracy	Precision	Recall	F1-score
SVM	87%	88%	65%	78%
RNN	88.1%	88%	80%	84%
LSTM	88%	88%	77%	83%
GPT2	59%	60%	60%	60%
BERT	**89%**	**81%**	**80%**	**78%**

6 Conclusion and Future Work

In this research, we presented a model designed to detect cyberbullying and offensive language in Arabic effectively. The model comprises four main elements: text preprocessing, feature extraction and representation, classification, and prediction. Current cyberbullying detection models are mainly designed with English or similar languages in mind, which poses a challenge when dealing with the distinct nuances of the Arabic language especially in political domain. To address this issue, our proposed model capitalized on diversity and incorporates advanced decision-making techniques, thereby enhancing its accuracy in detecting cyberbullying. Performance evaluations demonstrate that our TL approach with gpt2 outperforms existing models in accurately identifying instances of cyberbullying. While our research has concentrated on a particular domain, it's important to acknowledge the potential for further investigations in other domains. Each domain presents unique challenges and opportunities, warranting dedicated research efforts and yielding distinct outcomes. Future work could explore these diverse domains, with Tunisian Dialect, independently to uncover their respective results and implications. Linguistic expressions in the Arabic language may be subject to various interpretations due to the cultural nuances specific to each region and social context. What is considered innocuous in one culture may be seen as offensive in another, due to factors such as local proverbs or accent variations. Therefore, for accurate detection of cyberbullying in Arabic-speaking environments, it is essential to construct an annotated dataset that takes into account these regional and cultural diversities. This approach will capture the linguistic subtleties specific to each Arabic-speaking community, thereby improving the ability of detection models to identify online behaviors.

References

1. Ahmed, M.S., Maher, S.M., Khudhur, M.E.: Arabic cyberbullying detecting using ensemble deep learning. Indon. J. Electr. Eng. Comput. Sci. **32**(2), 1031–1041 (2023)
2. Aldhyani, T.H., Al-Adhaileh, M.H., Alsubari, S.N.: Cyberbullying identification system based deep learning algorithms. Electronics **11**(20), 3273 (2022)
3. AlFarah, M.E., Kamel, I., Al Aghbari, Z., Mouheb, D.: Arabic cyberbullying detection from imbalanced dataset using machine learning. In: International Conference on Soft Computing and its Engineering Applications, pp. 397–409. Springer, Heidelberg (2021). https://doi.org/10.1007/978-3-031-05767-0_31
4. Almutairi, A.R., Al-Hagery, M.A.: Cyberbullying detection by sentiment analysis of tweets' contents written in Arabic in Saudi Arabia society. Int. J. Comput. Sci. Netw. Secur. IJCSNS **21**(3), 112–119 (2021)
5. Alrowais, F., et al.: Robertanet: enhanced roberta transformer based model for cyberbullying detection with glove features. IEEE Access (2024)
6. Arabia, I.: Cyber safety report. Research into the online behavior of arab youth and risks they face (2015)
7. Bechir, S.B., Mekki, A., Ellouze, M.: Transfer learning model for cyberbullying detection in Tunisian social networks. In: International Conference on Model and Data Engineering, pp. 62–75. Springer, Heidelberg (2023). https://doi.org/10.1007/978-3-031-55729-3_6
8. Kanan, T., Aldaaja, A., Hawashin, B.: Cyber-bullying and cyber-harassment detection using supervised machine learning techniques in Arabic social media contents. J. Internet Technol. **21**(5), 1409–1421 (2020)
9. Khairy, M., Mahmoud, T.M., Omar, A., Abd El-Hafeez, T.: Comparative performance of ensemble machine learning for Arabic cyberbullying and offensive language detection. In: Language Resources and Evaluation, pp. 1–18 (2023)
10. Mazari, A.C., Kheddar, H.: Deep learning-based analysis of Adlgerian dialect dataset targeted hate speech, offensive language and cyberbullying. Int. J. Comput. Digital Syst. (2023)
11. Patchin, J.W., Hinduja, S.: Measuring cyberbullying: implications for research. Aggress. Violent. Beh. **23**, 69–74 (2015)
12. Saadi, M.Q., Dhannoon, B.N.: Arabic cyberbullying detection using support vector machine with cuckoo search. Iraqi J. Sci., 5322–5330 (2023)
13. Schmidt, A., Wiegand, M.: A survey on hate speech detection using natural language processing. In: Proceedings of the Fifth International Workshop on Natural Language Processing for Social Media, pp. 1–10 (2017)
14. Shannag, F., Hammo, B.H., Faris, H.: The design, construction and evaluation of annotated Arabic cyberbullying corpus. Educ. Inf. Technol. **27**(8), 10977–11023 (2022)

Assessing BERT Models for Arabic Named Entity Recognition in a Multi-dialectal Context

Bassem Kessentini, Tahar Alimi, Rahma Boujelben(✉), and Lamia Hadrich Belguith

ANLP Research Group, MIRACL Laboratory, FSEGS, University of Sfax, Sfax, Tunisia
rahma.boujelbane@gmail.com, lamia.belguith@fsegs.usf.tn

Abstract. Named Entity Recognition (NER) is a crucial task in Natural Language Processing (NLP), enabling the identification of key entities within texts such as names, locations, and organizations. However, for Arabic, this task is particularly challenging due to its rich morphology, complex syntax, and diverse dialects. This study addresses these challenges by fine-tuning AraBERTv2, a BERT-based model specifically designed for Arabic. We prepare a diverse dataset that covers both Modern Standard Arabic (MSA) and various regional dialects, ensuring broad linguistic coverage. Our results show that the fine-tuned model achieves an F1 Score of 0.859, demonstrating its robustness across different Arabic text forms.

Keywords: Arabic Named Entity Recognition (NER) · Transformer Models · BERT for NLP · Cross-Dialectal NER

1 Introduction

Named Entity Recognition (NER) is a fundamental task in Natural Language Processing (NLP), crucial for extracting entities like names, locations, and organizations from unstructured text. It plays a key role in applications such as information retrieval, question-answering systems, and conversational agents, where accurate entity identification significantly enhances user interaction and system efficiency. However, achieving high performance in NER is particularly challenging for languages with complex linguistic structures, such as Arabic.

The Arabic language presents unique difficulties for NER due to its rich morphology, which involves extensive inflectional forms and complex syntactic structures. Additionally, the widespread use of regional dialects introduces significant variability, as these dialects often diverge substantially from Modern Standard Arabic (MSA) in vocabulary, syntax, and phonology. This linguistic diversity poses challenges for generalizing NER models across different Arabic varieties. Furthermore, the limited availability of annotated datasets for both MSA and dialectal Arabic hampers the development of robust and comprehensive models.

L. H. Belguith and K. Shaalan (Eds.): LPKM 2024, LNNS 1303, pp. 194–203, 2025.
https://doi.org/10.1007/978-3-031-85067-7_19

While transformer-based models like BERT have set new benchmarks in various NLP tasks, their success heavily depends on large, well-annotated datasets. AraBERT, a BERT variant pre-trained on Arabic, has demonstrated promising results in MSA, but its potential for handling dialectal variations remains underexplored. This study seeks to address this gap by fine-tuning AraBERTv2 on a carefully curated multi-dialectal corpus, aiming to improve NER performance across diverse Arabic text types.

2 Literature Review

Arabic Named Entity Recognition (NER) has seen significant advancements, driven largely by the development of key annotated corpora. These corpora address the intricate linguistic and dialectal diversity of the Arabic language, providing essential resources to refine and enhance NER technologies [8].

2.1 Review of Arabic NER Corpora

The evolution of NER systems for Arabic has been bolstered by several notable corpora, each tailored to specific linguistic challenges:

- **ANERcorp:** This corpus consists of approximately 150,000 tokens derived from formal news texts. It serves as a foundational resource for studying NER within Modern Standard Arabic (MSA), particularly in highly formal contexts [3].
- **CADENE:** Comprising 258,000 tokens across four classical Arabic dialects, CADENE facilitates the study of regional linguistic variations, enhancing understanding and recognition of dialectal diversity in NER tasks [7].
- **OntoNotes 5:** A comprehensive dataset with 250,000 tokens annotated from various sources, OntoNotes 5 is widely utilized for model training and benchmarking, offering a rich resource for exploring syntactic and semantic nuances in both formal and informal texts [5].
- **WojoodNER:** The largest corpus under review, containing 550,000 tokens, WojoodNER spans both MSA and several regional dialects. It leverages advanced transformer-based models like BERT to manage complex entity types, including nested and overlapping entities [4].
- **Antoun Corpus:** Encompassing 300,000 tokens from both formal and informal sources, this corpus has been instrumental in fine-tuning the AraBERTv2 model. It demonstrates the adaptability of transformer models to varied text types, achieving notable improvements in precision, recall, and F1 score [2].

2.2 Review of Approaches in Arabic NER

The methodologies applied to Arabic NER have evolved from traditional rule-based systems to advanced machine learning techniques, reflecting the complexity of the task:

- **ANERcorp:** Combines rule-based methods with Support Vector Machines (SVM) to enhance entity recognition in formal Arabic texts. This hybrid approach has shown significant improvements in precision and recall [3].
- **CADENE:** Employs Conditional Random Fields (CRF) integrated with rule-based preprocessing to handle the nuanced differences in classical Arabic dialects. This method improves accuracy in dialect-specific entity recognition [7].
- **OntoNotes 5:** Utilizes BiLSTM networks combined with CRF layers, providing robust syntactic and semantic annotation. This combination enhances entity detection across diverse text forms [5].
- **WojoodNER:** Implements transformer-based models like BERT, capable of recognizing nested and overlapping entities. This approach significantly boosts performance in both MSA and dialectal Arabic [4].
- **Antoun Corpus:** Fine-tunes AraBERTv2 with adjustments to learning rates and batch sizes, yielding improved performance metrics: precision of 0.905, recall of 0.890, and an F1 score of 0.897. This demonstrates the effectiveness of tailored configurations for Arabic NER [2].

2.3 Current Challenges and Research Motivation

Despite these advancements, several challenges remain in Arabic NER. Dialectal variability continues to pose significant difficulties, affecting the robustness and generalizability of NER systems. Inconsistent annotation standards across corpora further complicate model training, leading to performance discrepancies. Additionally, the computational demands of deep learning models, particularly for low-resource languages like Arabic, raise concerns about scalability and efficiency.

This study seeks to address these challenges by developing an integrated framework that leverages advanced deep learning techniques. By fine-tuning transformer-based models on a diverse, multi-dialectal corpus, we aim to improve the accuracy and robustness of Arabic NER across various text forms and linguistic contexts.

3 Arabic Multi-Dialect NER Corpus

This section introduces the Arabic multi-dialect NER corpus, meticulously designed to cover a diverse array of entity types and dialects. The corpus includes both formal and informal Arabic texts, ensuring a comprehensive linguistic resource. It employs the BIO annotation scheme, with examples provided to demonstrate the annotation quality and consistency.

3.1 Sources

The dataset for this Named Entity Recognition (NER) study has been carefully curated to provide a comprehensive representation of various entity types in the

Arabic language, such as person names, locations, and organizations. Each token is annotated using the BIO labeling scheme, which identifies the role and type of each entity. The dataset consists of 130,000 tokens for training and 12,000 tokens for validation, facilitating robust model training and evaluation.

To ensure linguistic diversity, the dataset includes Arabic dialects from Algeria, Morocco, Palestine, and Lebanon, alongside Modern Standard Arabic (MSA). It incorporates multiple text sources to balance formal and informal language use:

- **News Articles**: Providing structured, formal MSA text.
- **Social Media Posts**: Capturing informal expressions and colloquial terms from various regions, reflecting everyday language use.
- **Literary Works**: Covering classical and modern literary Arabic to ensure the inclusion of varied stylistic elements.

Our dataset combines newly collected data with existing annotated corpora, offering a rich and varied linguistic resource. Specifically, the dataset composition is as follows:

- **WojoodNER (20%)**: Includes Palestinian and Lebanese dialects, offering extensive coverage of nested and overlapping entities [4].
- **CADENE (20%)**: Focuses on classical Arabic and Algerian dialects, capturing key dialectal features [7].
- **OntoNotes 5 (20%)**: Provides diverse annotations of syntactic and semantic features, enhancing linguistic depth [5].
- **Newly Collected Data (40%)**: Ensures balance and representativeness by incorporating varied and recent sources (Table 1).

Table 1. Data Representation by Source and Dialect

Source/Dialect	Training Tokens	Validation Tokens	Total Tokens
News (MSA)	50,000	5,000	55,000
Social Media (Algeria)	20,000	2,000	22,000
Social Media (Morocco)	20,000	2,000	22,000
Social Media (Palestine)	20,000	2,000	22,000
Literature (Classical)	20,000	1,000	21,000

This table categorizes the dataset by source and dialect, highlighting the token distribution across training and validation sets. The inclusion of both formal texts (news articles) and informal sources (social media posts) ensures a balanced and comprehensive dataset. This diversity strengthens the generalizability of NER models across various Arabic contexts.

3.2 Annotation Tags

The BIO tagging schema employed in this study includes multiple entity types relevant to the Arabic context. Below is a detailed description of the tags (Table 2):

Table 2. Annotation Tags and Descriptions

Tag	Description
B-PER	Beginning of a person's name entity
I-PER	Inside of a person's name entity
B-LOC	Beginning of a location entity
I-LOC	Inside of a location entity
B-ORG	Beginning of an organization entity
I-ORG	Inside of an organization entity
B-MISC	Beginning of a miscellaneous entity category
I-MISC	Inside of a miscellaneous entity category
O	Outside of any named entity

3.3 Data Standardization and Annotation Methods

To ensure high-quality annotations and robust model training, a series of preprocessing steps were applied:

- **Text Normalization:** Diacritics were removed, and variant forms of identical letters were standardized to reduce variability across the corpus.
- **Ambiguity Resolution:** Ambiguous entities were meticulously reviewed and resolved to ensure annotation consistency and reliability.

Following preprocessing, each token was annotated using the BIO schema. This systematic annotation process helps the model accurately detect and distinguish between different entity types.

As illustrated in Fig. 1, we present an example of annotated data from Moroccan social media texts, showcasing the informal nature of dialectal Arabic and the diverse entity types identified. Similarly, Fig. 2 highlights annotated data from Algerian social media posts, providing insight into the unique linguistic characteristics of this dialect.

Fig. 1. Example of Annotated Data from Social Media (Morocco)

Fig. 2. Example of Annotated Data from Social Media (Algeria)

4 Model Configuration

The Arabic NER model configuration involves several critical steps to optimize training and ensure accurate predictions.

4.1 Data Preparation and Tokenization

- **Tokenization:** An Arabic-specific pretrained tokenizer from AraBERT was used, adept at handling morphological variations and diacritics.
- **Tag Encoding:** Tokens were assigned BIO tags to represent their entity roles, enabling the model to learn entity boundaries effectively.
- **DataLoader Implementation:** Efficient data loading and batching were achieved through a DataLoader, optimizing computational resource usage during training.
- **Tensor Alignment:** Ensured consistent input-output tensor alignment within the transformer architecture for accurate sequence processing.

4.2 Training and Evaluation Techniques

To refine model performance, the following techniques were employed:

- **Optimizer:** AdamW was chosen for its adaptive learning rate adjustment, crucial for transformer-based models. Weight decay was used to mitigate overfitting.
- **Learning Rate Scheduler:** A linear scheduler with a warmup phase was applied to stabilize training, gradually increasing the learning rate from a low initial value.

- **Performance Metrics:** Precision, recall, and F1 scores were calculated after each epoch to monitor model performance and detect potential overfitting.

5 Arabic Dialect NER Model

Our Arabic NER framework leverages three advanced BERT-based models, each designed to address specific linguistic challenges in Arabic. By combining their strengths through an ensemble approach, we enhance the accuracy and robustness of entity recognition across diverse Arabic dialects and contexts.

5.1 Model Descriptions

This section details the three BERT-based models used in our NER framework, highlighting their individual strengths and contributions to the system.

AraBERT V2. AraBERT V2 is specifically pre-trained on Modern Standard Arabic (MSA) and excels in processing formal texts, such as news articles and official documents. Its training on MSA allows it to capture complex linguistic nuances, making it ideal for tasks involving structured and formal Arabic content [1].

Multilingual BERT. Designed for multilingual tasks, this model is particularly effective in handling code-switching and transliteration, which are common in social media and other informal text sources. Its training on datasets like WojoodNER ensures its adaptability to mixed-language content, enhancing its performance in informal Arabic settings [9].

HooshvareLab BERT. Tailored for regional Arabic dialects, HooshvareLab BERT specializes in informal texts where dialectal variations are prevalent. Trained on corpora such as CADENE and WojoodNER, it effectively captures the syntactic and lexical peculiarities of various dialects, improving entity recognition in dialect-specific contexts [6].

5.2 Model Ensemble and Challenge Resolution

To overcome the limitations of individual models, we implemented an ensemble approach that integrates the outputs of AraBERT V2, Multilingual BERT, and HooshvareLab BERT. This strategy leverages their complementary strengths, ensuring robust performance across different text types and dialects.

We addressed the issue of data sparsity in certain dialects through synthetic data augmentation, enriching the dataset to provide balanced learning opportunities. Distributed computing techniques were employed to manage the high computational demands of training multiple models, enhancing system efficiency and scalability. These optimizations significantly improved the NER system's ability to handle complex, multi-dialectal Arabic texts.

6 Experimental Study and Comparative Analysis

We conducted a comprehensive evaluation of our NER framework to compare the performance of three BERT-based models: AraBERT V2, Multilingual BERT, and HooshvareLab BERT. The evaluation focused on key metrics such as precision, recall, and F1 score to assess each model's effectiveness in recognizing named entities across diverse Arabic texts. Additionally, we compared our results with those of previous studies to highlight the improvements achieved through our approach.

6.1 Performance Evaluation

The performance metrics for each model are summarized in Table 3. These metrics provide a detailed view of how each model performs in terms of precision, recall, and F1 score.

Table 3. Performance of NER Models

Model	Precision	Recall	F1 Score
AraBERT V2 (Our Study)	0.9215	0.9024	0.8599
Multilingual BERT	0.8748	0.8533	0.7957
HooshvareLab BERT	0.8081	0.8133	0.7857

AraBERT V2 demonstrated the highest precision among the three models, making it particularly effective for formal texts such as news articles. However, its slightly lower recall suggests that it may overlook some entities in informal contexts. Multilingual BERT provided a well-balanced performance, offering competitive precision and recall, which makes it versatile for mixed-language environments. On the other hand, HooshvareLab BERT excelled in recall, capturing a broader range of entities in informal dialectal texts, but at the cost of lower precision, indicating a higher rate of false positives.

6.2 Comparative Analysis with Previous Studies

Our study shows significant improvements over the results reported by Antoun et al. (2020). As shown in Table 4, our fine-tuned AraBERT V2 model achieved higher precision (0.9215 vs. 0.905) and recall (0.9024 vs. 0.890), demonstrating its enhanced capability in accurately identifying entities. Although the F1 score of our model is slightly lower (0.8599 vs. 0.897), this trade-off is attributed to the increased emphasis on precision and recall, which are critical in applications requiring high confidence in entity recognition.

These improvements can be attributed to our refined preprocessing methods and the use of a more diverse training dataset, which collectively enhance the model's robustness in handling both formal and informal Arabic texts. This comparison highlights the critical role of data quality and preprocessing in optimizing model performance for Named Entity Recognition tasks.

Table 4. Comparison of AraBERT V2 Performance (Our Study vs. Antoun et al. 2020)

Study	Precision	Recall	F1 Score
Our Study	0.9215	0.9024	0.8599
Antoun et al. 2020	0.905	0.890	0.897

6.3 Discussion

The results underline the complementary strengths of the three models. AraBERT V2 is optimal for high-precision tasks in formal settings, while Multilingual BERT offers a balance suitable for a variety of contexts. HooshvareLab BERT is ideal for applications requiring high recall in informal and dialectal texts. By integrating these models in an ensemble framework, we can further improve the robustness and adaptability of NER systems across diverse Arabic contexts.

7 Conclusion and Future Work

This study presents a robust NER framework for Arabic, leveraging an ensemble of BERT-based models to address the linguistic and dialectal complexities of the language. By fine-tuning these models on a diverse multi-dialectal corpus, we achieved significant improvements in precision and recall.

Future work will focus on expanding the dataset to include additional dialects and exploring transfer learning techniques to further enhance model adaptability. We also aim to investigate the integration of other NLP tasks, such as sentiment analysis and machine translation, to build a comprehensive language understanding framework for Arabic.

References

1. Antoun, W., Baly, F., Hajj, H.: Arabert: transformer-based model for Arabic language understanding (2020)
2. Antoun, W., Baly, R., Hajj, H.: Arabert: transformer-based model for Arabic language understanding. In: Proceedings of the 2020 Conference on Empirical Methods in Natural Language Processing (EMNLP), pp. 9–14. Association for Computational Linguistics (2020)
3. Author, F., Author, S.: Arabic and Arabizi named entity recognition using transformer-based approach. arXiv preprint arXiv:2308.14669 (2023)
4. Elkhbir, N., Zaratiana, U., Tomeh, N., Charnois, T.: Lipn at wojoodner shared task: a span-based approach for flat and nested Arabic named entity recognition, pp. 789–796 (2023). https://doi.org/10.18653/v1/2023.arabicnlp-1.89
5. Hovy, E., Marcus, M., Palmer, M., Ramshaw, L., Weischedel, R.: Ontonotes: a large training corpus for enhanced processing. In: Studies in Natural Language Processing, pp. 123–456 (2013)

6. Mansour, M., Tohamy, M., Ezzat, Z., Torki, M.: Arabic dialect identification using bert fine-tuning (2020)
7. Name, A.: Exploring regional linguistic variations with cadene. J. Comput. Linguist.(2023)
8. Quimbaya, A.P., Shinyama, Y., Morwal, S.: Comprehensive overview of named entity recognition: models, domain-specific applications and challenges. J. Adv. Comput. Linguist. **1**(1), 1–30 (2024)
9. Rönnqvist, S., Kanerva, J., Salakoski, T., Ginter, F.: Is multilingual bert fluent in language generation? (2019)

Tunisian Normalized Pronunciation

Houcemeddine Turki[1](✉), Maik Gibson[2], Rafik Zribi[3], Imed Adel[4], Mohamed Ali Hadj Taieb[1], Mohamed Ben Aouicha[1], Kawthar Ellouze[5], Hager Ben Ammar[6], and Abderrezak Bennour[7]

[1] Data Engineering and Semantics Research Unit, Faculty of Sciences of Sfax, University of Sfax, Sfax, Tunisia
turkiabdelwaheb@hotmail.fr
[2] Redcliffe College, Gloucester, UK
[3] IÉSEG School of Management, Lille, France
[4] École supérieure des sciences et de la technologie de Hammam Sousse, Université de Sousse, Sousse, Tunisia
[5] Concordia University, Montréal, Canada
[6] Lycée Gustave-Flaubert, La Marsa, Ariana, Tunisia
[7] Institut Supérieur des Langues de Gabès, Université de Gabès, Gabès, Tunisia

Abstract. Tunisian (ISO 693-3: aeb) is a set of linguistic varieties spoken in Tunisia and evolved from the Arabic language. In this research work, we attempt to establish a standardized pronunciation that can serve as a communication tool for the diverse Tunisian population. This paper draws on prior linguistic research and uses a rich dataset maintained by the Linguistic Data Consortium to identify common phonological features across various Tunisian dialects. This research creates a harmonized Tunisian pronunciation that transcends regional differences, with a focus on consonants, vowels, and assimilation among other patterns. This effort is driven by the need for accurate communication, education, and media in a linguistically diverse nation.

Keywords: Tunisian · Phonology · Normalized · Pronunciation

1 Introduction

Tunisia, a North African nation steeped in cultural, historical, and linguistic diversity, serves as a captivating mosaic of traditions and dialects [1]. At the crossroads of the Mediterranean, Tunisia has witnessed the interplay of various civilizations and linguistic influences, resulting in a landscape where Arabic, Berber, and other languages have converged and evolved [1]. In this context, the study of Tunisian Arabic (ISO 693-3: aeb), the vernacular spoken by most Tunisians, is pivotal for understanding the nation's cultural development [1]. Tunisian Arabic, classified as a dialect of the broader Arabic language family, falls under the Maghrebi Arabic subgroup, which includes the Arabic dialects spoken in North Africa [2]. Within this subgroup, Tunisian Arabic is recognized as a distinct dialect with its own phonological, lexical, and grammatical features [2]. It shares a closer linguistic kinship with Maghrebi dialects such as

L. H. Belguith and K. Shaalan (Eds.): LPKM 2024, LNNS 1303, pp. 204–212, 2025.
https://doi.org/10.1007/978-3-031-85067-7_20

Algerian Arabic, Libyan Arabic, and Maltese [3]. These dialects share common features and vocabulary, enhancing mutual intelligibility [3]. However, intelligibility between Tunisian Arabic and non-Maghrebi dialects, such as Levantine or Gulf Arabic, and Modern Standard Arabic, is generally lower due to significant linguistic differences [4]. phonology is essential in computational linguistics, impacting applications like text-to-speech, diacritization, and speech recognition [5]. Understanding Tunisian Arabic's phonological intricacies has practical implications for communication [1], education [6], and media [7,8]. Standardizing pronunciation can bridge linguistic gaps and promote cultural cohesion. Exploring common phonological features across Tunisian dialects fosters unity and transcends regional differences [9].

This research presents a normalized pronunciation for Tunisian, derived from the shared phonological features across all Tunisian dialects, using an evidence-based approach. The paper is structured as follows: Sect. 2 details our methods. Section 3 delves into the consonantal aspects of Tunisian, followed by an exploration of the vowel system in Sect. 4. Sections 5 to 9 elucidate specific phonotactic and morphophonemic phenomena in Tunisian, including Assimilation, Stress, Epenthesis and Elision, Syllable Structure, and Metathesis. Section 10 concludes by providing the pronunciation of a brief excerpt in Tunisian.

2 Methods

In the realm of linguistic research, Tunisian dialects have long been a subject of scholarly exploration, yielding a collection of phonological descriptions that span various regions and variations [9]. In this research work, we take a novel approach, one rooted in the principles of Evidence-Based Linguistics as championed by Belyaeva and Chernyavskaya (2016) [10]. Our primary aim is to craft a standardized pronunciation of Tunisian that transcends regional biases, making it a viable communication tool for mass media addressing the diverse Tunisian population, as advocated by Achour Kallel (2011) [8] and La Rosa (2022) [7]. Central to our methodology is the identification of common phonological features shared among the various Tunisian dialects, a foundation laid by prior phonological descriptions (Table 1). These features serve as the building blocks upon which we construct a normalized, pan-Tunisian pronunciation.

Crucially, this research hinges on the utilization of the IWSLT Tunisian Arabic Shared Task Training Data (LDC2022E01), a rich resource curated by the Linguistic Data Consortium [11]. By scrutinizing this extensive corpus, we fortify our methodology with empirical evidence, enabling us to fine-tune the adopted rules and make necessary adjustments as we progress toward our goal of establishing a harmonized Tunisian pronunciation that transcends regional boundaries. This endeavor stands as a testament to the evolution of linguistic research, merging the traditions of past scholarship with the precision of modern data analysis to serve the needs of a diverse and dynamic linguistic landscape.

Table 1. Reference excerpts from which the phonological rules are derived

Dialect	Reference	Consonants	Vowels	Assimilation	Syllable structure	Elision and Epenthesis	Stress
Tunis	Maamouri (1967) [12]	39–65, 95–123	5–12, 66–94	124–149	13–18, 124–167	147–153	19–25
Tunis	Chekili (1982) [13]	11–16	13	20	16–17	18–20	17–18
Sahil	Talmoudi (1979) [14]	20–41, 43–49	17–19, 57–59	50–56	42, 60	N/A	42–43
Sahil	La Rosa (2021) [15]	Research Paper					
Sfax	Lajmi (2009) [16]	Research Paper					
Sfax	Mizouri (2021) [17]	Research Paper					
North-Western	Cherni et al. (2007) [18]	13–14	14–15	N/A	15–16	16	N/A
North-Western	Zarb (2023) [9]	42–48	49–53	57	58	54–56	N/A
South-eastern	Benkato et al. (2023) [19]	Research Paper					
South-Western	Saada (1984)	23–31	32–38	38–39	42–45	N/A	N/A
South-Western	Cherni et al. (2007) [18]	13–14	14–15	N/A	15–16	16	N/A
Judeo Tunisian	Gebski (2023) [20]	Research Paper					

3 Consonants

Tunisian generally preserves the majority of Arabic consonantal sounds [21], including interdental fricatives, as outlined in Table 2. There is, however, an exception in the case of [dˤ], which is pronounced as [ðˤ]. This exception is common in other Arabic languages such as Moroccan [23, p. 2], and Khuzestani Arabic [22]. But, it is kept in South Levantine [23]. Similar to Standard Arabic [21], Tunisian frequently employs "gemination" (commonly known as Shaddah), as seen in the word /haddad/, which means "to threaten". Tunisian Arabic also exhibits a slight use of the Hamza (glottal stop), a feature that aligns with its roots in Standard Arabic, although Tunisian re-places most of the glottal stops inherited from Standard Arabic with /h/.

Moreover, Tunisian exhibits variations in the pronunciation of the /q/ sound across different dialects. In sedentary varieties, it is pronounced as [q], while in nomadic varieties, it takes the form of [g]. For example, the word "he said" is articulated as /qa:l/ in sedentary dialects but as /gɛ:l/ in nomadic ones.This feature is a Hilalian influence and can be found in other Maghrebi Arabic languages such as Moroccan [23, p. 2]. Nevertheless, certain words consistently maintain the [g] sound, particularly those associated with agriculture, irrespective of the dialect. For instance, the word for "cow"' is consistently pronounced as /bɛgra/, with the /g/ sound originating from the original Arabic [q]. In our situation, we consider the sedentary pronunciation of both sounds as the normalized one given that it is the one adopted by most of the dialects of Tunisian.

In addition to the consonants inherited from Arabic and those that are commonly found, Tunisian Arabic also encompasses a range of less common consonant sounds. The sounds /p/ and /v/ are present in borrowed words but are often replaced by /b/, as in the case of /ba:bu:r/ (boat, originally borrowed from French vapeur). However, these sounds are preserved in some words. These two foreign sounds also exist in other Maghrebi Arabic dialects, influenced by French and Spanish occupation [24]. Besides, we find several emphatic letters [bˤ], [ɫ], [mˤ], [nˤ], [rˤ], and [zˤ] that substitute their non-emphatic equivalents where they are adjacent to an [ɐ] sound. This phenomenon is not common in the

Table 2. Tunisian Arabic Consonant Phonemes and Features. Uncommon consonants are put between parentheses (Please see Supplementary Table S1 for practical examples)

		Labial		Interdental		Dental/Alveolar		Palatal	Velar	Uvular	Pharyngeal	Glottal
		plain	emphatic	plain	emphatic	plain	emphatic					
Nasal		m	(mˤ)			n	(nˤ)					
Stop	voiceless	(p)				t	tˤ		k	q		(ʔ)
	voiced	b	(bˤ)			d			ɡ			
Fricative	voiceless	f		θ		s	sˤ	ʃ		χ	ħ	h
	voiced	(v)		ð	ðˤ	z	(zˤ)	ʒ	ɣ		ʕ	
Trill						r	(rˤ)					
Approximant						c	(ɫ)	j	w			

phonology of Modern Standard Arabic [21] but exists in other Maghrebi Arabic languages, particularly Moroccan [23, p. 2].

4 Vowels

Tunisian Arabic, as a rich and diverse dialect, encompasses a phonological landscape of seven distinct vowel sounds. These vowels are predominantly categorized as front or close-mid to closed. These seven vowel sounds correspond to four fundamental vowel qualities that form the core of the Tunisian vowel system. These qualities include /a/, /ɛ/, /u/, and /i/, each playing a unique role in the articulation and pronunciation of words, imbuing the language with a distinctive phonetic character. The distribution of short /a/ typically occurs in proximity to emphatic, uvular, and pharyngeal consonants, while long /a/ tends to appear adjacent to emphatic consonants or following uvular or pharyngeal consonants, primarily influenced by the phonological phenomenon known as Imalah, as in several other Arabic languages [15]. That is why /a/ and /ɛ/ are considered the same phoneme in Modern Standard Arabic [21], Egyptian [26, pp. 20–22], and South Levantine Arabic [23]. However, because there are exceptions to this rule in Tunisian such as /vispa/ (big motorcycle) and as the duality of /a/ and /ɛ/ allows to differentiate between [bˤ], [ð], [mˤ], [nˤ], [rˤ], and [zˤ] and their non-emphatic minimal pairs, we have chosen to consider /a/ and /ɛ/ as separate vowel qualities. Another distinctive feature of Tunisian is the disappearance of the diphthongs /aj/ and /aw/ that are respectively changed into /u:/ and /i:/. Egyptian [26, pp. 20–22] and South Levantine [23] exclusively use /o:/ and /e:/ to substitute /aj/ and /aw/ while Maltese [25], Lebanese [29, p. 23], and Judeo-Tripolitanian [19] just keep them. Table 3 meticulously outlines these vowel qualities, providing valuable insights into their specific phonetic properties and usage in Tunisian Arabic. For a deeper understanding of how these vowel qualities are applied in everyday language and speech within Tunisia, Supplementary Table S2 offers a comprehensive collection of examples. These examples shed light on the versatile and dynamic nature of Tunisian Arabic, showcasing the rich interplay of vowel sounds that underpin its unique phonological structure.

Table 3. Tunisian Vowel Quality in Different Lengths and Syllable Positions

Vowel Quality	Short		Long	
	Open syllable	Closed Syllable	Not Final	Final
/a/	[ɐ]	[ɐ]	[ɐ:]	[ɐ]
/ɛ/	[æ]	[æ]	[ɛ:]	[æ]
/u/	[u]	[o]	[u:]	[u]
/i/	[i]	[ɪ]	[i:]	[i]

5 Assimilation

In Tunisian, there are many situations when consonant sounds are substituted in particular conditions by other sounds that are simpler to pronounce (So-called assimilation). In the first situation of assimilation, the semi-vowels /w/ and /j/ change when they occur at the beginning of a word and are followed by a consonant.

- When the semi-vowel /w/ appears at the word-initial position followed by a consonant, it assimilates and becomes [u:] in pronunciation. Example: /wqif/ ß [u:qɪf] (He stopped)
- Similarly, the semi-vowel /j/ at the beginning of a word followed by a consonant undergoes assimilation and is pronounced as [i:]. Example: /**j**qu:l/ →[**i**:qu:l] (He says)

The conversion of semi-vowels into vowels is a common fact and exists in Berber languages like Ghomari [30, pp. 36–37] and even in several European languages such as English [26].

The second situation of assimilation occurs when adjacent consonants influence each other's pronunciation unless an epenthetic [ɪ] separates them, resulting in changes in their phonetic realization. This fact can alter the pronunciation of /ɛ/ as /a/ when a consonant next to it is converted into an emphatic, a pharyngeal, or a uvular one following the Imalah rule. All the common adjacent consonants that are assimilated are outlined in Supplementary Table S3. This phenomenon also occurs in Moroccan [23, pp. 19–20] and Levantine Arabic [27] among other Arabic languages [28]. For instance, in the case of the phoneme /ttˤ/, it undergoes assimilation, resulting in a phonetic form of [tˤ:] as seen in the example [tˤtˤɐww ɪl] ([She] lengthens). Similarly, various other phonemes like /χh/, /χɣ/, /tg/, and more, exhibit distinctive assimilation patterns, showcasing the richness and complexity of phonological processes within the language. The third and final situation of assimilation happens for the /l/ in the /il/ definite article (meaning The).

The /l/ is pronounced as the letter following it if that letter is a sun consonant. In Tunisia, sun consonants are /t/, /θ/, /Z/, /d/, /ð/, /r/, /z/, /ʃ/, /sˤ/, /ðˤ/, /tˤ/, /l/, and /n/. An example is [ɪ**n**næhr] (The river).

This stands in contrast to Modern Standard Arabic and other Arabic languages that do not consider /ʒ/ as a sun consonant [29].

6 Stress

Stress in Tunisian is determined by the word's syllable structure rather than phonological distinctions. When there are two consecutive closed syllables, the stress typically falls on the final syllable, as exemplified by the word [mɪf**tɛ:ħ**] (meaning "key"). Conversely, when there are two syllables and one of them is open, the stress falls on the first syllable, as in [bɛ:tɪt] ("she stayed"). Moreover, in one syllable words, the stress naturally falls on that single syllable, as in [mrˤɐ], meaning "woman." For words with more than one syllable, stress generally falls on the penultimate syllable, as seen in [ʒæ**ri**:dæ], which means "newspaper." Furthermore, when dealing with affixes, they are considered an integral part of the word. For example: [bɛ:tɪt] - "She stayed", or [mbɛ:tɪtʃ] - "She did not stay". These instances provide a clear demonstration of how stress allocation in Tunisian is influenced by syllable composition, affecting both isolated words and words with affixes. Other Maghrebi Arabic languages like Moroccan follow quite the same stress rules as Tunisian [23, pp. 8–10].

7 Epenthesis and Elision

In Tunisian Arabic, elision is a common phonological phenomenon where the first short vowel of a word is omitted when preceded by a word ending with a vowel. This creates a smooth and fluid pronunciation between words. For example, [bdæɪlklɛ:m] (The talking began) is pronounced as [bdælklɛ:m] because the first vowel [ɪ] of the definite article [ɪl] is omitted due to the preceding word, [bdæ], ending with a vowel [æ]. These elisions and epenthetic additions are essential aspects of Tunisian Arabic phonology, contributing to the language's sound and rhythm. They exist in other Arabic dialects like Moroccan [23, pp. 22–23], South Levantine [23], and Egyptian [26, pp. 25–30]. Yet, these two phonological alternations rarely exist in Modern Standard Arabic [26, pp. 25–30].

8 Syllable Structure

The Tunisian syllables can be classified into three categories according to the position of their final vowels (V, Bold in Supplementary Table S4): Open syllable if it finishes with a vowel sound, Simply closed syllable if the final vowel is short and only fol-lowed by a single ungeminated consonant, and Double closed syllable if the other situations. Regardless of their type, all the syllables in Tunisian follow $\mathrm{C}_{(0-3)}(:)\mathrm{V}(:)C_{(0-2)}(:)$ as a pattern. This syllable structure seems to be similar to the one of other Maghrebi Arabic languages, particularly Moroccan [23, pp. 6, 14–15] but seems to be different from the one of Modern Standard Arabic [23, pp. 6, 14–15] and Eastern Arabic languages like South Levantine [23]. This is mainly explained by the significant elimination of short vowels from words inherited from Classical Arabic in Maghrebi Arabic languages [23, pp. 6, 14–15]. Supplementary Table S4 shows practical examples of all the possible syllable structures for Tunisian.

9 Metathesis

Metathesis is the phenomenon in which the initial vowel of a word undergoes a positional shift. This typically occurs when dealing with unconjugated verbs or nouns without suffixes, specifically when they start with a consonant-consonant-vowel-consonant (CCVC) structure. In such cases, where the consonants are not geminated and the vowel is short (V), the initial vowel undergoes metathesis. However, when these nouns are conjugated or when suffixes are added, the first vowel is repositioned, resulting in a consonant-vowel-consonant-consonant (CVCC) structure. For example, in Tunisian Arabic, for the verb [ktɪb] "(he) wrote", "(she) wrote" becomes [kɪtbɪt] after metathesis. Similarly, for [dbæʃ] "some stuff", "my stuff" becomes [dæbʃi] in Tunisian Arabic, illustrating how metathesis impacts the pronunciation and structure of words in this language.

10 Conclusion

This paper tried to create a standardized pronunciation framework that captures the common phonological features across various Tunisian dialects. Our investigation into the consonantal and vowel systems, along with phonotactic and morphophonemic phenomena such as assimilation, stress patterns, epenthesis, and elision, has provided a comprehensive understanding of the phonetic characteristics of Tunisian Arabic. By normalizing these features, we offer a unified pronunciation model that transcends regional variations, facilitating clearer communication and enhancing the representation of Tunisian Arabic in education and media. This standardized model is crucial for bridging linguistic gaps and fostering unity among speakers from different parts of Tunisia. Future research can build on this foundation by exploring the sociolinguistic implications of phonological standardization and its impact on language use in various social contexts. Additionally, further studies could investigate the application of these phonological rules in natural language processing tools, potentially improving the accuracy and usability of speech recognition and text-to-speech systems in Tunisian.

11 Supplementary Material

All supplementary tables, along with recordings of the examples within the supplementary tables and of a brief excerpt, can be accessed at https://doi.org/10.6084/m9.figshare.25765332. Additionally, a Python code for converting Tunisian texts written in Diacritized Arabic Script, following the NOTA Guidelines [30], into IPA, is also provided at the same link. An extended edition of this research paper is available at http://doi.org/10.2139/ssrn.4732831.

Acknowledgments. We thank Mohamed Maamouri (University of Pennsylvania, USA), FethiBougares (Le Mans University, France, and Elyadata, Tunisia), Dominique Caubet (Alliance Sorbonne-Paris-Cité, France), Ramzi Hachani (United Nations Office for Project Services, USA), Tariq Daouda (Université de Montréal, Canada), Nassim Regragui (Copenhagen Business School, Denmark), LameenSouag (Alliance Sorbonne-Paris-Cité, France), and Lucas Werkmeister (Wikimedia Deutschland, Germany) for their valuable comments and discussions. This work is supported by the Derja Association, a non-profit organization dedicated to promoting the use of Tunisian in various fields. It was developed from an initial attempt to describe Tunisian Phonology on English Wikipedia. We used ChatGPT, a large language model by OpenAI, for typesetting, proofreading, and copyediting.

References

1. Daoud, M.: The sociolinguistic situation in Tunisia: language rivalry or accommodation? Int. J. Sociol. Lang. **2011**(211), 9–33 (2011)
2. Sayahi, L.: Diglossia and Language Contact: Language Variation and Change in North Africa. Cambridge University Press, Cambridge (2014)
3. çéplö, S., Bátora, J., Benkato, A., Milička, J., Pereira, C., Zemánek, P.: Mutual intelligibility of spoken Maltese, Libyan Arabic, and Tunisian Arabic functionally tested: a pilot study. Folia Linguistica **50**(2), 583–628 (2016)
4. S'hiri, S.: Speak Arabic please!: Tunisian Arabic speakers' linguistic accommodation to middle Easterners. In : Language Contact and Language Conflict in Arabic: Variations on a Sociolinguistic Theme, pp. 149–174 (2002)
5. Masmoudi, A., Bougares, F., Ellouze, M., Estève, Y., Belguith, L.: Automatic speech recognition system for Tunisian dialect. Lang. Resour. Eval. **52**, 249–267 (2018)
6. Maamouri, M.: Language education and human development: Arabic diglossia and its impact on the quality of education in the Arab region. In: ERIC Clearinghouse (1998)
7. La Rosa, C.: Linguistic varieties and social radios in the Tunisian sahel: the case of news broadcasting. Annali di Ca' Foscari - Serie orientale **58**, 81–112 (2022)
8. Achour Kallel, M.: Choix langagiers sur la radio Mosaïque FM. Dispositifs d'invisibilité et de normalisation sociales. Langage société (4), 77–96 (2011)
9. Zarb, M.: The Arabic Dialect of Testour (Northwestern Tunisia): A Phonological and Morphological Analysis. University of Vienna, Vienna (2023)
10. Belyaeva, L., Chernyavskaya, V.: Evidence-based linguistics: methods in cognitive paradigm. Voprosy kognitivnoy lingvistiki **3**, 77–84 (2016)
11. Anastasopoulos, A., Barrault, L., Bentivogli, L., Boito, M., Bojar, O., Cattoni, R., et al.: Findings of the IWSLT 2022 evaluation campaign. In : Proceedings of the 19th International Conference on Spoken Language Translation (IWSLT 2022), pp. 98–157 (2022)
12. Maamouri, M.: The Phonology of Tunisian Arabic. Cornell University, Ithaca (1967)
13. Chekili, F.: The Morphology of the Arabic Dialect of Tunis. University of London, London (1982)
14. Talmoudi, F.: The Arabic Dialect of Susa (Tunisia). University of Gothenburg, Gothenburg (1979)

15. La Rosa, C.: Mahdia dialect: an urban vernacular in the Tunisian sahel cdontext. Languages **6**(3), 145 (2021)
16. Lajmi, D.: Spécificités du dialecte Sfaxien. Synergies Tunisie **1**, 135–142 (2009)
17. Mizouri, I.: Les parlers sfaxiens: spécificités linguistiques et culturelles. Langue(s) Parole **6**, 139–159 (2021)
18. Cherni, T., Lowe, I.: Tunisian Phonological and Morphological Varieties: Differences between the Regional Dialects of El Kef and Tozeur. University of Tunis, Tunis (2007)
19. Benkato, A.: The Arabic dialect of Benghazi, Libya: Historical and comparative notes. Zeitschrift für Arabische Linguistik (59), 57–102 (2014)
20. Grbski, W.: The Phonology of the Judaeo-Arabic Dialect of Gabes (Southern Tunisia). J. Semitic Stud. **68**(1), 165–197 (2023)
21. Thelwall, R., Sa'adeddin, M.: Arabic. J. Int. Phon. Assoc. **20**(2), 37–39 (1990)
22. Bahrani, N., Ghavami, G.: Khuzestani Arabic. J. Int. Phon. Assoc. **51**(2), 299–313 (2021)
23. Cotter, W.: The Arabic dialect of Gaza City. J. Int. Phon. Assoc. **52**(1), 122–134 (2022)
24. Sayahi, L.: Phonological adaptation of Spanish loanwords in Northern Moroccan Arabic. Univ. Pennsylvania Work. Papers Linguist. **11**(1), 253–263 (2005)
25. Krier, F.: Analyse phonologique du maltais. Phonetica **32**(2), 103–129 (1975)
26. Mahdi, A.: The Role of Semi-Vowels. J. Iraqi Univ. **29**(3), 409–420 (2012)
27. Zuraiq, W., Zhang, J.: Phonological assimilation in urban Jordanian Arabic. Kansas Work. Papers Linguist. **28**, 33–64 (2006)
28. Alfozan, A.: Assimilation in Classical Arabic: A Phonological Study. University of Glasgow, Glasgow (1989)
29. Kambuziya, A.: Arabic definite prefix: an autosegmental analysis. South Asian Lang. Rev. **17**(2), 63–73 (2007)
30. Turki, H., et al.: Normalized Orthography for Tunisian Arabic. Association Derja (2024)
31. Yoda, S.: "Sifflant" and "chuintant;; in the Arabic dialect of the Jews of Gabes (south Tunisia). Zeitschrift für arabische Linguistik (46), 7–25 (2006)
32. Sullivan, N.: Writing Arabizi: Orthographic Variation in Romanized Lebanese Arabic on Twitter. University of Texas at Austin (2017)
33. Ennaji, M., Makhoukh, A., Es-saiydi, H., Moubtassime, M., Slaoui, S.: A grammar of Moroccan Arabic. Université Sidi Mohamed Ben Abdellah (2004)
34. Mourigh, K.: A Grammar of Ghomara Berber. Leiden University, Leiden (2015)
35. Mitterer, H., Csépe, V., Blomert, L.: The role of perceptual integration in the recognition of assimilated word forms. Q. J. Exp. Psychol. **59**(8), 1395–1424 (2006)
36. Johnson, C.: Vowel-consonant metathesis in a Palestinian dialect. Acta Linguistica Hafniensia **17**(1), 61–77 (1982)
37. Idrissi, O.: Metathesis in Moroccan Arabic: optimality-theoretic account. Int. J. Linguist. Stud. **3**(1), 62–67 (2023)
38. Banjar, S.: A Synchronic study of metathesis in Eastern Arabic with reference to Makkan and Cairene dialects. King Saud's Bull. Arts Fac. **14**, 1–44 (2003)

Natural Language Processing

TraffScOnto: Ontology for Traffic Management in the Context of Smart City Domain

Nesrine Saafi(✉) and Karima Dhouib

MIRACL Laboratory, University of Sfax, 2 BP 242, 3021 Sfax, Tunisia
Saafi.nesrine1@gmail.com, Karima.dhouib@isims.usf.tn

Abstract. The increasing complexity of urban traffic management within the smart city context necessitates effective tools for data integration and decision-making. In response to this need, this paper introduces an ontology specifically designed for traffic management within the smart city domain. The ontology named "TraffScOnto" aims to address the challenge of semantic heterogeneity and data interoperability by providing an enriched vocabulary representing traffic related concepts and relationships. The formalization of this model has been carried out using OWL language and Protégé tool. "TraffScOtno" encompasses a comprehensive range of traffic phenomena including traffic flow dynamics, congestion factors, road infrastructure and various transportation modes.

Keywords: Knowledge management (KM) · Smart city (SC) · Ontology · Transportation

1 Introduction

As the global population continues to grow, cities face numerous daunting challenges, such as traffic congestion, air pollution, and the need for efficient transportation systems. To tackle these issues, many cities are turning to smart city technologies, leveraging data and connectivity to improve urban mobility and enhance residents' quality of life. Smart city technologies hold the promise of optimizing traffic flow, easing congestion, and enhancing road safety. However, their implementation comes with significant challenges. One major obstacle is the integration of diverse systems and data sources. Smart city technologies rely on data streams from various origins, including traffic sensors, GPS devices, and mobile applications. Integrating these disparate systems requires complex solutions and substantial investments in software development. To overcome this interoperability barrier we aim to exploit the contribution of the knowledge management domain by proposing an ontological model for representing traffic related concepts in order to mitigate semantic heterogeneity and facilitate data integration across diverse urban systems. The structure of this paper is as follows Sect. 2 will be an insight into knowledge management and smart cities,

L. H. Belguith and K. Shaalan (Eds.): LPKM 2024, LNNS 1303, pp. 215–224, 2025.
https://doi.org/10.1007/978-3-031-85067-7_21

Sect. 3 will be dedicated to present some knowledge management tools and platforms for smart cities, Sect. 4 will represents related works about existing traffic ontologies followed by a discussion in the fourth section, next in Sect. 5 we will describe the ontology construction steps and we finish by a conclusion and by giving our future works.

2 An Insight into Knowledge Management and Smart Cities

In [1], the author defines knowledge management as a comprehensive process encompassing individuals, procedures, culture, and the necessary technologies to capture, manage, share, and retrieve information. Additionally, [2] emphasizes the pivotal role of knowledge in the establishment and development of cities, influencing their economic, societal, and cultural aspects significantly. Within the context of smart cities, knowledge management assumes paramount importance for their advancement and effective operation. The success of smart cities heavily relies on the adept utilization of technology to collect and analyze data from diverse sources, aimed at enhancing the well-being of their residents. This necessitates the adoption of a meticulously structured approach to efficiently manage knowledge and information, ensuring the accuracy, reliability, and relevance of the amassed data, tailored to the unique requirements of each city. Furthermore, key aspects of knowledge management in smart cities, as delineated by Abdallah et al. (2019) in [3] such as data governance establishing a framework for data collection, storage, and sharing, including ownership, standards, and access protocols, data analytics utilizing tools to analyze vast datasets from sensors, IoT devices, and social media to identify patterns and insights and collaborative platforms facilitating collaboration among stakeholders through decision-making tools and knowledge-sharing platforms to enhance knowledge management.

3 Knowledge Management Tools and Platforms for Smart Cities

This section discusses the importance of using ontologies to address the challenges of heterogeneity in the smart city domain, particularly in the traffic domain. It highlights several innovative smart city platforms [4], including SmartSantander, CiDAP, DIMMER, CityPulse, and FIWARE, each of which focuses on different aspects such as data collection, big data analytics, energy efficiency, real-time data processing, and interoperability. The discussion then turns to the common problem of heterogeneity among these platforms, which results from differences in data sources, analytical capabilities and scalability. To address this challenge comprehensively, participants emphasize the need to use ontologies. Ontologies serve as standardized frameworks for defining concepts and relationships within smart city data, enabling seamless integration and interoperability across platforms. Our work focuses on the traffic domain and analyses several traffic platforms that use ontologies. Several studies are cited that highlight the critical role of ontology-based architectures in addressing heterogeneity issues in the traffic domain. These studies emphasize the importance of ontologies for data integration, predicting traffic congestion severity, managing different data formats, and addressing semantic inconsistencies across IoT devices and platforms. The section concludes by stating that the following section will be a survey of existing traffic ontologies, which will serve as the basis for developing the authors' own traffic ontology called "TraffScOnto".

4 Related Works About Existing Traffic Ontologies

Traffic management is a critical aspect of urban planning and development, particularly in the rapidly growing landscape of smart cities. Effective management of traffic flow, congestion mitigation, and road safety requires comprehensive frameworks for representing and analyzing traffic-related data. Ontologies, as formalized representations of knowledge within specific domains, play a crucial role in facilitating the integration, sharing, and utilization of heterogeneous traffic data in smart city environments. In this section, we explore existing works in the realm of traffic ontologies, examining their contributions, methodologies, and limitations in addressing the complexities of urban traffic management. (Susel et al., 2016) [5] proposes an ontology-driven architecture aimed at enhancing traffic safety and improving driver comfort through various automated tasks. The ontology layer of this architecture comprises three interconnected groups of concepts: vehicle-related concepts, road-related concepts, and sensor-related concepts. The vehicle-related concepts encompass a taxonomy of different vehicle types, including commercial, public (such as buses and taxis), private (like cars, bicycles, and motorbikes), and priority vehicles (such as ambulances, police cars, and fire trucks). Additionally, these concepts allow for the representation of information pertaining to vehicle routes and locations. The road-related concepts include a taxonomy of various road types (e.g., local roads, prefectural roads, national highways, and expressways), as well as the representation of other infrastructure components such as road segments, traffic lights, traffic signs, lanes, road markings (such as painted arrows), and other elements like tunnels, parking lots, roundabouts, bridges, gas stations, and toll booths. Lastly, the sensor-related concepts are based on the SSN (Semantic Sensor Network) ontology. A mapping schema is proposed to translate sensor data into semantic data. This allows sensor data to be automatically represented as instances of the SSN ontology, with the observed property being the car flow property. In their publication (Tan et al., 2021) [6] have created a detailed knowledge graph of urban traffic systems using a top-down approach. Their work has shown that this model can uncover hidden connections between different parts of traffic systems, helping to find important traffic insights. This new method not only shows how different traffic elements interact in cities but also shows how it can be a useful tool for understanding complex traffic patterns and finding better solutions for city mobility. The KM4City ontology proposed by (Bellini et al., 2014) [7], developed by the University of Florence (Italy), serves as a foundational framework for smart city applications, particularly tailored to support data collection and integration efforts within the Tuscany region of Italy. This ontology encompasses a wide array of concepts related to urban infrastructure and services, including streets (such as road elements, junctions, and lanes), local public transportation (including routes, bus stops, and ride information), and various traffic sensors and event types (such as traffic observations, speed measurements, and traffic concentration data). Through its comprehensive coverage of urban data domains, the KM4City ontology facilitates the organization and utilization of heterogeneous data sources for urban planning, management, and analysis purposes. In their publication [8], Syzdykbayev et al. present an ontology specifically tailored for collaborative navigation within the context of autonomous cars, drivers, and pedestrians in smart cities. This ontology serves as a comprehensive representation of the pertinent concepts, relationships, and rules associated with navigation in a collaborative setting. It encompasses a wide range of concepts, including traffic rules, road conditions, vehicle and pedestrian behavior, as well as communication protocols. By employing this ontology, effective communication and coordination among diverse actors within the collaborative navigation environment can be achieved. Moreover, it can serve as the foundation for the development of intelligent navigation systems that aim to alleviate traffic congestion, enhance safety measures, and improve overall transportation efficiency within smart cities. The article emphasizes the significant role of ontologies in fostering interoperability and standardization across smart city systems. The proposed ontology for collaborative navigation is poised to facilitate the creation of advanced transportation systems that benefit both drivers and pedestrians alike. Through its structured representation and incorporation of relevant knowledge, the ontology empowers the development of intelligent transportation solutions capable of addressing the unique challenges and complexities of navigation in smart cities.

The existing body of work on traffic-related ontologies highlights the increasing use of Semantic Web technologies in traffic management and smart cities. However, each approach has its limitations. (Susel et al., 2016) developed an ontology-based architecture for intelligent transport systems using a traffic sensor network, but their taxonomy may not fully cover all aspects, potentially leading to semantic gaps. (Tan et al., 2021) focused on urban traffic ontologies, but their scope may be too narrow for larger metropolitan areas with complex infrastructures. (Bellini et al., 2014) aimed for comprehensive urban knowledge management, but their ontology lacks the granularity required for effective urban mobility management. Syzdykbayev et al.[8] also presented an ontology, but it may miss critical aspects of collaborative navigation in urban environments.

Building on these insights, we propose "TrafScOnto", an enriched traffic ontology tailored for smart cities. Our approach includes a thorough literature review, expert consultation, and the use of semantic web technologies. By addressing the limitations of existing ontologies, "TrafScOnto" aims to provide a comprehensive taxonomy of diverse traffic phenomena essential for smart city traffic management. We will rigorously evaluate our ontology through stakeholder feedback to ensure its effectiveness and applicability in improving traffic conditions in smart city environments.

5 Ontology Construction

5.1 Motivation

Initially, our research objective was to find how to propose a contribution to the smart city domain and more precisely to the traffic management through the ontological approach and especially how to address the challenge of semantic heterogeneity and data interoperability by providing an enriched vocabulary representing concepts related to traffic. Afterwards, as we progressed in our study, we asked ourselves the question if this innovative field could really be applied in Tunisia. And from there we did research in this direction and we discovered that there is already a national smart cities program which has already been launched since 2019. This program [9] has the ultimate objective of contributing to the optimization of the development of wealth, the guarantee of universal access to rights to basic services, and the monitoring of environmental protection. Thus, we have oriented our research especially during the identification of concepts to the exploration of resources which can be a support for the definition of concepts linked mainly to infrastructure hoping that this ontology will be deployed through the platforms that will eventually be developed a part of this program.

5.2 Adopted Construction Method

For the construction of our ontology we have adopted METHONOLOGY methodology [10]. We have chosen this methodology over other approaches because it provides a structured framework designed for the comprehensive development of ontologies. Also, it emphasis on iterative development and its comprehensive set of tools and techniques for each phase make it suitable for creating complex ontologies such as traffic management ontologies. We Begin with the domain analysis which facilitates a comprehensive understanding of the domain by delineating its key characteristics and functionalities. Drawing from validated documentation endorsed by domain experts, this phase lays the groundwork for subsequent stages. Knowledge acquisition follows, wherein domain-specific knowledge is identified and extracted, forming the basis for the ontology. Subsequently, the acquired knowledge undergoes modeling, where it is structured and represented using ontology modeling languages. This modeling step ensures that the ontology accurately captures the intricacies of traffic systems, including vehicles, infrastructure and sensors. The methodology then progresses to an interrogation phase, enabling validation of the ontology's responsiveness to user queries and requirements. Finally, verification and validation steps for the ontology are carried out. Overall, this method integrates both automated tools

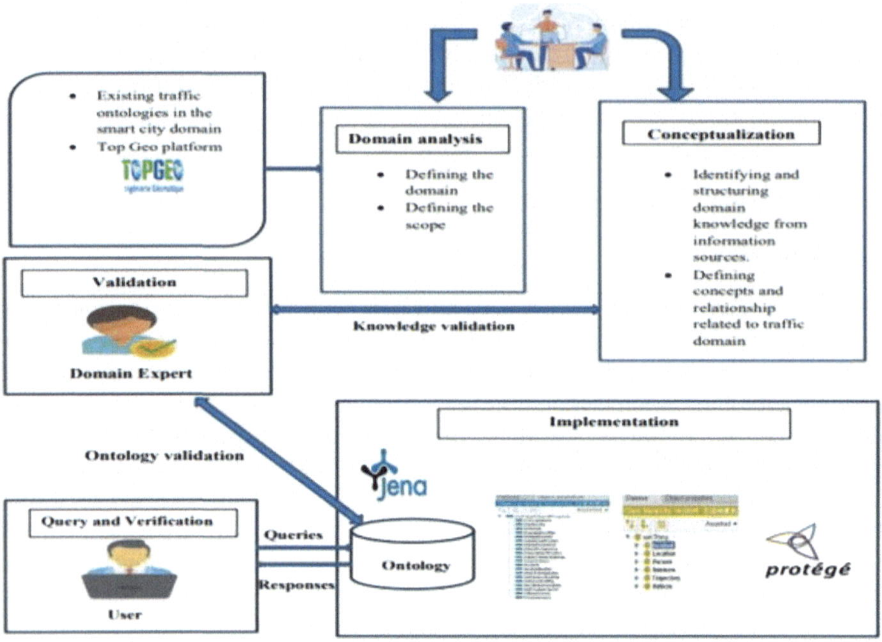

Fig. 1. Construction approach of "TraffScOnto" ontology

used in the modeling phase and manual expert interventions justifying its semi-automated category. The following figure describes the different steps (Fig. 1).

The proposed approach relies on related works dealing with traffic ontologies in the field of smart cities, which are thoroughly discussed in the previous section. Additionally, we draw upon documentation such as the Atlas Sfax [11] and interviews with domain experts from the Ministry of Equipment and Housing of Sfax. This approach will be detailed in the next section which englobes the different construction steps.

5.3 Construction Steps

- Domain analysis: During this stage, our focus lies on understanding the traffic domain. Following interviews with a domain experts (civil engineers) affiliated with the Ministry of Equipment and Housing in Sfax, and gaining insight into the proposed project to be carried out in the years 2023–2024 titled "Development of a platform for managing municipal road occupancy and advertising, as well as data collection and entry operations." Indeed, the diversity of traffic-related data is vast and characterized by heterogeneity, making the development of such a platform quite complex, as well as its operation. In fact, incorporating traffic ontology into this kind of project can deal with this gap. Then it offers numerous advantages such as providing a semantic understanding of traffic related data, enabling more sophisticated processing and analysis. Added to that, it facilitates the integration of diverse data sources, harmonizing disparate data sets for a comprehensive view of traffic conditions. Furthermore, leveraging the ontology enables enhanced decision support capabilities, aiding in optimal road occupancy management and advertising placement decisions. By exploring the Atlas Sfax documentation providing informations typically includes data on road networks, traffic flow patterns, congestion hotspots, accident statistics, transportation infrastructure such as traffic signals and signage, public transportation routes, parking facilities, and any ongoing or planned traffic management initiatives. Additionally, the documentation may contain analysis and insights into traffic trends, seasonal variations, peak hours, and factors influencing traffic dynamics in Sfax. Maps, diagrams, and statistical data may be included to visualize the information effectively and focusing in existing works dedicated to traffic in the smart city domain using ontologies as a solution to deal with the gap of semantic heterogeneity we have proposed an ontological model in order to provide a common and shared vocabulary that can be used in the development of traffic platforms. This modeling relies on data acquired from domain expert and other different documentations that will be used for the next step of our approach.

- Conceptualization: During the conceptualization phase, our focus lies on extracting and organizing domain knowledge from various sources, delineating key concepts, and establishing relationships pertinent to the traffic domain. In fact, this phase consists to identify the taxonomy of our ontology.

"TraffScOnto" englobes six principal modules which are infrastructure, incident, person, sensor, vehicle and trajectory that will be detailed along this section. Infrastructure class is inspired from existing traffic ontologies [5,8] then the concept of the road is updated based on the Atlas Sfax documentation [10], with a classification of road typology tailored to suit the specific characteristics of roads in Sfax aiming to improve its adaptability.

Incident class within the traffic ontology is structuring following existing ontology model [6] while incorporating additional elements based on [12], particularly those related to incidents leading to traffic congestion may allow for a more focused response to traffic incidents by distinguishing between those that frequently occur (recurrent) and those that are less common (non-recurrent), enabling resources to be allocated more efficiently.

Person class a preexisting concept in traffic-related works [5–7], it has been enhanced by subdividing the pedestrian subclass into adult, child, old, and vulnerable categories. This classification aims to address and categorize individuals at higher risk or requiring special attention in traffic scenarios.

Sensor class is defined building upon the existing concept of sensor [5,8], recent advancements have led to its classification into two distinct categories based on [13]: vehicle sensors and infrastructure sensors, further subdivided into intrusive and non-intrusive types. This classification may facilitates a more nuanced understanding of sensor functionalities and applications, enabling tailored deployment strategies and improved integration into traffic management systems.

Vehicle class is a preexisting concept used in pervious works [5,8]. Based on [14] we are trying to classify vehicles according to their functionalities and characteristics. In this aim we have adding the subclass of "HeavyWeightVehicle".

In fact, weight classification in traffic management domain aids in setting and assigning priority lanes based on vehicle weight, also real time systems utilize weight data to optimize signal timings, implement toll pricing and ensure smoother traffic management. In our ontology model we establish several fundamental relationships between key concepts to represent the complex dynamics of traffic management systems. Vehicles are intricately linked to their locations, either by circulating in specific areas or by occupying specific positions at any given time. Events and sensors play a crucial role, with sensors detecting events and being detected by sensors in the environment. The relationship between drivers and vehicles is direct, as drivers are responsible for driving vehicles, establishing a fundamental link between the two entities. Infrastructure elements such as roads and traffic lights are equipped with sensors that monitor various parameters, creating an equipped-by or installedin relationship. In addition, vehicles host sensors to monitor their surroundings, while specific sensors are located within vehicles to collect data effectively.

- Implementation: our ontology was created using Protégé tool [15] which is an open source for editing and building. It also supports a variety of ontology languages including RDF, OWL and XML and can be customized with plugins to extend its functionality. The following figure represents the global view of "TraffScOnto" ontology which englobes 105 concepts and 21 semantic relation ships (Fig. 2).

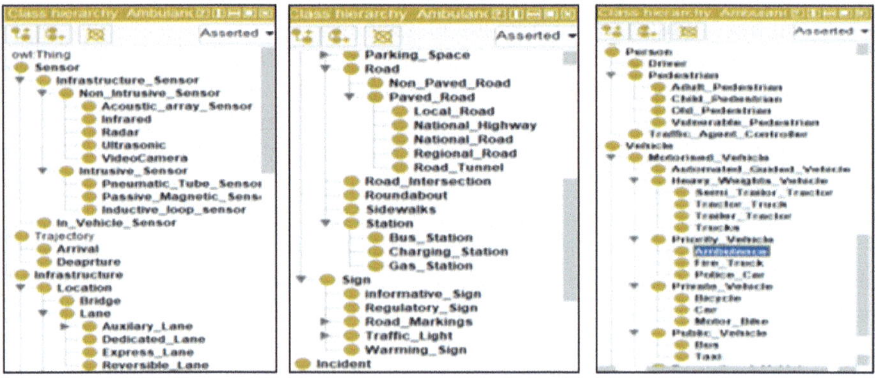

Fig. 2. "TraffScOnto" ontology global view

The resultant ontology is now ready for use. It can be queried using interrogation languages like SPARQL [16] and SQWRL, and it can be enhanced with inference rules such as SWRL.

- Query and verification: Querying the ontology enables smoother interaction between users and machines, allowing machines to accurately and clearly respond to user inquiries. After we have enriched our ontology model with real instances we are trying to execute some queries using Protégé tool and the following represents a simple query syntax using the SPARQL language at the aim of determining the list of incidents that happens in a special location (Fig. 3).

Fig. 3. SPARQL query example

- Validation:

The validation phase of our developed ontology, TraffScOnto involves rigorous testing to ensure its accuracy and effectiveness in real-world traffic management scenarios such as urban congestion management, incident response coordination, public transportation optimization and event traffic management. This includes evaluating its responsiveness to user queries and assessing its performance against actual traffic data. Expert review and feedback is essential to refine and validate the practical applicability of the ontology.

6 Conclusion and Future Works

Our paper introduces our ontology for traffic management in smart cities, laying the foundation for effective data integration and knowledge representation. Future works includes integrating heterogeneous data sources, refining mapping techniques, and developing a user interface for intuitive interaction. Additionally, leveraging the ontology for data analytics will provide valuable insights for urban traffic management. Through these efforts, we aim to enhance smart city initiatives and optimize urban mobility systems.

References

1. Wahl, Z.: What is the knowledge management and why is it important? (2018). https://www.kminstitute.org/blog/what-knowledge-management-and-why-it-important. Accessed 28 Feb 2023
2. Chang, D.L., Sabatini-Marques, J., Da Costa, E.M., Selig, P.M., Yigitcanlar, T.: Knowledge-based, smart and sustainable cities: a provocation for a conceptual framework. J. Open Innovation: Technol. Market, and Complex. **4**(1), 5 (2018)
3. Abdalla, W., Renukappa, S., Suresh, S., Al-Janabi, R.: Challenges for managing smart cities initiatives: an empirical study. In: 3rd International Conference on Smart Grid and Smart Cities (ICSGSC), pp. 10-17. IEEE (2019). https://doi.org/10.1109/ICSGSC.2019.00-26
4. Cheng, B., Longo, S., Cirillo, F., Bauer, M., Kovacs, E.: Building a big data platform for smart cities: experience and lessons from santander. In: 2015 IEEE International Congress on Big Data, pp. 592–599. IEEE (2015). https://doi.org/10.1109/bigdatacongress.2015.91
5. Fernandez, S., Hadfi, R., Ito, T., Marsa-Maestre, I., Velasco, J.R.: Ontology-based architecture for intelligent transportation systems using a traffic sensor network
6. Tan, J., Qiu, Q., Guo, W., Li, T.: Research on the construction of a knowledge graph and knowledge reasoning model in the field of urban traffic. Sustainability **13**(6), 3191 (2021). https://doi.org/10.3390/su13063191
7. Bellini, P., et al.: Km4City ontology building vs data harvesting and cleaning for smart-city services. J. Vis. Lang. Comput. **25**(6), 827–839 (2014)
8. Syzdykbayev, M., Hajari, H., Karimi, H.A.: An ontology for collaborative navigation among autonomous cars, drivers, and pedestrians in smart cities. In: 2019 4th International Conference on Smart and Sustainable Technologies (SpliTech). https://doi.org/10.23919/splitech.2019.8783045
9. Tunisian Smart Cities. (n.d.).Mission. https://www.tunisiansmartcities.com/mission.php
10. Fernández-López, M., Gómez-Pérez, A.: Overview and analysis of methodologies for building ontologies. Knowl. Eng. Rev. **17**(02) (2002). https://doi.org/10.1017/s0269888902000462
11. Topgeo. (n.d.). Atlas Sfax. https://www.topgeo.net/projects/atlas-sfax
12. Mejri, H.: Un système d'aide à la régulation d'un réseau de transport multimodal perturbe: réponse au problème de congestion. Ecole Centrale de Lille (2012)
13. Yu, X., Prevedouros, P.D.: Performance and challenges in utilizing non-intrusive sensors for traffic data collection. ARS **02**, 45–50 (2013)
14. Académie de Poitiers. (n.d.). Cours sur la classification des véhicules. https://ww2.ac-poitiers.fr/vehicules-materiels/sites/vehiculesmateriels/
15. Protégé. (2024). Protégé: A free, open-source ontology editor and framework for building intelligent systems. https://protege.stanford.edu/. Accessed 26 May 2024
16. Wiśniewski, D., Potoniec, J., Ławrynowicz, A.: Analysis of ontology competency questions and their formalizations in SPARQL-OWL. J. Web (2019)

SERTUS Dataset Collection from Spontaneous Environments

Latifa Iben Nasr(✉), Abir Masmoudi, and Lamia Hadrich Belguith

ANLP RG, MIRACL Laboratory, FSEGS, Sfax University, Sfax, Tunisia
{latifa.ibennasr,lamia.belguith}@fsegs.usf.tn, masmoudiabir@gmail.com

Abstract. This paper introduces SERTUS (Speech Emotion Recognition TUnisian Spontaneous), an extensive dataset collection intended to propel research in Speech Emotion Recognition (SER), particularly within the realm of Tunisian Dialect (TD). SERTUS encompasses both registers of the Tunisian Dialect: the Popular (familiar) register and the intellectual register, capturing a diverse range of emotions in spontaneous environments and natural interactions across different regions of Tunisia. This work delineates the methodology utilized in crafting SERTUS, highlighting the challenges and strategies involved in capturing spontaneous interactions. Moreover, we underscore the importance of including TD and the multidomain nature of the dataset, illustrating its potential applications in various domains such as sports, politics, and culture. It's imperative to note that this corpus collection adhered to a rigorous protocol to ensure corpus quality, as it will undergo annotation in future research endeavors.

Keywords: Data collection · Spontaneous emotion speech · Tunisian Dialect

1 Introduction

Emotions encompass the psychological traits linked to an individual's thoughts, feelings, actions, and experiences of pleasure and pain [1]. The ability to express and recognize emotions has long been a functional objective for fully capable intelligent virtual agents [2]. This capability serves various purposes, including sentiment analysis [3,4] and mental health diagnostics [5]. Furthermore, humans utilize speech, body language, and facial expressions to communicate their emotional states [6]. Currently, SER has found widespread application in human-machine interaction (HMI), virtual reality, and user behavior analysis [7].

Expanding upon the linguistic complexities, Arabic SER is an emerging field of study. The primary factor hindering the progress of Arabic SER, when compared to other languages, is the scarcity of resources [8]. The Arabic language, renowned for its diverse array of dialects and intricate linguistic subtleties, poses a significant challenge in SER tasks. Among these dialects, the TD stands out due to its unique characteristics and complexities. Spoken by an estimated 12 million

L. H. Belguith and K. Shaalan (Eds.): LPKM 2024, LNNS 1303, pp. 225–233, 2025.
https://doi.org/10.1007/978-3-031-85067-7_22

people across various regions [19], TD comprises numerous regional variations, including the Tunis dialect (Capital), Sahil dialect, Sfax dialect, Northwestern TD, Southwestern TD, and Southeastern TD [20]. Furthermore, TD showcases a distinct linguistic amalgamation, incorporating vocabulary from diverse languages such as French, Turkish, and Berber [11]. For instance, Tunisians frequently integrate French phrases into their everyday conversations, such as "ça va," "désolé," "rendez-vous," "merci," etc. Indeed, Tunisia's linguistic landscape is characterized as "poly-glossic," where multiple languages and language varieties coexist [21], reflecting the nation's rich historical heritage influenced by French colonization and Ottoman Turkish rule. Consequently, comprehensively analyzing emotional expressions within TD necessitates a nuanced understanding of its linguistic intricacies, encompassing both intellectualized and popular dialects [22], each presenting its own set of challenges for SER tasks. Despite the linguistic diversity, research on SER in TD remains limited, with only a handful of studies, such as those conducted by [10,13], and [14], offering a modest amount of data.

Our primary objective is to address this research gap by presenting our data collection efforts, a portion of which has already been utilized in previous studies [10]. Additionally, we enrich our corpus by incorporating a new dataset. We present the SERTUS data collection methodology, which aims to gather diverse and naturalistic data for training and evaluating emotion recognition models. We describe our approach to selecting data sources based on the two registers of TD, which are two oral linguistic registers [22]: the intellectualized dialect found in the conversations of scholars and radio and television broadcasts. This level of dialect is defined by the use of a literal lexicon and borrowings but keeps all the structures of the dialect, and the popular register, which conveys daily needs. Also, we develop data collection protocols. Our work contributes to advancing the field of emotion recognition by providing researchers with a valuable resource to develop more robust and generalizable models. This research makes several significant contributions:

- Compilation of spontaneous multidomain corpora
- Inclusion of two registers of TD within the dataset,
- Application of speech preprocessing techniques to ensure data quality and consistency.

The structure of this paper is organized as follows: Sect. 2 outlines the related work. Section 3 provides the approach and data profile of our corpus, SERTUS. The findings are detailed in Sect. 4. Finally, Sect. 5 concludes the paper by outlining key insights and potential perspectives for future research.

2 Related Work

In the domain of SER, database collection plays a pivotal role in advancing research and development efforts. Arabic Speech Emotion Recognition (ASER) is an emerging field of study. There is a notable scarcity of available datasets for the Arabic language. Between 2000 and 2021, the emotional database size in the Arabic language only reached 4.76% of the total size of all languages [9].

Additionally, the majority of datasets used for Arabic Speech Emotion Recognition are acted or elicited and recorded [8]. Notable examples include KSUEmotions, an acted database in standard Arabic comprising 5 h and 5 min of data, and EYASE, a semi-natural dataset in the Egyptian dialect consisting of 579 utterances ranging from 1 to 6 s. To the best of our knowledge, only two datasets are available in the TD: the REGIM_TES dataset [13] , consisting of 720 acted emotional speech samples with lengths of up to 5 s, and another dataset TuniSER [14] containing 2771 induced speech utterances with durations ranging from 0.41 to 15.31 s, with an average duration of around 1 s.

When compared with other languages such as English and French, the Arabic language notably lacks in both the quality and quantity of available datasets for SER. For instance, datasets in English, such as the IEMOCAP dataset [15], consist of natural, spontaneous conversations between actors, totaling approximately 12 h of audio recordings. Similarly, the SAVEE dataset [16] in English comprises natural emotional speech from four male speakers, with a total duration of 4 h. Furthermore, in French, datasets like the RECOLA [17] dataset contain natural emotional speech recordings in various contexts, totaling around 8 h of audio data.

However, when compared to these well-established datasets in English and French, the Arabic language suffers from a dearth of comparable resources in terms of both quality and quantity. As previously mentioned, the available Arabic datasets predominantly consist of acted or elicited speech, lacking the naturalness and spontaneity characteristic of real-world emotional expressions. Additionally, the sizes of existing Arabic datasets are significantly smaller in comparison, further exacerbating the challenges faced in developing robust SER systems for Arabic. This disparity underscores the urgent need for additional efforts to address the shortcomings in Arabic speech emotion recognition research, particularly in terms of dataset quality and size.

3 Approach and Data Profile

The SERTUS methodology is tailored to acquire data from spontaneous environments for emotion recognition research, with the goal of capturing naturalistic vocal expressions in real-world settings spanning various domains, capturing the two registers of TD. The SERTUS collection process is depicted in Fig. 1, and we will elaborate on it subsequently.

3.1 Data Source Selection

In Tunisia, there is the right to make fair use of resources that are copyrighted according to Article 23, as in our case within a research framework. Additionally, as we mentioned above, there are two oral linguistic registers [22]. For the first register, we drew inspiration from Tunisian television programs, particularly

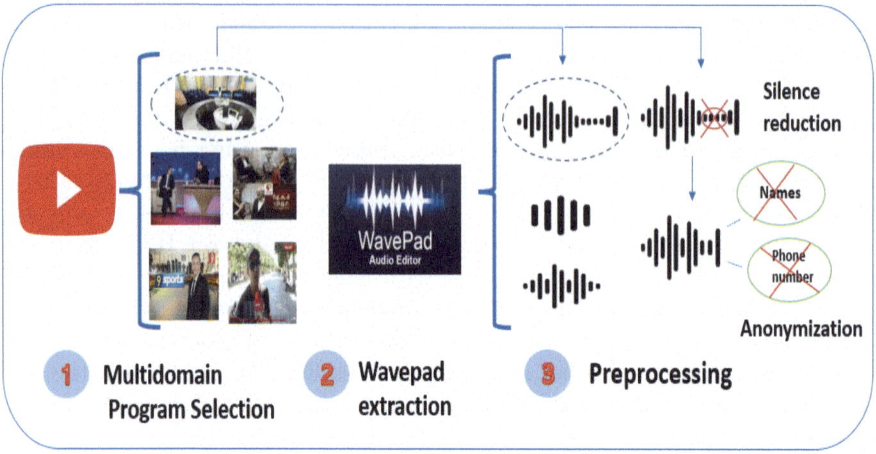

Fig. 1. SERTUS collection process

talk shows available on YouTube. Our aim was to analyze the range of emotions expressed by guests when sharing their viewpoints on various topics impacting the social, political, cultural, and sporting realms. As for the second register, we selected data from the 'Tunisian reality' series also accessible on YouTube. This series consists of interviews conducted on the streets using sidewalk microphones in the capital city of Tunisia. This series captures spontaneous emotional responses and comments from individuals in public spaces, addressing a wide range of topics such as politics, the economy, etc. We also made sure to cover different regions, such as the south and the north, in order to capture dialectical differences.

By focusing on a variety of talk shows and the "Tunisian reality" series, with an emphasis on the various domains and the two registers of TD, and by including diverse public regions, we are able to provide a comprehensive representation of spontaneous emotional reactions within the Tunisian population, ultimately leading to a better understanding. These sources offer a wide range of emotional expressions and linguistic variations (north and south) to better capture the characteristics of TD, including multilingualism, morphological, syntactical, and lexical differences, as well as a variety of word types [10,11].

3.2 Collection Protocol

Rigorous protocol is established to guide the data collection process and ensure the consistency and quality of recordings. These protocols encompass procedures such as software selection, participant recruitment, as well as ethical and legal considerations in data collection from public sources.

We have chosen the wavpad software for selecting suitable audio segments, which is a tool for editing audio files [12]. Participant selection involves individuals who are accustomed to working with this software and speech processing.

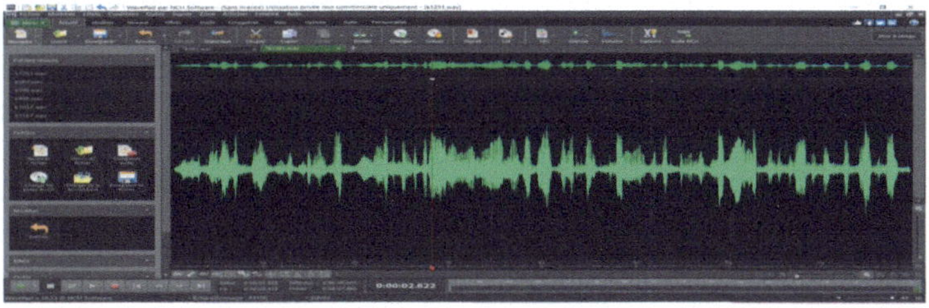

Fig. 2. Interface of "WavePad" software

The number of participants is two, one with a computer science bachelor's degree and the other with a master's degree in computer science, one female and one male.

Our vocal data collection originates from authentic conversations, potentially involving personal information such as names, professions, locations, and other details. To ensure confidentiality and comply with data protection regulations, we consider it imperative to meticulously oversee and adhere to the legal framework regarding the management of personal data. Under the 2006 Good Practice Guide, we emphasize the importance of anonymizing this data to prevent the identification of individuals. We have taken this obligation into account in the preprocessing stage.

3.3 Collection Process

In the data collection process, as previously mentioned, we have identified six programs in the Tunisian dialect, along with sequences from street interviews. To achieve this, as mentioned, we utilize WavePad software to extract only the responses of the speakers, eliminating any accompanying questions, whether from TV program presenters or journalists conducting street interviews, as well as extraneous noise. Each speaker's response is treated as a distinct audio segment to ensure clarity and coherence. This process involves meticulous attention to detail, ensuring that only the relevant speech segments are retained for further analysis. Fig. 2 provides a visual representation of this process, illustrating an example of a speech segment being extracted from the audio data.

Throughout the collection process, we maintain a constant sampling rate of 44.1 kHz to ensure consistency across all audio recordings. This uniformity enhances the reliability of the corpus and facilitates in-depth analyses of the emotional content present within the dataset.

3.4 Preprocessing

As stated earlier, our approach involves identifying sensitive components in recordings that may reveal personal details. Anonymization is achieved by substituting any direct speaker identification cues with periods of silence, achieved by reducing the frequency of speech where personal information is present. We acknowledge that certain combinations of cues may still potentially disclose identity, such as a speaker mentioning a first name without a last name, which may not be classified as identifying information. Professional roles, such as journalist, are generally not considered identifiers unless specific details are disclosed.

Upon evaluation of the corpus, instances allowing direct identification, such as specific job titles and locations, are rare but require careful handling. Anonymization is a manual process, a thorough review of all audio segments to safeguard privacy and comply with legal standards.

Another Preprocessing step is necessary. Given the spontaneous nature of our corpus, silent periods are common, whether for reflection or before a speaker's response. To ensure uniformity, any silent period exceeding 2 s has been standardized to a 2-s silence.

4 Findings

Our recently compiled database, SERTUS, represents a substantial collection of spontaneous speech in the TD, totaling over 20 h of audio recordings. These recordings encapsulate a diverse array of speech segments, ranging from brief snippets lasting a mere 1.15 s to more extensive segments extending up to 10.97 min in duration. On average, each segment spans approximately 20 s, capturing a broad spectrum of linguistic expressions and emotional nuances. The database comprises contributions from 1259 speakers, further enriching its diversity and comprehensiveness.

The meticulous process of compiling this corpus involved dedicated efforts spanning a period of 5 months. Throughout this time, we meticulously sourced, curated, and organized a vast array of audio data to ensure the richness and diversity of the dataset.

As previously noted, our dataset encompasses speech segments from various domains, reflecting the diverse nature of human communication. For example, in the sports domain, we have:

الترجي الرياضي شيخنا كيف العادة

/Altrjy AlryA.dy ŝyxnA kyf Al‘ATD/Attaragi team, our joy, as usual.

Similarly, in the cultural domain, we observe:

الأرقانيزاصون متع الحفلة هبال

/Al.AwrVAnyzA.swn mt‘ Al.hflT hbAl/The organization of the party was excellent.

Additionally, our dataset encompasses two registers: intellectual and popular. For example, in the intellectual register:

بورقيبة قاعد يخطب قدام أربعمية تابوت أربعمية من الشهداء من الناس إلي جندلتهم القوات الفرنسية

/bwrqybT qAd yx.tb qdAm ‘rbmyT tAbwt ’rbmyT mn AlŝhdA’ mn AlnAs Aly jndlthm AlqwAt AlfrnsyT/Bourguiba was giving a speech in front of a gathering of four hundred coffins, four hundred martyrs, among the people whose forces were defeated by the French forces

And in the popular register, we have:

أي كان حاجة في اطارها بوركوا با أصلا حاجة كيف هكا تنجم تحسن وضع البلاد و تزها الدنيا
أما كان فيها منا ولا منا وقتها حاجة مرفوضة قطعا ملا شنوة يحبو يشردون و عينينا تغزر

/'ay kAn .hAjT fy A.tArihA bwurkwA bA 'a.slA .hAjT kyf hkA tnjm t.hsn w.d' Al-blAd w tzhA Ad-dnyA 'mmA kAn fyhA mnA wlA mnA wqtuhA .hAjT mrfw.TD q.t'A mlaA ŝnwT y.hbw yŝrdwn w 'ynynA t.gzr/Right, if there is something in its framework, why not. In the first place, something like that can improve the country's situation and adorn the world. But, if there is any division or discord, it is absolutely rejected. Blimey! they want us to disperse while having us watching.

Furthermore, to offer insights into the distribution and composition of our dataset, the accompanying Table 1 categorizes the collected data by domain, providing a detailed breakdown of the dataset's size and composition within each domain. This comprehensive approach ensures that researchers have access to a well-rounded dataset that encompasses a wide range of linguistic, cultural, and thematic elements, facilitating diverse avenues of research and analysis in the field of speech emotion recognition.

Table 1. Statistics on the number of minutes collected per domain

Domain	Program	Number of minutes
Culture	Kahwa arbi & Labess show	180
Policy	Wahech checha	60
Sport	Attesia sport	60
Society	Safi kalbek & Maa ala	180
Multi-domain	Tunisian reality	1225

5 Conclusion and Future Perspectives

Arabic stands as one of the oldest languages still in use today. It remains among the earliest widely spoken languages, as noted by [18]. Within the realm of language study, Arabic holds a unique position, boasting a rich history and a vast linguistic heritage. Its influence spans continents, shaping cultures, literature, and intellectual discourse for centuries.

The compilation of the SERTUS corpus stands as a pivotal advancement in emotion recognition research, particularly within the context of Tunisian spontaneous speech. It represents a concerted effort to capture and analyze authentic emotional expressions in natural, unscripted interactions, shedding light on the intricate dynamics of human communication within the Tunisian sociocultural landscape. Assembled with meticulous care, this corpus serves as a rich repository of emotional data, offering researchers and practitioners a valuable resource for conducting comprehensive analyses across diverse sociocultural contexts.

Looking ahead, the potential of the SERTUS corpus extends far beyond its current scope. In the years to come, we envision leveraging this corpus to explore new frontiers in emotion recognition research, pioneering the development of more sophisticated models capable of discerning subtle nuances and cultural variations in emotional expression within Tunisian speech. Moreover, the practical applications of the SERTUS corpus are vast, holding promise for fields such as mental health diagnostics and HMI technology. By harnessing insights gleaned from this corpus, we aim to contribute to the development of innovative solutions that enhance user experiences, improve mental health diagnostics, and foster greater understanding and empathy in cross-cultural communication scenarios.

In essence, the SERTUS corpus not only signifies a significant achievement in emotion recognition research but also serves as a catalyst for ongoing exploration and innovation in understanding and harnessing the power of emotions within the Tunisian context and beyond. Initiatives like the compilation of the SERTUS corpus play a crucial role in bridging gaps in Arabic emotion recognition research, offering researchers a valuable tool to analyze emotional expressions within a specific sociocultural framework. By leveraging language and emotion, initiatives such as the SERTUS corpus pave the way for deeper insights into human communication across diverse sociocultural contexts, thereby overcoming challenges and facilitating advancements in Arabic emotion recognition research.

References

1. Li, C., Bao, Z., Li, L., Zhao, Z.: Exploring temporal representations by leveraging attention-based bidirectional LSTM-RNNs for multi-modal emotion recognition. Inf. Process. Manage. **57**(3), 102185 (2020)
2. Irfan, B., Narayanan, A., Kennedy, J.: Dynamic emotional language adaptation in multiparty interactions with agents. In: Proceedings of the 20th ACM International Conference on Intelligent Virtual Agents, pp. 1–8 (2020)
3. Islam, M.N., Khan, N.I., Roy, A., Rahman, M.M., Mukta, S.H., Islam, A.N.: Sentiment analysis of bangladesh-specific covid-19 tweets using deep neural network. In: 2021 62nd International Scientific Conference on Information Technology and Management Science of Riga Technical University (ITMS), pp. 1–6. IEEE (2021)
4. Mukta, M.S.H., et al.: A comprehensive guideline for Bengali sentiment annotation. Trans. Asian Low-Resour. Lang. Inf. Process. **21**(2), 1–19 (2021)
5. Fei, Z., et al.: Deep convolution network based emotion analysis towards mental health care. Neurocomputing **388**, 212–227 (2020)
6. Chatterjee, A., Narahari, K.N., Joshi, M., Agrawal, P.: SemEval-2019 task 3: EmoContext contextual emotion detection in text. In: Proceedings of the 13th International Workshop on Semantic Evaluation, pp. 39–48 (2019)
7. Chen, J., Sun, C., Zhang, S., Zeng, J.: Cross-modal dynamic sentiment annotation for speech sentiment analysis. Comput. Electr. Eng. **106**, 108598 (2023)

8. Alamri, H.: Emotion recognition in Arabic speech from Saudi dialect corpus using machine learning and deep learning algorithms (2023)
9. Iben Nasr, L., Masmoudi, A., Hadrich Belguith, L.: Survey on Arabic speech emotion recognition. Int. J. Speech Technol., 1–16 (2024)
10. Nasr, L.I., Masmoudi, A., Belguith, L.H.: Natural Tunisian speech preprocessing for features extraction. In 2023 IEEE/ACIS 23rd International Conference on Computer and Information Science (ICIS), pp. 73–78. IEEE (2023)
11. Masmoudi, A., Bougares, F., Ellouze, M., Estève, Y., Belguith, L.: Automatic speech recognition system for Tunisian dialect. Lang. Resour. Eval. **52**, 249–267 (2018)
12. Mustafa, H.H., Darwish, N.R., Hefny, H.A.: Automatic speech emotion recognition: a systematic literature review. Int. J. Speech Technol. 1–19 (2024)
13. Meddeb, M., Karray, H., Alimi, A.M.: Automated extraction of features from Arabic emotional speech corpus. Int. J. Comput. Inf. Syst. Industr. Manage. Appl. **8**, 11–11 (2016)
14. Messaoudi, A., Haddad, H., Hmida, M.B., Graiet, M.: TuniSER: toward a tunisian speech emotion recognition system. In: Proceedings of the 5th International Conference on Natural Language and Speech Processing (ICNLSP 2022), pp. 234–241 (2022)
15. Busso, C., et al.: IEMOCAP: interactive emotional dyadic motion capture database. Lang. Resour. Eval. **42**, 335–359 (2008)
16. Jackson, P., Haq, S.: Surrey audio-visual expressed emotion (SAVEE) database. University of Surrey, Guildford, UK (2014)
17. Ringeval, F., Sonderegger, A., Sauer, J., Lalanne, D.: Introducing the RECOLA multimodal corpus of remote collaborative and affective interactions. In 2013 10th IEEE International Conference and Workshops on Automatic Face and Gesture Recognition (FG), pp. 1–8. IEEE (2013)
18. Dahmani, H., Hussein, H., Meyer-Sickendiek, B., Jokisch, O.: Natural Arabic language resources for emotion recognition in Algerian dialect. In: Arabic Language Processing: From Theory To Practice: 7th International Conference, ICALP 2019, Nancy, France, October 16-17, 2019, Proceedings 7, pp. 18–33. Springer International Publishing (2019)
19. Zribi, I., Boujelbane, R., Masmoudi, A., Ellouze, M., Belguith, L.H., Habash, N.: A conventional orthography for Tunisian Arabic. In: LREC, pp. 2355–2361 (2014)
20. Gibson, M.L.: Dialect contact in Tunisian Arabic: sociolinguistic and structural aspects (Doctoral dissertation, University of Reading) (1999)
21. Lawson, S., Sachdev, I.: Codeswitching in Tunisia: attitudinal and behavioural dimensions. J. Pragmat. **32**(9), 1343–1361 (2000)
22. Boukadida, N.: 'Connaissances phonologiques et morphologiques derivationnelles et apprentissage de la lecture en arabe (Etude longitudinale) (2008). https://tel.archives-ouvertes.fr/tel00300544/document

An Agricultural Sentiment Dataset for Pest Control and Crop Diseases

Ameni Chamekh[1,2(✉)], Mariem Mahfoudh[1,3], and Khouloud Boukadi[1,2]

[1] MIRACL Laboratory, Sfax, Tunisia
[2] Faculty of Economics and Management of Sfax, Sfax, Tunisia
ameni.chamekh@fsegs.usf.tn
[3] Higher Institute of Computer Science and Multimedia of Sfax, Sakiet Ezzit, Tunisia

Abstract. This paper focuses on creating a dataset in an agricultural domain to understand and analyze sentiments related to pesticides, pests and crop disease. Collecting data allows us to gain insights into the opinions, attitudes, and experiences of farmers, agricultural experts, and consumers. Therefore, we collected data from social media and online platforms based on a list of keywords extracted from the AGROVOC multilingual controlled thesaurus. We have obtained two datasets: the first comprises 1617 tweets regarding pests and crop diseases from Twitter, while the second consists of 10 181 reviews on pesticide products from Amazon. Aspect extraction has been employed, identifying and extracting specific features or aspects discussed in reviews or tweets. This process enables a more granular and detailed data analysis by allowing us to go deeper into the data and understand the sentiment associated with each aspect rather than the overall sentiment. We identified eight aspects within the Amazon dataset, including "price", "package", "toxicity", etc. Similarly, we extracted relevant aspects from our Twitter dataset, resulting in five aspects such as: "pests", "safety", "toxicity", etc. Then, the data will be pre-annotated in collaboration with agricultural domain experts from INAT. This data will be used for future analyses and will inform the purchasing decisions of various decision-makers.

Keywords: Sentiment Analysis · Agriculture · Dataset · TALN · Pest Control · Deep Learning

1 Introduction

Sentiment analysis, also known as opinion mining or text sentiment analysis, is a field that examines people's opinions, feelings, and attitudes toward various topics of interest [6]. Typically, it categorises text into three primary sentiments: positive, negative, and neutral [10]. Sentiment analysis has emerged as a significant research area with applications across various professions and industries, including e-commerce [4], finance [7], the film industry [8,9] and agriculture

L. H. Belguith and K. Shaalan (Eds.): LPKM 2024, LNNS 1303, pp. 234–243, 2025.
https://doi.org/10.1007/978-3-031-85067-7_23

[2,6,11] and numerous others [1,5]. Our research focuses on the agricultural sector, aiming to provide valuable insights to farmers and individuals involved in this field through sentiment analysis of agricultural data. The objective is to understand and analyze sentiments about pesticides, pests, and crop diseases in agriculture. Analyzing this data reveals benefits and useful information that can enhance agricultural practices. Despite sentiment analysis's promising opportunities, its application in the agricultural context remains unexplored. Some studies have examined it [3,11], but their focus primarily concerns identifying sentiments expressed in online commerce reviews of agricultural products. Furthermore, following a literature review, it is evident that there is a significant lack of data corpus for sentiment analysis in agriculture. In [2], researchers used a manually scraped corpus consisting of 300 tweets and Facebook posts, which can be considered very limited.

Therefore, we have constructed our datasets by collecting data from various sources, including social media platforms like Twitter and online commerce platforms like Amazon. We focus on reviews and tweets from both farmers and the general public, specifically those related to pesticide products, parasites, and crop diseases. The datasets were collected based on a list of keywords extracted from the AGROVOC multilingual controlled thesaurus. This thesaurus is characterized by its hierarchical organization, which allows us to navigate efficiently through different levels of term specificity, starting from more general categories and moving towards more precise and specific concepts. This structure not only facilitates the search for relevant terms but also enhances the understanding of relationships between different agricultural concepts. Furthermore, our choice is based on AGROVOC's specialization in the agricultural field, its compliance with international standards, its wide terminological coverage, its open-access availability, and its evolving and constantly updated nature. Then, aspect extraction is utilized to identify and extract specific features discussed in reviews or tweets, enhancing detailed data analysis and comprehension of sentiment associated with each aspect rather than focusing on the overall sentiment. This process was automated through the application of natural language processing (NLP) techniques and predefined lexicons. Our aim is to create an informative and diverse data corpus for future efficient sentiment analysis. The rest of this paper is organized as follows: Sect. 2 explores related work. Section 3 outlines the proposed method for constructing our datasets. The conclusion and future endeavours are presented in Sect. 4.

2 Related Work

In this section, we present some pertinent works from the literature that offer insights into the current state of research in the field. The authors in [11] collected Chinese agricultural product reviews from an e-commerce platform using the R language. They selected agricultural products such as bananas, peaches, pitaya, potatoes, tomatoes, cucumbers, green vegetables, and other agricultural products to build a dataset of 23,418 agricultural product reviews. This dataset includes 14,888 sentences expressing positive emotions and 8,530 sentences expressing negative emotions. In [3], researchers utilized a dataset consisting of manually annotated reviews categorized into positive, negative, and neutral sentiments sourced from various channels, including social media, news sites, e-commerce platforms, and offline surveys. The data labelling is performed twice to cross-verifying the results. This process yielded 6152 positive, 1873 neutral, and 7925 negative sentiment data points. The results demonstrate efficient information extraction and visualization of emotional trends in online evaluations, offering valuable insights for those invested in the agricultural product sector. In their study [6], researchers highlighted e-commerce's role in improving sales strategies by leveraging user feedback. They extracted comments from the "fruits" category of the publicly available "online_shopping_10_cats" dataset from GitHub. This dataset comprised over 60,000 comments across various categories, with a balanced distribution of positive and negative sentiments. The comments extracted from the "fruits" category of the dataset totaled 10,000 comments, including 5,000 positive and 5,000 negative comments. Given the limitations of the public dataset, [6] employed web crawling techniques to gather additional review data from e-commerce platforms such as "Jing-Dong". They carefully organized a private dataset by specifying search keywords and filtering conditions to ensure relevance to agricultural products. Then annotated the sentiment polarity of the product reviews, resulting in a dataset of over 14,000 reviews with a substantial number of positive and negative sentiments. Overall, this study offers valuable insights for vendors in the agricultural sector to improve product quality and enhance customer satisfaction. The authors in [2] have utilized the Facebook Graph API and the Twitter API to extract posts and tweets from Facebook and Twitter, respectively. They extracted information regarding 1) Insects affecting sugar cane, rice, soybean, and cocoa crops; 2) Chemicals used in crop management; and 3) Diseases, symptoms, treatments, and recommendations for crop control and management. The corpus comprises 300 tweets and Facebook posts (100 positive, 100 negative, and 100 neutral). They were manually labelled by a group of three individuals, two of whom were in the agriculture domain. The labeling process involved reading the tweets and identifying all aspects and associated polarities (positive, negative, or neutral). Table 1 offers a comparison of the datasets employed in these studies, highlighting classes, annotation, aspect, data source, and data size.

Table 1. Comparison of agriculture datasets for sentiment analysis

Dataset	Classes	Annotation	Aspect	Data Source	Dataset Size
[2]	3 (Positive, Negative, Neutral)	Manual	None	Twitter and Facebook	300
[11]	2 (Positive, Negative)	Manual	Yes	E-commerce Platform	23,418
[3]	3 (Positive, Negative, Neutral)	Manual	None	Social Media, News Sites, E-commerce Platforms, Offline Surveys	15,950
[6]	2 (Positive, Negative)	Pre-annotated	None	"Fruits" category from "online shopping 10 cats" on Github	10,000
	2 (Positive, Negative)	Manual	None	Private dataset on agricultural products	14,222

Most studies in the agricultural field often focus on e-commerce products, leaving a significant gap in data related to agricultural anti-parasitic products. To address this gap, we propose adopting a data collection approach to leverage the vast information available on e-commerce platforms and social media networks. Using these sources, we have created a dataset specifically focusing on agricultural anti-parasitic products, parasites and crop disease. This dataset will serve as a fundamental resource for conducting thorough sentiment analysis.

3 A Proposed Agricultural Sentiment Dataset

In this work, we propose a dataset focused on agricultural anti-parasitic products, parasites, and crop diseases. We address the lack of datasets in this field and offers a method for constructing such datasets. We initiated our work with the following steps:

- Collect data from social media and online platforms, including Amazon and Twitter, comprising reviews and tweets related to pesticides, pests control and crop diseases.
- Preprocess the datasets to ensure that the text is clean and uniform by employing various actions such as normalization, removing stopwords, etc.
- Extract aspects from our data corpus by identifying specific features discussed in reviews or tweets.
- Annotate the datasets in collaboration with agricultural experts from the National Agronomic Institute of Tunisia (INAT). This process involves reading the data and identifying associated polarities.

3.1 Data Collection

With the rapid expansion of social media, sentiment analysis has become vital, especially in agriculture, as online opinions and concerns offer valuable insights. Following a literature review, it is evident that there is a significant lack of datasets for sentiment analysis within the agricultural domain. Therefore, we collected reviews on agricultural pesticide products and tweets about pests and diseases affecting crops from various sources, including Amazon and Twitter, based on a list of keywords extracted from the AGROVOC multilingual controlled thesaurus[1]. This enabled us to construct an informative and varied dataset. We choose to use this thesaurus instead of other tools due to its hierarchical structure of terms, which facilitates the search and classification of agricultural concepts. This hierarchical organization allows us to navigate efficiently through the different levels of specificity of terms, starting from more general categories and moving towards more precise and specific concepts. Then, we utilized web scraping, a technique derived from the English term "to scrape". This method, implemented in Python, provides an automated, rapid, and efficient solution for collecting a large amount of data.

[1] https://agrovoc.fao.org/browse/agrovoc/en/.

Twitter Dataset. We chose Twitter as our data collection platform because it is widely used for discussing current and general topics. Facebook is less suitable for our purposes as it focuses more on personal social interactions. We employed the AGROVOC thesaurus to identify keywords associated with pest control and crop diseases. Our keywords list includes terms such as "pest control", "insecticides", etc. Then, we developed a web extension to extract tweets automatically using our list of keywords. We collected 1617 tweets categorized by types of pests, with columns including "class", "user", "userLink", "tweet", and "date". The collection period extends from May 19, 2009, to February 27, 2024. In summary, through a web extension, we employed "Twitter" and "AGROVOC" to efficiently collect data on pest control and crop diseases in agriculture. 1a shows the Twitter data categorization.

Amazon Dataset. We utilized web scraping to gather various opinions on agricultural pest control products sold on Amazon. Amazon was chosen due to its extensive product range and global reputation. This platform offers advanced filtering and sorting features, thus facilitating the collection of relevant data. We extracted data using a list of keywords from the AGROVOC thesaurus. Subsequently, we developed Python scripts using the BeautifulSoup and Selectorlib libraries to navigate web pages and extract necessary information. This approach enabled us to retrieve details such as titles, dates, and customer comments, providing a comprehensive insight into users' concerns and experiences regarding pest control products. Our Python script was designed to extract information on pest control products from Amazon using specific search queries. Reviews were categorized into six categories, each corresponding to different types of pests. In total, 10181 reviews were collected and distributed across six CSV files, with the following breakdowns: "Insects: 3824 reviews", "Fungi: 2421 reviews", "Herbs: 3296 reviews", "Bacteria: 133 reviews", "Nematodes: 178 reviews", and "Viruses: 323 reviews". The data collection period spans from June 17, 2009, to February 17, 2024. The Amazon dataset is represented by the following columns: "class_id," "treatment_type," "product_id," "review_author_name," "review_date," "review_content," "review_rating," and "review_is_verified. Figure 1b displays Amazon data categorization.

3.2 Data Preprocessing

The data preprocessing stage involves preparing textual data. It is crucial before employing aspect extraction as it ensures that the data is clean, uniform, and ready to be used by specific aspect extraction techniques. Effective data preparation directly enhances the quality of extraction results and the reliability of insights generated from aspect analysis. We utilized several actions such as :

- **Conversion to lowercase.** We normalized the text by converting it to lowercase, which allows treating words uniformly. For example, *"Price"* and *"price"* are considered identical after this transformation.

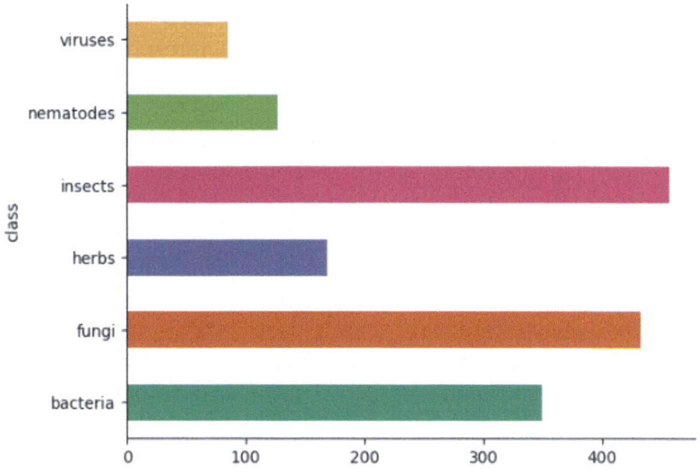

(a) Twitter Data Categorization

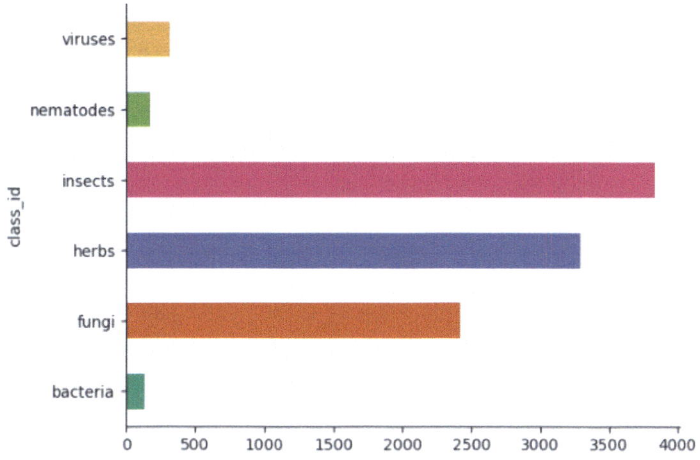

(b) Amazon Data Categorization

Fig. 1. Categorization of Datasets

- **Removal of HTML tags and links.** We removed HTML tags *"¡ ¿"*, as well as links like: *https://www.example.com*, which is no useful information for analysis, in order to clean the text.
- **Removal of numbers, symbols, and punctuation.** We eliminated noise introduced by numbers *[0 .. 9]*, symbols : *"$, %, &, etc."* and punctuation *"?, !, ', ", (), [], etc."*. to simplify the text, enabling us to focus on meaningful words in the reviews.
- **Removal of emojis.** We removed emojis as they do not always contribute to the text's meaning. This allowed us to focus specifically on analyzing textual

data rather than incorporating visual elements such as emojis, which can introduce ambiguity and variability in sentiment interpretation due to their subjective meanings and cultural differences.

- **Removal of stopwords.**
  - The removal of stopwords, done using the stopwords from the NLTK library, eliminates common words such as connectors, pronouns, prepositions, and auxiliary words, thus reducing noise in the text to focus on the most informative words for subsequent analysis.
  - This process enhances the quality of the analysis by removing text elements that do not contribute to the textual content's meaning. Example stopwords include: *"i", "my", "we", "our", "it", "its", "they", "them", "their", "what", "which", "can", "will", "just", "should", "now", etc.*
- **Normalization.** We normalized the text using a technique called lemmatization. Lemmatization is a natural language processing technique that involves reducing words to their base form, called a lemma. We employed it using the language model loaded by spaCy. For example, the words *"price", "pricey", and "pricing"*, lemmatization with spaCy will produce the following results:
  - *"price"* will be converted to its lemma *"price"*.
  - *"pricing"* will also be converted to its lemma *"price"*.
  - *"pricey"* will also be "price" because *"pricey"* is a variation of *"price"* that means something expensive or costly, but in lemmatization, we aim to obtain the base form of the word, which is *"price"*.

3.3 Aspect Extraction

Aspect extraction is crucial in sentiment analysis as it enables the determination of sentiment towards specific aspects rather than the overall sentiment of a review [1]. It provides a finer understanding of customer opinions towards features such as effectiveness against pests and diseases, method of use, and safety for use on crops. The process involves identifying specific topics or themes mentioned in the text that are relevant to the domain. We employed Python libraries such as spaCy to conduct automatic aspect extraction based on natural language processing (NLP) techniques and predefined lexicons, efficiently extracting and analyzing relevant aspects from our datasets. Predefined lexicons were motivated by a thorough analysis of our datasets and research in specialized thesauruses where these characteristics are frequently mentioned. Each aspect, such as "price," "package," and "effectiveness," has associated keywords indicative of that aspect. Each token in the text is compared against the keywords associated with each aspect in the lexicon. If a token matches any of the keywords for a particular aspect, that aspect is considered relevant to the text. This matching process is case-insensitive to ensure flexibility in identifying aspects. After identifying aspects mentioned in the text, a Counter object from Python's collections module is used to count the occurrences of each aspect, facilitating determination of the most commonly mentioned aspects in the reviews. This approach enabled us to identify the following key aspects in the Amazon datasets: "price", "package", "shipping", "effectiveness", "usage", "pests", "safety", and "toxicity". Similarly, for the Twitter corpus, we conducted the extraction of relevant aspects. Agricultural tweets, unlike reviews on commercial pesticide products, do not directly address sales-related aspects such as "price", "package" and "shipping". Consequently, we excluded these aspects from our analysis. However, we opted to retain other critical aspects pertinent to the agricultural context, including "effectiveness", "usage", "pests", "safety", and "toxicity". The selected aspects, such as "price", "package", and "shipping", are criteria inherited from the commerce domain since pesticides are sold online on e-commerce platforms like Amazon. In contrast, aspects such as "effectiveness," "usage," "pests," "safety," and "toxicity" are specifically related to our application domain, agriculture. The approach ensures coverage of both commercial and agricultural aspects. The extracted aspects provide crucial information for future sentiment analysis, representing key features that influence opinions regarding anti-parasitic products, crop pests, and diseases.

3.4 Data Annotation

Data annotation is an essential step for sentiment analysis. It converts raw data into actionable insights by assigning each review a specific sentiment, such as positive, negative, or neutral. This annotation forms the basis of sentiment analysis models, enabling automated classification of customer opinions on products or services. For annotating the Amazon corpus, we utilized a Python script to process a dataset comprising review ratings, converting these numerical ratings into textual sentiments. Specifically, it defines a function that maps a rating string *(e.g., "5 étoiles")* to a corresponding sentiment *(e.g., "very positive")* by extracting the numerical value and using conditional statements to determine the sentiment. The function is applied to the "review_rating" column of the dataset using pandas' "apply" method, and the results are stored in a new column named "sentiment". The annotation schema used categorizes ratings from 1 to 5 stars into sentiments: "very negative", "negative", "neutral", "positive", or "very positive". The process includes extracting the numerical rating, mapping it to a sentiment based on predefined conditions, and adding these sentiments to the DataFrame. Finally, the modified DataFrame, now with a new "sentiment" column, is printed to display the results. The annotation of the Amazon dataset reveals 6574 reviews of "very positive" sentiments, 1647 reviews of "positive" sentiments, 779 reviews of "very negative" sentiments, 386 reviews of "negative" sentiments, and 795 reviews of "neutral" sentiments. This annotation makes the data more informative and facilitates subsequent analysis of customer sentiments.

Moreover, the tweets will be manually labeled in collaboration with INAT's agricultural domain experts. This process involves reading the tweets and assigning associated polarities (positive, negative, or neutral). The involvement of agricultural experts from INAT will add scientific rigor to our research by ensuring accurate and relevant annotated data, enhancing the study's reliability. To maintain consistency and reliability in sentiment annotations, we will develop detailed guidelines that clearly define the sentiment categories and provide specific examples to minimize subjective interpretation differences. We will label each dataset twice independently and cross-verify the results for consistency to ensure the objectivity and accuracy of the annotated data.

4 Conclusion

In conclusion, our study involved the collection of reviews and tweets related to pesticides, pests, and crop diseases from Amazon and Twitter. We implemented data preprocessing steps to ensure that the data is clean and uniform. Subsequently, we employed aspect extraction to identify key aspects. Within the Amazon dataset, we identified eight aspects, including "price", "package", and "toxicity". Similarly, relevant aspects were extracted from our Twitter dataset, resulting in five aspects such as "pests", "safety", and "toxicity". Additionally, we have applied a data pre-annotation on the Amazon corpus. In the upcoming phases of our research, we aim to validate the corpus and annotations. Following this, we will develop our sentiment analysis model and evaluate the performance of the aspect detection technique, presenting precision and accuracy measures. Subsequently, we intend to enhance the accuracy and effectiveness of our model by incorporating advanced machine learning and deep learning techniques. Through our model, we aim to gain a deeper understanding of the sentiments related to pesticides, pests and crop diseases in agriculture, ultimately contributing to improved agricultural practices.

References

1. Belguith, M., Aloulou, C., Gargouri, B.: Aspect level sentiment analysis based on deep learning and ontologies. SN Comput. Sci. **5**(1), 58 (2023)
2. Bermeo-Almeida, O., del Cioppo-Morstadt, J., Cardenas-Rodriguez, M., Cabezas-Cabezas, R., Bazán-Vera, W.: Sentiment analysis in social networks for agricultural pests. In: ICT for Agriculture and Environment: Second International Conference, CITAMA 2019, Guayaquil, Ecuador, January 22-25, 2019, Proceedings 2, pp. 122–129. Springer (2019)
3. Cao, Y., Sun, Z., Li, L., Mo, W.: A study of sentiment analysis algorithms for agricultural product reviews based on improved Bert model. Symmetry **14**(8), 1604 (2022)
4. Chamekh, A., Mahfoudh, M., Forestier, G.: Sentiment analysis based on deep learning in e-commerce. In: International Conference on Knowledge Science, Engineering and Management, pp. 498–507. Springer (2022)
5. Edara, D.C., Vanukuri, L.P., Sistla, V., Kolli, V.K.K.: Sentiment analysis and text categorization of cancer medical records with LSTM. J. Ambient. Intell. Humaniz. Comput. **14**(5), 5309–5325 (2023)
6. Liu, R., Wang, H., Li, Y.: AgriMFLN: mixing features LSTM networks for sentiment analysis of agricultural product reviews. Appl. Sci. **13**(10), 6262 (2023)
7. Mejbri, H., Mahfoudh, M., Forestier, G.: Deep learning-based sentiment analysis for predicting financial movements. In: International Conference on Knowledge Science, Engineering and Management, pp. 586–596. Springer (2022)

8. Mutinda, J., Mwangi, W., Okeyo, G.: Sentiment analysis of text reviews using lexicon-enhanced BERT embedding (LeBERT) model with convolutional neural network. Appl. Sci. **13**(3), 1445 (2023)
9. Tan, K.L., Lee, C.P., Anbananthen, K.S.M., Lim, K.M.: Roberta-LSTM: a hybrid model for sentiment analysis with transformer and recurrent neural network. IEEE Access **10**, 21517–21525 (2022)
10. Tan, K.L., Lee, C.P., Lim, K.M.: A survey of sentiment analysis: approaches, datasets, and future research. Appl. Sci. **13**(7), 4550 (2023)
11. Zikang, H., Yong, Y., Guofeng, Y., Xinyu, Z.: Sentiment analysis of agricultural product ecommerce review data based on deep learning. In: 2020 International Conference on Internet of Things and Intelligent Applications (ITIA), pp. 1–7. IEEE (2020)

Deep Learning Approach to Identify and Classify Arabic Verbal Multi-word Expressions

Meniara Bsissa[1(✉)], Iskandar Keskes[1], and Najet Hadj Mohamed[1,2]

[1] Faculty of Economics and Management of Sfax (FSEGS), MIRACL Laboratory, University of Sfax, Sfax, Tunisia
bsissa.meniara@gmail.com, iskandarkeskes@gmail.com, hadjmed.dhouha@gmail.com

[2] University of Tours, LIFAT, Tours, France

Abstract. This paper focuses on the identification and classification of Arabic Verbal Multi-Word Expressions (VMWE), which are combinations of words with a unitary meaning and containing at least one verb. We describe our contribution to annotating the Arabic-annotated-conll17 corpus with VMWEs, following the annotation guide of the PARSEME framework, and using the ChatGPT model to enrich the corpus with sentences that contain more linguistic phenomena. We propose a method for the identification and classification of Arabic VMWEs based on machine learning (Random Forest, Decision trees and Support Vector Machine) and deep learning (Conv1D+BiLSTM) models. The results show that our Random Forest model outperformed others in identification, while the Decision Tree model excelled in classification. Additionally, our deep learning model exhibited significant performance improvement after data enrichment with ChatGPT. These findings underscore the effectiveness of our approach in addressing this complex problem.

Keywords: Multi-word expression · Arabic language · VMWE identification · VMWE classification · deep learning

1 Introduction

A multi-word expression is a lexical unit that contains more than one word and has an idiosyncratic property, it does not adhere to the compositional logic of its individual elements. For example, the Arabic expression قتل الوقت ("qatala al-waqt") does not literally mean 'kill time' but rather 'pass time'. MWEs are very common in natural languages and represent challenges for automatic natural language processing (NLP), as they necessitate specialized techniques for detection and interpretation. According to [3,15] there are different main categories of MWEs. For example, the most common category is Nominal MWEs (NMWE) which results from the combination of two or more names. Verbal

L. H. Belguith and K. Shaalan (Eds.): LPKM 2024, LNNS 1303, pp. 244–252, 2025.
https://doi.org/10.1007/978-3-031-85067-7_24

MWEs (VMWEs) are a category that includes at least one verb associated with another lexicalized element. Within this category, there are several subtypes, including Light Verb Constructions (LVC), Verbal Idioms (VID), Inherently Reflexive Verbs (IRV), Verb-Particle Constructions (VPC), Multi-Verb Constructions (MVC), and Inherently Adpositional Verbs (IAV) [12].

VMWEs pose significant challenges to both linguists and technologists due to their capacity to morph across different contexts (variability) and incorporate disruptive elements while retaining meaning (discontinuity). These challenges hinder the development of annotated resources, which are essential for automated linguistic processing and annotation. Prepositional MWEs are linguistic constructs consisting of a preposition and other elements forming a coherent set in terms of semantics and syntax. Additionally, the authors note the existence of Adjectival MWEs.

2 Related Work

2.1 Annotation and Language Resources of Arabic MWEs

Arabic is a complex and diverse language with limited specific resources such as lexicons and corpora annotated with MWEs, due to factors like the lack of standards and tools for MWEs annotation and minimal researcher and institutional interest, although some efforts have been made to create or utilize lexical resources for Arabic MWEs using manual, automatic, or semi-automatic methods. [11] created a list of 5,000 manually extracted Arabic multiword expressions from dictionaries categorized by syntax, and used a pattern-matching algorithm to label them in a large Arabic text corpus for better understanding and natural language processing applications. Next, [6] worked with a large Arabic-English text database LDC GALE[1] and a list of multi-word expressions [5] to improve translation accuracy in a computer program called Moses[2]. They focused on four types of expressions, including verbs, nouns, adjectives, and adverbs. They introduced four new methods for handling these expressions and found over 69,000 examples. Then, [1] developed a semi-automatic resource called SAMER for Arabic multi-word expressions. This resource includes all possible flexible forms and syntactic structures of a list of Arabic verbal expressions in their canonical form. They searched for all occurrences of these forms in the Arabic Treebanks (ATB) and Arabic Gigawords. They effectively built a resource containing 1,884 Arabic verbal expressions. More recently, [7] successfully modeled Arabic verbal expressions within the PARSEME[3] multilingual framework. They manually annotated an Arabic corpus with 1,252 verbal expressions from 1,062 sentences of the Prague Arabic Dependency Treebank [10]. Lastly, [8] constructed an Arabic version of the AlphaMWE parallel corpus, which is referred to as AlphaMWE-Arabic, enriched with annotations of Multiword Expressions. This

[1] https://www.ldc.upenn.edu/collaborations/past-projects/gale/data/gale-pubs.
[2] http://www2.statmt.org/moses/.
[3] https://parsemefr.lis-lab.fr/doku.php.

effort encompassed the utilization of machine translation (MT), post-editing, and the annotation of diverse Arabic dialects, including both standard and regional varieties such as Tunisian and Egyptian Arabic.

2.2 Identification of Multi-word Expressions

In natural language processing, the automated identification of MWEs is crucial as they constitute significant components of lexicons and language semantics. In this context, [16] studied multi-word expressions in the Irish language. They used both multilingual and monolingual pre-trained language models (mBERT and gaBERT) to identify these expressions. Focusing on verbal expressions, they showed that high-quality language-specific resources could compensate for a lack of linguistic data in certain natural language processing tasks. Their approach achieved an F-score of 45.66.

Next, [13] focus on a semantic model called BioLORD for identifying idiomatic expressions in medical terms. They aim to assist in the translation of complex biomedical entities by scoring their idiomaticity, using BioLORD's ability to compare entity definitions with their semantic representations. After that, [2] work on enhancing the automatic identification of Romanian multi-word expressions using multilingual language models. They introduce a lateral inhibition layer and adversarial training to improve these models' performance, achieving a significant increase in the F-score and setting a new state-of-the-art for this task. Finally, [9] presented a method for improving the identification of MWEs in Arabic by using lexicons with basic morphosyntactic information. This is the first study to concentrate on both seen and unseen of VMWEs for Arabic.

3 Annotation of the Arabic-Annotated-Conll17 Corpus with VMWEs

We utilized the data from the shared task of CoNLL[4] (The Conference on Computational Natural Language Learning) 2017, which pertains to the Arabic corpus, as the source corpus for annotation with Arabic MWEs, following the guidelines of the PARSEME network. We performed the annotation task with three types of VMWEs; Light Verb Construction (LVC) which combines a weak or auxiliary verb with a predicative noun that carries the main meaning of the action for example, أجرى تحقيق ("'ağra'a taḥqīq"; means conducting an investigation"): The verb أجرى ("'ağra'a"; means conducting") is light and the name تحقيق ("taḥqīq"; means investigation) refers to an event, Verbal Idiom (VID), they have at least two lexicalized elements, including a head verb and one of its complements. Idioms typically have both a literal meaning and an idiomatic meaning such as ألقى الضوء ("alqā al-ḍaw'"; literally: he shed light, means: he

[4] https://lindat.mff.cuni.cz/repository/xmlui/handle/11234/1-1989.

focused attention on something) and Inherently Adpositional Verb (IAV) which are verbs or expressions that are associated with an idiomatic preposition or postposition, which is essential or significantly alters the meaning of the verb or expression if it is not present for instance, رغب في ("raġiba fī"; means to desire).

3.1 Annotation Protocol

To conduct the annotation task, we implemented a structured four-step approach. Initially, we engaged in a training phase, where annotators collaboratively worked to classify and categorize expressions within 50 sentences. This process utilized decision trees and tests from the annotation guide to ensure precision. Subsequently, we entered a phase dedicated to updating the annotation guidelines. This involved refining the guidelines used for annotating Arabic VMWEs to incorporate distinctive Arabic terms, such as the MASDAR a verbal noun that does not convey tense or quantity. We integrated a decision tree tailored to Light Verb Constructions (LVCs) that accounts for the MASDAR, as shown in Fig. 1. For example the expression اجراء دراسة ("ajrā' dirāsah"; means to conduct a study) is considered as an LVC.full despite the term اجراء ("ajrā"') is a masdar. The testing phase followed, wherein two annotators independently marked sentences post-training. They assessed the consistency of their annotations using a recognized standard measure [4]. A high consistency level allowed them to proceed; otherwise, additional training was undertaken. Finally, in the annotation phase, the annotators, equipped with training and testing experience, were prepared to label the entire corpus accurately which contains 740 examples of AVMWEs.

3.2 Enrichment of the Corpus Using ChatGPT API

We have enhanced our system to better identify and sort complex Arabic phrases, using ChatGPT to create more Arabic examples. According to [14], ChatGPT produces high-quality data based on the quality of the prompts. Therefore, we have tested various scenarios and commands to enable ChatGPT to generate relevant Arabic VMWEs. The Table 1 illustrates the results of our data Enrichment efforts. It indicates a total of 197 VMWEs generated. Among these, there are 84 instances of LVC.full, 46 instances of LVC.cause, 47 instances of VID, and 20 instances of IAV.

4 Proposed Method

In this study, we focus on analyzing Arabic verbal expressions. We look at two processes: identification (deciding if an expression is a verbal poly-lexical one) and classification (categorizing expressions). Despite trying various machine learning models, our results were unsatisfactory. So, we proposed a more effective deep learning model combining ConvNet and BiLSTM. To implement this method, we followed these steps: data preparation, feature extraction, and finally, the identification and classification phase of VMWEs.

Apply test LVC.0 - [**N-ABS** : *Is the noun abstract ?*]

- **NO** ⇒ It is not an LVC, exit
- **YES or UNSURE** ⇒ Apply test LVC.1 - [**N-PRED** : *Is the noun predicative ?*]
  - **NO** ⇒ It is not an LVC, exit
  - **YES or UNSURE** ⇒ Apply test masdar - [**V - MASDAR** : Is the verb a verbal noun ?]
    - **YES or UNSURE** ⇒ Apply test LVC.2 - [**V-SUBJ-N-ARG** : *Is the subject of the verb a semantic argument of the noun ?*]
    - **YES or UNSURE** ⇒ Apply test LVC.3 - [**V-LIGHT** : *The verb only adds meaning expressed as morphological features ?*]
      - **NO** ⇒ It is not an LVC, exit
      - **YES** ⇒ Apply test LVC.4 - [**V-REDUC** : *Can a verbless NP-reduction refer to the same event/state ?*]
        - **NO** ⇒ It is not an LVC, exit
        - **YES** ⇒ It is an **LVC.full**
    - **NO** ⇒ Apply test LVC.5 - [**V-SUBJ-N-CAUSE** : *Is the subject of the verb the cause of the noun ?*]
      - **NO** ⇒ It is not an LVC, exit
      - **YES** => It is an **LVC.cause**

Fig. 1. LVC-specific decision tree after update.

4.1 Data Preparation

In data science, to get good results, we need good data. If the data is messy, we have to fix it first. This step is very important and is called data preparation. Thus, we combined our annotated dataset with the one from [7] and then split the final dataset into training, development, and test data using a method suggested by [12] in version 2.1 of PARSEME. Thereafter, we pre-processed the three datasets by removing punctuation, special characters like §, @, -, etc., and non-Arabic words.

4.2 Word Embeddings

We started our approach by creating vector representations of our data using Word2Vec embedding methods trained on our raw data and pre-trained AraVec

Table 1. Number of Arabic VMWEs generated by ChatGPT.

Phrases	LVC.full	LVC.cause	VID	IAV	Total
591	84	46	47	20	197

which has developed models based on various Arabic text sources, such as tweets and Wikipedia articles.

4.3 Identification and Classification

We initially tested a wide range of machine learning models to determine the most suitable ones for our problem. Based on their performance and relevance, we selected Decision Trees, Random Forest, SVM, Conv1D, and BiLSTM to solve the problem of identifying and classifying Arabic VMWEs. We presented two versions of our deep learning model with the same architecture except for the first layer (embedding layer), see Fig. 2. The key layers of the proposed model are:

- Embedding layer (Word2Vec or AraVec): this layer turns words into dense vectors, meaningfully representing them for the model. We implemented the Word2Vec (Skip-gram and CBOW) model on our raw data with a vector size equal to 100 and a window equal to 5.
- 1D Convolution layer: the Conv1D layer applies filters across the data's temporal dimension, extracting important local features from sequential data. The key parameters are; filters: set to 64, kernel size: set to 3 and activation function: uses ReLU to introduce non-linearity, enhancing the model's ability to learn complex relationships.
- Pooling layer: this global pooling layer reduces the dimensionality of features from the Conv1D layer with pool size equal to 2, retaining the most important information for easier learning and interpretation by the neural network's subsequent layers.
- Two Bidirectional LSTM layers: BiLSTM layer process both past and future information for data sequence modeling, unlike unidirectional LSTM layer. It helps detect long-distance dependencies and improves temporal relations representation. the second layer of BiLSTM with dropout equal to 0.4 improves the results.
- Global pooling layer: it synthesizes information from the BiLSTM layer, keeping only the highest value for each feature across the sequence.
- Dense output layer: a fully connected layer that produces the model's final predictions using features from previous layers.

To enrich word embedding, we used the word's lemma as additional information.

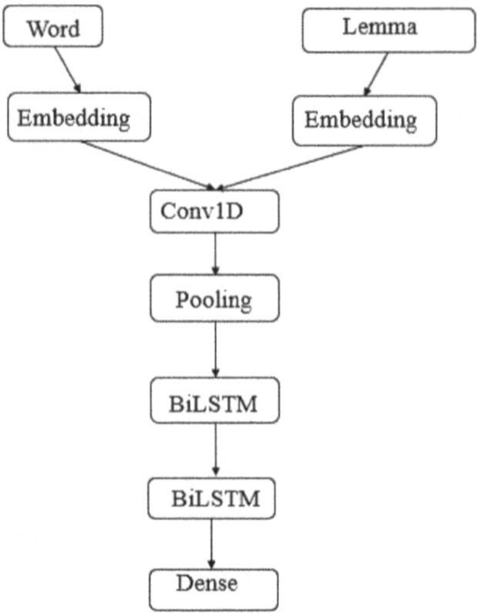

Fig. 2. Architecture of our model: ConvNet + BiLSTM.

5 Implementation and Evaluation

We compared the performance of various machine learning models on development data, then evaluated them on test data using Accuracy and F-score.

5.1 Results of Machine Learning Models

- Identification: after various training sessions on development data with different machine learning models, the best result was achieved using the Random Forest model. It yielded an accuracy of 84.65 and an F-score of 29.62, surpassing other tested models.
- Classification: the Random Forest model had the highest accuracy at 82.50. The Decision Tree model excelled in macro F-score 30.80. For micro F-score, Random Forest was superior 81.40.

5.2 Results of the Proposed Deep Learning Model

We set our model's parameters to use ReLU for convolution, sigmoid and softmax activations for identification and classification, respectively. We used binary and categorical crossentropy for loss functions, and the adam optimizer to adjust the model.

- Identification: to evaluate our proposed model on test data, we used the best-performing parameters from development data: Word2Vec for embedding type, 100 for embedding dimension, 200 epochs, and a batch size of 64. We also demonstrated the training data enrichment impact with ChatGPT-generated data on our model's performance, as shown in Table 2.

Table 2. Performance of the proposed model for identification on test data before and after data enrichment.

	Accuracy	F-micro	F-macro
Before	0.93	0.94	0.63
After	0.93	0.92	0.65

Table 3. Performance of the proposed model for classification on test data before and after data enrichment.

	Accuracy	F-micro	F-macro
Before	0.93	0.93	0.36
After	0.93	0.93	0.37

- Classification: to evaluate the performance of our model in classifying different VMWE categories on test data, we used the following parameters: Word2Vec for embedding type, 200 for embedding dimension, 100 for the number of epochs, and 64 for batch size. Table 3 shows the results before and after data enrichment. We observed a slight increase in F-macro.

6 Conclusion and Future Works

This research delves into the complexities of Arabic verbal multi-word expressions, which are intricate word sequences with distinct characteristics that cannot be deduced from their individual components. Recognizing the scarcity of studies in this linguistic area, particularly for the Arabic language, the project aimed to develop a systematic approach for the automatic identification and classification of VMWEs. The methodology involved the annotation of the Arabic-annotated-conll17 corpus, guided by the PARSEME network's standards and enhanced with examples generated via ChatGPT. Adjustments were made to the annotation framework to align with Arabic linguistic nuances, including the MASDAR form. The study introduced a machine learning and deep learning-based method, which was rigorously tested against various models to ascertain its efficacy in recognizing and categorizing Arabic VMWEs. The outcome is a significant contribution to the field, presenting the second corpus annotated with Arabic VMWEs and a robust method for their automated processing. It paves the way for further exploration in extending annotations to other EP types, refining the method with advanced language models, and broadening its application across different textual genres.

References

1. Al-Badrashiny, M., Hawwari, A., Ghoneim, M., Diab, M.: SAMER: a semi-automatically created lexical resource for Arabic verbal multiword expressions tokens paradigm and their morphosyntactic features. In: Proceedings of the 12th Workshop on Asian Language Resources (ALR12), pp. 113–122 (2016)
2. Avram, A., Barbu Mititelu, V., Cercel, D.C.: détection d'expressions multimots roumaines à l'aide d'un entraînement contradictoire multilingue et d'une inhibition latérale. In: Actes du 19e atelier sur les expressions multi-mots (MWE 2023), pp. 7–13 (2023)
3. Baldwin, T., Kim, S.N.: Multiword expressions, vol. 2, pp. 267–292 (2010)
4. Cohen, J.: A coefficient of agreement for nominal scales, vol. 20, pp. 37–46 (1960)
5. Fellbaum, C.: Wordnet: An Electronic Lexical Database. MIT Press, Cambridge (1998)
6. Ghoneim, M., Diab, M.: Multiword expressions in the context of statistical machine translation. In: Proceedings of the Sixth International Joint Conference on Natural Language Processing, pp. 1181–1187 (2013)
7. Hadj Mohamed, N., Khelil, C.B., Savary, A., Keskes, I., Antoine, J.Y., Belguith, L.H.: Annotation d'expressions polylexicales verbales en arabe: validation d'une procédure d'annotation multilingue. In: Traitement Automatique des Langues Naturelles, pp. 280–286 (2022)
8. Hadj Mohamed, N., Rassem, M., Han, Lifeng et Nénadic, G.: AlphaMWE-Arabic : Arabic edition de corpus parallèles multilingues avec annotations d'expressions multimots. In: Actes de la 14e Conférence internationale sur les progrès récents dans le traitement du langage naturel, pp. 448–457 (2023)
9. Hadj Mohamed, N., Savary, A., Ben Khelil, C., Antoine, J.Y., Keskes, I., Belguith Hadrich, L.: Lexicons gain the upper hand in Arabic MWE identification. In: The 2024 Joint International Conference on Computational Linguistics, Language Resources and Evaluation (LREC-COLING 2024), pp. 1–10 (2024)
10. Hajic, J., Smrz, O., Zemánek, P., Šnaidauf, J., Beška, E.: Prague arabic dependency treebank: development in data and tools. In: Proceedings of the NEMLAR International Conference on Arabic Language Resources and Tools, vol. 1 (2004)
11. Hawwari, A., Bar, K., Diab, M.: Building an Arabic multiword expressions repository. In: Proceedings of the ACL 2012 Joint Workshop on Statistical Parsing and Semantic Processing of Morphologically Rich Languages, pp. 24–29 (2012)
12. Ramisch, C., et al.: Édition 1.2 de la tâche partagée PARSEME sur l'identification semi-supervisée d'expressions verbales multimots. In: Actes de l'atelier conjoint sur les expressions multimots et les lexiques électroniques, pp. 107–118 (2020)
13. Rémy, F., Khabibullina, A., Demeester, T.: Détection d'expressions idiomatiques multimots dans la terminologie clinique à l'aide de l'apprentissage par représentation basé sur la définition. In: Actes du 19e atelier sur les expressions multi-mots (MWE 2023), pp. 73–80 (2023)
14. Ubani, S., Polat, S.O., Nielsen, R.: Zeroshotdataaug: generating and augmenting training data with chatgpt (2023)
15. Victoria, R., Koenraad, D.S., Gyri, S.L., Eduard, B., Agata, S., Petya, O.: Mwe dans les bancs d'arbres : de l'enquête aux lignes directrices (2016)
16. Walsh, A., Lynn, T., Foster, J.: A BERT's eye view: identification of Irish multiword expressions using pre-trained language models. In: Proceedings of the 18th Workshop on Multiword Expressions@ LREC2022, pp. 89–99 (2022)

Big Data, AI Impact, and Rule-Based Systems

A Design Pattern-Based Approach for Analyzing MapReduce Applications

Mayssa Bessifi[1(✉)], Ahlem Ben Younes[2], and Leila Ben Ayed[3]

[1] FSEGS, University of Sfax, Sfax, Tunisia
mayssabessifi@gmail.com
[2] ENSIT, University of Tunis, Tunis, Tunisia
[3] ENSI, University of la Manouba, Manouba, Tunisia

Abstract. MapReduce is a widespread programming paradigm used to develop scalable parallel applications in various fields. Despite the simplicity of the MapReduce model, using it to effectively solve real-life problems can be challenging for developers and designers. Furthermore, ensuring the reliability and correct requirements of any system is fundamental at the early stage of development, as late corrections are estimated to be more than 200 times greater than corrections during requirement engineering. Therefore, a systematic approach is required to assist designers in developing reliable MapReduce applications. In this paper, we present a design pattern-based approach to the specification and formal verification of MapReduce applications using the BPMN notation and the Event B method.

Keywords: MapReduce · Design pattern · Fromal verification · BPMN · Event B

1 Introduction

Nowadays, zettabytes of data are generated every day. This flood of data is accessible in all formats, i.e. structured, unstructured and semi-structured. The large volume, the variety and the rising speed of generation of this huge amount of data give rise to the term Big Data [6]. In order to store, process and analyze such data new technologies and storage mechanism are required. One of the most successful paradigms is MapReduce. It has been proposed by google and emerged as the core of several Big Data frameworks such as Hadoop, spark and Storm, etc.

Hadoop [6] is developed by Apache Software Foundation as an open-source framework. It is the leader implementation of the MapReduce paradigm to solve Big Data problems. Hadoop cluster is mainly composed by two layers: a Hadoop distributed file system called HDFS [6] for storing data and Hadoop MapReduce paradigm for parallel processing. The Hadoop system is highly scalable; i.e., it can combine the computing and the storage power of thousands of computers to stock and process the massive data in a distributed way.

L. H. Belguith and K. Shaalan (Eds.): LPKM 2024, LNNS 1303, pp. 255–260, 2025.
https://doi.org/10.1007/978-3-031-85067-7_25

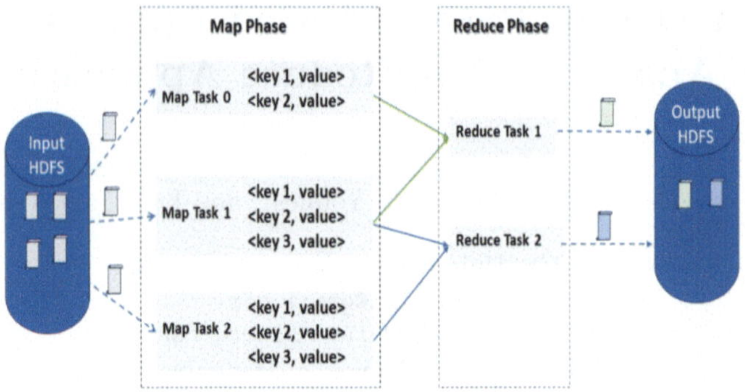

Fig. 1. Procedure of a MapReduce job.

A MapReduce workflow processes terabytes of data through MapReduce jobs connected through producer-consumer relationships. Each MapReduce job, as depicted in Fig. 1, is created in two phases: map and reduce. The map phase, composed of parallel map tasks, receives raw data as input and generates a set of key-value pairs. Any given "key" can be used by several pairs. Then, the reducer, which can be one or many parallel reduce tasks, processes pairs from mappers and aggregates values with the same key. The output of the reducer can be written in HDFS or send to another Mapper, in the case of MapReduce workflow.

Although the Hadoop MapReduce framework is easy to grasp, the development of complex MapReduce workflow can be a tedious task and require the collaboration of many developers. The intervention of different developers raises the possibility of mistakes and bugs, in the map and reduce programs, that can interrupt the MapReduce execution or produce inaccurate output. Also, as MapReduce applications are like every other application, from the design phase a set of requirements are envisaged and need to be verified from an early stage to reduce the possibilities of failure or inconsistency. Therefore, we propose in this paper an approach driven by MapReduce design patterns, for the modeling and formal verification of MapReduce workflow using both the standard BPMN2 [4] and the Event B method [1].

The objective of the proposed approach is twofold. First, the approach is developed to help designers to easily design their MapReduce workflow using the graphical tool of a well-known and rich standard which is BPMN, and based on a set of predefined MapReduce patterns. Second, by automatically transformed the MapReduce workflow to a formal notation, Event B, the designer can make further analysis and consequently detect any errors at earlier stage.

The rest of the paper is organized as follows: Sect. 2 presents the proposed approach. In Sect. 3, we validate the proposed approach by a case study. In Sect. 4, we present the related work. The last section concludes the paper.

2 Proposed Approach

This section describes the proposed approach of the specification and verification of MapReduce workflow. As depicted in Fig. 2, the approach starts by the modelling of the MapReduce workflow as a sequence of MapReduce patterns using BPMN2.0 notations. After the graphical modelling, a sequence of transformations implemented using Kermeta [3] language are automatically executed; First the generated BPMN2 model is transformed into an XML Metadata Interchange (XMI) file conforming to the proposed BPMN meta-model. Second a model-to-model transformation is executed to generate an Event B model conforms to the Event B meta-model. Finally, the generated Event B model is transformed into a RODIN project. After transformation phase, the RODIN platform is used to validate and verify the RODIN project. A set of proof obligations is generated. If a proof obligation is not proven, then the designer uses the trace file generated from the transformation phase to locate the error in the BPMN specification and fix it.

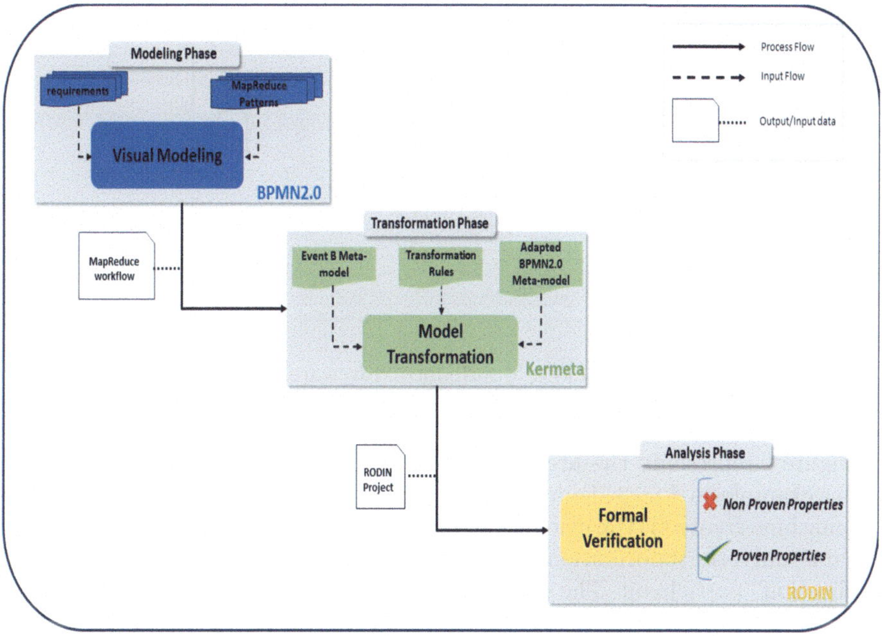

Fig. 2. Overview of the proposed approach.

3 Case Study

The "Fireware" application is the solution to the "Denial-of-Service" DOS attack problem. Figure 3 illustrates the MapReduce workflow of the Fireware application, composed by three patterns: occurrence calculation pattern, sort pattern and filter pattern. The first pattern "Occurrence Calculation" processes a very large log file that contains all the IP addresses that have accessed a given URL to calculate the number of hits for each IP address. The list of key-value pairs produced by this pattern will be sorted in an array by the following pattern "Sort". The last pattern "Filter" will take the sorted array and apply a filter (maximum number of accesses allowed) on all key (IP address), value (number of accesses) pairs.

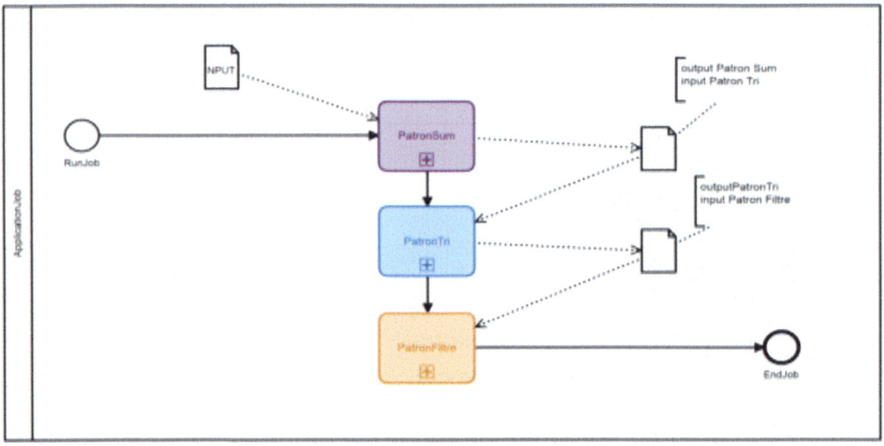

Fig. 3. Graphical modelling of a MapReduce "Fireware" application with BPMN

Figure 4 represents the architecture of the Event B project corresponding to the example in Fig. 3. Four machines linked by refinement are created: the first machine created is the machine named "MACHINE_APPLICATION" which contains only the information concerning the control flows of the "Fireware Application" participant. The second machine, "MachinePatron1", preserves the previous machine and contains the information concerning the control flows of the first sub-Process "Pattern calculation of occurrence" and its internal functioning. The third "MachinePatron2" and the fourth machine "Machine Patron3" contain information concerning the other two sub-processes "Sorting Pattern" and "Filtering Pattern". Each machine preserves its previous one.

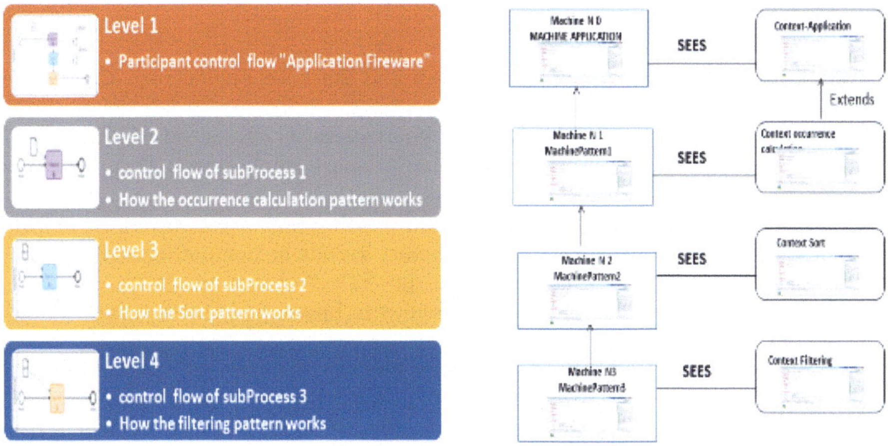

Fig. 4. The architecture of the RODIN project of the "firewire" application

4 Related Works

In recent years, the validation of the correctness of big data applications, using formal methods, has gained an increased interest. Research like [5,9], adopt a run time verification approach, at code level, based on the MSVL (Modelling, Simulation, and Verification Language). Others works like [2,8], use Petri net modelling to verify different kinds of properties, such as rationality of MapReduce applications and performance analysis of Spark applications. Quattrocchi et al. in [2,7], propose a formal definition of Spark programs based on the CLTLoc [7] logic for the validation of the required resources given a deadline.

All the above cited works are based on the model checking technique which suffer from the explosion number of states, compared to the theorem proving technique used by the Event B method.

5 Conclusion

In this paper, a model-driven approach is presented to ensure the reliability of MapReduce applications. The user can easily use the Camunda graphical tool of BPMN to construct his MapReduce workflow by using a chosen set of MapReduce design patterns. The graphical model is then transformed automatically to an Event B project composed by a set of contexts and machines linked by refinement. The use of refinement in the proposed approach improve the readability and enhance the provability of the generated formal specification.

In the future, we will provide more real-world examples of MapReduce applications and we will test the feasibity of the approach with real-time processing frameworks such as Spark [5].

References

1. Abrial, J.R.: Modeling in Event-B: System and Software Engineering. Cambridge University Press, Cambridge (2010)
2. Baresi, L., Bersani, M.M., Marconi, F., Quattrocchi, G., Rossi, M.: Using formal verification to evaluate the execution time of spark applications. Form. Aspects Comput. **32**, 33–70 (2020)
3. Bessifi, M., Younes, A.B., Ayed, L.B.: Bpmn2eventb supporting transformation from bpmn2. 0 to event b using kermeta. In: Smart Trends in Computing and Communications: Proceedings of SmartCom 2021, pp. 247–255. Springer (2022)
4. Correia, A., e Abreu, F.B.: Adding preciseness to bpmn models. Procedia Technol. **5**, 407–417 (2012)
5. Fan, K., Wang, M.: Dag-based formal modeling of spark applications with MSVL. Information **14**(12), 658 (2023)
6. Ghazi, M.R., Gangodkar, D.: Hadoop, MapReduce and HDFS: a developers perspective. Procedia Comput. Sci. **48**, 45–50 (2015)
7. Marconi, F., Quattrocchi, G., Baresi, L., Bersani, M.M., Rossi, M.: On the timed analysis of big-data applications. In: NASA Formal Methods: 10th International Symposium, NFM 2018, Newport News, VA, USA, 17–19 April 2018, Proceedings 10, pp. 315–332. Springer (2018)
8. Zeliu, D., Deke, G., Xi, C., Jin, C.: Mapreduce rationality verification based on object petri net. J. Syst. Eng. Electron. **30**(5), 861–874 (2019)
9. Zhang, N., Wang, M., Duan, Z., Tian, C.: Verifying properties of mapreduce-based big data processing. IEEE Trans. Reliab. **71**(1), 321–338 (2020)

Analyzing the Impact of Big Data in Mental Health

Sonda Rekik[1(✉)], Mourad Ellouze[2], and Lamia Hadrich Belguith[1]

[1] ANLP Group MIRACL Laboratory, FSEGS, University of Sfax, Sfax, Tunisia
sandou.rekik@yahoo.com, lamia.belguith@fsegs.usf.tn
[2] ICL, Junia, Catholic University of Lille, LITL, 59000 Lille, France
mourad.ellouze@univ-catholille.fr

Abstract. This paper provides an overview of the application of big data in psychology, exploring its relevance, potential applications, and associated challenges. It begins with a presentation of the concept of big data, including its architectures and technologies, emphasizing their importance in psychological research. After that, the paper delves into a set of existing data in the literature, taking into account high-level criteria such as volume, variety, and velocity. These criteria are expanded to include additional standards like ethical considerations, acquisition velocity, and value. Following that, the paper explores various studies about the analysis and processing of massive datasets in psychology, with an emphasis on study objectives and techniques employed, such as data warehouses and parallel processing. Finally, the paper concludes with a discussion on the criticisms of the use of big data in psychology and the significance of integrating other techniques such as artificial intelligence (AI) and natural language processing (NLP). Overall, the present study highlights the potential of big data in advancing psychological research, while acknowledging the need to address associated challenges and adopt complementary methodologies.

Keywords: Big Data · Psychology · Data quality · Optimized and intelligent processing · Mental Health

1 Introduction

The term "Big Data" has surged in popularity, as indicated by its doubling frequency in common search engines annually [15]. Wearable devices and mobile apps have facilitated the collection of extensive behavioral data for real-world psychological research. For instance, a survey conducted by the Association for Psychological Science (APS) found that over 70% of psychologists use or plan to use mobile apps for data collection purposes. In fact, monitoring patient conditions requires real-time observation, as the psychological state is particularly delicate and susceptible to influence by various circumstances. At this level, systems based on mobile applications or electronic components enable real-time data generation. However, managing these resources poses challenges in terms of

L. H. Belguith and K. Shaalan (Eds.): LPKM 2024, LNNS 1303, pp. 261–270, 2025.
https://doi.org/10.1007/978-3-031-85067-7_26

storage and processing. To address these issues, many researchers have adopted Big Data technologies such as Apache Spark and AI techniques like machine learning. For example, to establish a remote instant health prediction system for depression and anxiety in pregnant women based on the Apache Spark big data processing engine, researchers of [12] used the Naive Bayes method to train a prediction model on the characteristics of mental health crisis conversations. This approach resulted in accurate classification of user behaviors and demographics during mental health crisis conversations, with an accuracy of 92.95%. Advanced analytics like machine learning can help identify patterns in large datasets, aiding in personalized care [17]. As demonstrated, Big Data and AI can play a crucial role in uncovering hidden trends, personalizing treatments, and providing real-time support to enhance outcomes. These functionalities can address the requirement for comprehensive knowledge and full interventions in the fields of mental health and psychology. For that, this paper seeks to give an overview of the current applications of Big Data in mental health, aiming to offer a comprehensive understanding and explore potential research avenues.

The paper comprises five main sections. Section 2 defines the concept of big data and explores its relevance in psychology, discussing various architectures and technologies associated with big data. In Sect. 3, the application of big data in psychology is examined, focusing on data collection criteria such as volume, variety, and velocity, as well as data quality considerations. This section is further divided into two subsections: data collection and data analysis techniques, including the use of data warehouses, parallel processing, and storage of data from multiple sources. Section 4 critiques the application of big data in psychology, emphasizing the need to integrate other techniques such as artificial intelligence, natural language processing, and the Internet of Things.

2 Exploring the Intersection of Big Data and Psychology

The incorporation of big data technology into psychology creates new pathways for researching social dynamics, brain processes, and behavior, emphasizing the significance of thorough understanding of architectures and technologies. In essence, the integration of Big Data holds promise for revolutionizing personalized medicine by harnessing the latest technological advancements in data collection, management, and analysis within the medical and healthcare systems. Figure 1 illustrates the increasing interest and research activity in the intersection of big data and healthcare over time, as depicted by the number of publications associated with big data in healthcare plotted by year.

Leveraging these advancements, researchers can address real-time monitoring needs, especially for individuals with mental disorders. This involves developing systems capable of managing and analyzing large datasets in real time to provide instant insights and tailored interventions. In fact, among key aspect of this approach is parallel processing, which allows tasks to be divided across multiple computing nodes, thereby speeding up processing times and enabling rapid

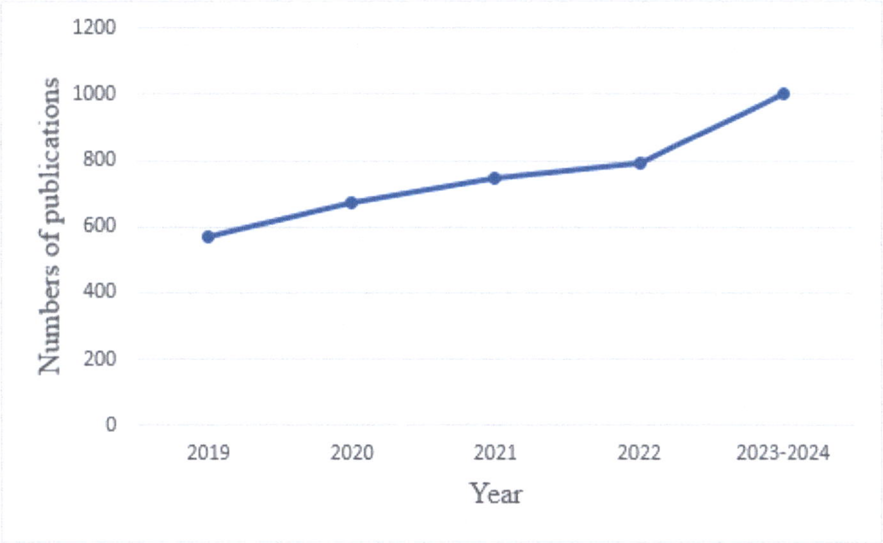

Fig. 1. Publications associated with big data in healthcare are plotted by year according to PubMed

analysis of massive datasets. Another key aspect is distributed architectures[1], such as Hadoop and Spark, which utilize networked computers to collaborate on shared tasks, distributing computational work across multiple machines instead of a central server. These keys are essential for processing large datasets in psychology. In this context, the study of [12] was dedicated to identifying anxiety and depression among pregnant women by employing performance-optimized algorithms for efficient feature extraction. Its primary objective is to develop a real-time remote health prediction system utilizing the Apache Spark Big Data processing engine alongside machine learning models. This system collects data from pregnant women to forecast their health conditions, employing the most effective machine learning algorithm available. This study underscores the pressing importance of addressing mental health challenges during the perinatal period by highlighting the critical need for timely diagnosis and treatment interventions. This process was achieved by pioneering a hybrid big data analytics platform tailored to swiftly identify and support pregnant women grappling with anxiety and/or depression. Moreover, the study of [11] outlines a technical vision for determining a person's mental state by analyzing behavioral patterns, emphasizing the benefits over traditional psychometrics. It suggests using smartphones to track user behavior, enabling more efficient study of human minds. The research of this study aims to dissect human conduct using large-scale data to estimate the severity of depression and internet addiction. By recording daily activities via smartphones or computers and storing data in Big Data Storage,

[1] https://estuary.dev/distributed-architecture/.

this approach offers a less time-consuming and cheaper alternative to traditional methods for both specialists and patients. The main challenge lies in dealing with streaming data, addressed using Apache Spark, an open-source real-time platform for large data sets.

On the other hand, parallel processing facilitates the division of tasks across multiple computing nodes, hastening the analysis of massive datasets in psychology to identify significant patterns and trends. In the study discussed in [8], the transformation process involves converting the format of the available data into a format suitable for processing. This conversion is necessary to ensure that big data can be efficiently processed. The proposed architecture adopts the Hadoop Distributed File System (HDFS) as a distributed storage mechanism for storing large datasets. HDFS is capable of handling vast amounts of data and making it easy to access. In addition, the data is distributed across multiple nodes, allowing for parallel processing to enhance efficiency. Furthermore, in the study discussed in [11], the authors highlight the importance of communication between nodes in execution graphs. In their framework, communication is facilitated by queues of messages managed by Apache Kafka. They emphasize the flexibility of their approach, where nodes can be replicated to handle high workloads, thus preventing bottlenecks in the execution graph. In contrast to batch processing solutions that allocate hardware resources to the entire execution graph, their approach, called CATENAE, allocates resources to node instances as needed.

3 Importance of Big Data in Mental Health

Big data in psychology offers new insights into human behavior with technological advancements, but also raises ethical and privacy issues. Nevertheless, it has revolutionized psychological research and treatment approaches.

3.1 Data Collection

In the age of big data, psychology utilizes vast datasets enabled by technological advancements to understand human behavior, offering new research and intervention opportunities. Analyzing large datasets can help create new models and therapies to improve psychological well-being, as depicted in Table 1. The data in Table 1 serve multiple objectives in mental health research, drawing from crisis hotlines, genetic studies, social media, and clinical subreddits[2]. The study of [2] examines crisis conversation trends, [5] investigates genetic influences, [6] analyzes social media sentiment, [9] studies clinical subreddit trends, and [18] explores online discussions on epilepsy and suicide. Through these efforts, we improve our understanding of mental health issues and guide targeted interventions.

In psychology, data quality is crucial for reliable research. Authors like [2] stress the importance of relevance and acquisition speed, while [5] highlights ethical considerations. Transparency is key, as emphasized by [6], and [9] emphasizes

[2] https://www.reddit.com/r/ListOfSubreddits/wiki/listofsubreddits/.

Table 1. Comparison of Data Sources for Mental Health Analysis.

Data	Criterion				Justification details
	Volume	Variety	Velocity	Veracity	
[2]	X	X		X	271,445 conversations consisting of 10,809/178 messages from various sources
[5]	X	X		X	271,445 conversations/10,809,178 messages/22 patient samples
[6]	X				Tweets related to a period of 24 weeks
[9]	X	X		X	515,374 publications downloaded from various clinical subreddits
[18]	X	X	X		515,374 publications downloaded from various sources including conversations
[20–22]	X		X	X	Large volume of data collected from Twitter in real-time validated by two experts
[13]		X		X	Data obtained from three sources was validated by a questionnaire.
[12]				X	Data obtained from hospital clinical records
[23]				X	Data obtained from testimonial groups
[24]				X	Data validated by two experts

reliability and precision. Innovative methods in [20, 21], and [22] address psychological disorders on social media in real-time, facilitating prompt identification and assistance. This aligns with [12] emphasis on diverse data sources for broader research, and [13] view on integrating multiple data sources for a holistic view of mental health. Moreover, veracity is an important criterion in psychological analysis to ensure the knowledge value. For that reason, different works, such as [12, 23, 24], validate the data obtained through expert annotation, recognized questionnaires, or reliable sources such as hospitals.

3.2 Data Storage

In this section, we explore data management and storage options in psychology, emphasizing important factors and practical methods.

NoSQL Databases. NoSQL databases like MongoDB are ideal for storing unstructured or semi-structured data prevalent in psychology, offering flexibility for diverse data types such as interview transcripts or biometric sensor data. The study of [19] emphasizes the urgency of understanding the impacts of mental health disorders amid their global prevalence and increasing complexity. To manage large datasets effectively, the researches of this study propose leveraging big data analytics tools like MongoDB and data mining techniques such as genetic algorithms. Similarly, [10] aims to predict mental conditions early; for that, it takes advantage of the MongoDB database for quick access, enhancing swift intervention in mental health crises.

Massively Parallel Storage. Massively parallel storage technologies like Hadoop Distributed File System (HDFS) and Spark are vital for storing and accessing vast amounts of data from psychological studies, ensuring reliability and efficient analysis access. In response to the surge in depression cases during the coronavirus pandemic, [13] proposes big data analytics on social networks for real-time depression detection, leveraging demographic characteristics and sentiment analysis from tweets. Throughout their workflow, it applies Big Data techniques, utilizing Hadoop clusters and HDFS for data storage and parallel processing. Similarly, [11] highlights the crucial role of HDFS within the HBase system, ensuring fault tolerance and efficient data replication across multiple nodes.

Data Warehouses. Data warehouses are used to store vast sets of data from multiple sources, facilitating access and analysis of data on a large scale. Similarly in psychology the data warehouse concept can be used to include clinical data for aggregating patient information. In this context, the researchers of [1] take advantages of diverse big data resources, influencing the scope of mental health inquiries, particularly in terms of data accessibility. These resources range from comprehensive clinical data sourced from electronic health records to aggregated service-level data like those from the European Observatory of Health Systems and Policies. Some studies also combine insights from various databases, such as the Psychiatric Genomics Consortium. In the same context, in [4], authors delve into NLP and machine learning methods to extract clinical notes. They convert unstructured text into vectors and utilize this data for tasks like medical concept extraction, disease classification, and phenotype extraction. Moreover, clinical data mining can be used to answer questions that cannot be addressed with randomized clinical trials [7].

Storage of Data from Multiple Sources. The efficiency of storing data from multiple sources in a single system allows for easier integration and holistic analysis, leading to a deeper understanding of human behavior and mental processes in the field of psychology. The following examples [2,5,6] demonstrate how data from different sources, such as anonymized conversations from mental health helplines, tweets, or publications from online clinical forums, can be integrated into a unified storage system for analysis in the field of psychology [9]. The consolidation of data provides the opportunity to conduct more in-depth analyses and uncover more comprehensive insights into various aspects of human psychology.

3.3 Data Analysis

In this section, we discuss the various approaches used to analyze and process massive datasets in psychology, focusing on study objectives and applied techniques. Real-time analytics technologies like Apache Kafka and Apache Storm play a crucial role in psychology for processing and analyzing data in real-time, especially for monitoring physiological responses or social media interactions. In this context, [14] emphasizes the significance of smartphones with embedded sensors in transforming healthcare delivery through platforms like RADAR-base, which utilizes Confluent and Apache Kafka for remote data collection. This platform ensures secure data transmission and scalability, currently utilized in studies like RADAR-CNS for conditions such as Multiple Sclerosis and depression. Similarly, [13] proposes a scalable platform for real-time social media data processing, facilitating efficient public health surveillance tasks. Their platform integrates various elements like content extraction crawlers and classifiers, demonstrated in a depression monitoring use case using Apache Kafka on a three-node cluster.

3.4 Machine Learning and Artificial Intelligence

In the realm of Big Data, Machine Learning (ML) and Artificial Intelligence (AI) techniques are pivotal, particularly in healthcare. Hybrid ML methods have been employed to identify pregnant women at risk for anxiety and depression, optimizing Big Data architecture for efficient processing of social media data [12]. Batch processing of Big Data was utilized to create a depression detection model; in fact, real-time data stream processing was crucial for immediate identification to mitigate depression-related suicide risks [10]. The proliferation of digital support tools, such as Shout, a mental health text messaging service in the UK, lacks systematic evaluation and understanding of user needs. For that, leveraging a large dataset of anonymized text message conversations, researchers aim to gain insights into mental health needs and assess the impact of crisis support interventions [2]. Similarly, in the UK, NLP platforms were utilized to swiftly mine and analyze health-related insights from tweets during the pandemic, providing real-time intelligence for healthcare decision-makers [6].

4 Discussion and Challenges

In the realm of mental health research, the quality of big data is influenced by several factors like accuracy and timeliness [1]. In general, research interviews and administrative data have a certain limitations, and understanding psychopathology is complex due to brain development and environmental factors [1,3]. In addition, some concerns exist about the reliability and transparency of technologies, such as NLP platforms and Twitter data [6]. For example, clinical subreddit posts do not guarantee definitive diagnoses, and predictive models for future mental disorders are limited, raising concerns about biases in big data collection [9]. For that integrating theory-driven probe development with big data and machine learning could enhance mental illness treatment [16].

Significant analytical challenges arise in the field of big data-based mental health research due to the quantity, diversity, and velocity of the data. As emphasized in [7], skewed representation, such as solely focusing on Twitter users, highlights the importance of avoiding biased conclusions. In addition, biases within data sources, such as electronic medical records and claims, can complicate research, potentially leading to erroneous correlations when applying traditional statistical models. Efforts to develop novel ways to solve these issues continue, although caution is advised to assure reliable causal inference. For example, Twitter can be considered as an intelligence source hinges on considering user demographics; however, challenges in retrospective tweet searchability and platform methodology disclosure require careful consideration during interpretation.

In light of these challenges, integrating AI, NLP, and internet Of Thing (IoT) techniques is vital to ensure data quality and meet the specific needs of psychology. Indeed, it makes it possible to exploit various data sources, including both textual data from conversations or testimonials, and sensor data collected from connected devices. This diversity of data sources helps meet the data variety requirements of Big Data. This enriched approach allows us to better understand psychological dynamics and provide more reliable and relevant results for mental health research.

5 Conclusion

For researchers, big data in psychology offers opportunities and challenges. In this paper, we explore various aspects of big data applications in psychology, highlighting its potential to enhance understanding of human behavior and enable thorough research results. In our study, we considered the different stages of the psychological data analysis process, starting with the quality of the data to be analyzed, moving to data storage and processing, and finishing with decision making. In our future work, we will develop an innovative approach based on advanced techniques to provide monitoring and assistance to people with psychological problems. We will implement methods for detecting and assessing the risks of psychological illnesses, as well as an in-depth analysis of the situation of each individual. In addition, we will focus on using generative artificial intelligence to provide personalized psychological assistance, generating tailored responses to various scenarios and providing recommendations for mental well-being. Finally, we will seek to improve the interaction to offer advice and personalized monitoring in an empathetic way, thus meeting the specific needs of users in terms of psychological support. To achieve that, we plan to use various types of big data, which will enable us to develop more comprehensive and accurate analytical models to support mental health.

References

1. Stewart, R., Davis, K.: 'Big data' in mental health research: current status and emerging possibilities. Soc. Psychiatry Psychiatr. Epidemiol. **51**(8), 1–13 (2016). https://doi.org/10.1007/s00127-016-1266-8
2. Liu, Z., Peach, R., Lawrance, E.L., Barahona, M., Ungless, M.A., Noble, A.: Listening to mental health crisis needs at scale: using natural language processing to understand and evaluate a mental health crisis text messaging service. Front. Digit. Health **3**, 1–10 (2021). https://doi.org/10.3389/fdgth.2021.779091
3. Rosenfeld, A., et al.: Big Data Analytics and AI in Mental Healthcare. Bar-Ilan University, Israel; McGill University, Canada; Aifred Health, 29 March 2019
4. Yu, Y., Li, M., Liu, L., Li, Y., Wang, J.: Clinical big data and deep learning: applications, challenges, and future outlooks. Big Data Min. Anal. **2**(4), 288–305 (2019). https://doi.org/10.26599/BDMA.2019.9020007
5. Cao, H., Meyer-Lindenberg, A., Schwarz, E.: Comparative Evaluation of Machine Learning Strategies for Analyzing Big Data in Psychiatry. Department of Psychiatry and Psychotherapy, Central Institute of Mental Health, Medical Faculty Mannheim, Heidelberg University, 68159 Mannheim, Germany. Published: 29 October 2018
6. Marshall, C., Lanyi, K., Green, R., Wilkins, G., Pearson, F., Craig, D.: Using natural language processing to explore mental health insights from UK tweets during the COVID-19 pandemic: infodemiology study. JMIR infodemiology, 28 July 2021. https://doi.org/10.2196/32449
7. Monteith, S., Glenn, T., Geddes, J., Bauer, M.: Big data are coming to psychiatry: a general introduction. Int. J. Bipolar Disord. (2015). https://doi.org/10.1186/s40345-015-0038-9
8. Tariq, M.U., Babar, M., Poulin, M., Khattak, A.S., Alshehri, M.D., Kaleem, S.: Human behavior analysis using intelligent big data analytics. Front. Psychol. **12**, 686610 (2021). https://doi.org/10.3389/fpsyg.2021.686610
9. Thorstad, R., Wolff, P.: Predicting future mental illness from social media: a big-data approach. Behav. Res. Methods **51**, 1586–1600 (2019). https://doi.org/10.3758/s13428-019-01235-z
10. Ul haq, A.K., Khattak, A., Jamil, N., Naeem, M.A., Mirza, F.: Data analytics in mental healthcare. Sci. Program. **2020**, paperID 2024160, 9. https://doi.org/10.1155/2020/2024160
11. Martínez-Castaño, R., Pichel, J.C., Losada, D.E.: A big data platform for real time analysis of signs of depression in social media. Int. J. Environ. Res. Public Health (IJERPH) **17**(13), 4752 (2020). https://doi.org/10.3390/ijerph17134752
12. Oğur, N.B., Çeken, C., Oğur, Y.S., Yuvacı, H.U., Yazıcı, A.B., Yazıcı, E.: Development of an artificial intelligence-supported hybrid data management platform for monitoring depression and anxiety symptoms in the perinatal period: pilot-scale study. IEEE Access **PP**(99), 1 (2023). https://doi.org/10.1109/ACCESS.2023.3262467
13. Angskun, J., Tipprasert, S., Angskun, T.: Big data analytics on social networks for real-time depression detection. J. Big Data **9**, 69 (2022). https://doi.org/10.1186/s40537-022-00622-2
14. Ranjan, Y., Böttcher, S., Kerz, M., Dobson, R.J.B., Rashid, Z., Folarin, A.A.: Poster: RADAR-base: A Novel Open Source m-Health Platform. Institute of Psychiatry, Psychology , Neuroscience, King's College London, London, SE5 8AF, UK

15. Andreu-Perez, J., Poon, C.C.Y., Merrifield, R.D., Wong, S.T.C., Yang, G.Z.: Big data for health. IEEE J. Biomed. Health Inform. **19**(4), 1193 (2015)
16. Rutledge, R.B., Chekroud, A.M., Huys, Q.J.M.: Machine learning and big data in psychiatry: toward clinical applications. Curr. Opin. Neurobiol. **55**, 152–159 (2019). https://doi.org/10.1016/j.conb.2019.02.006
17. Duffy, A., Faurholt-Jepsen, M., Ostacher, M.: Using big data to advance mental health research. Evid.-Based Ment. Health **23**(1), ebmental-2020-300143 (2020). https://doi.org/10.1136/ebmental-2020-300143
18. Falcone, T., et al.: Digital conversations about suicide among teenagers and adults with epilepsy: a big-data, machine learning analysis. Epilepsia. Accessed 4 March 2020, 25 March 2020, 26 March 2020. https://doi.org/10.1111/epi.16507
19. Dhaka, P., Johari, R.: Big data application: study and archival of mental health data, using MongoDB. Presented at the 2016 International Conference on Electrical, Electronics, and Optimization Techniques (ICEEOT), March 2016. https://doi.org/10.1109/ICEEOT.2016.7755300
20. Ellouze, M., Hadrich Belguith, L.: A hybrid approach for the detection and monitoring of people having personality disorders on social networks. Soc. Netw. Anal. Min. **12**(1), 67 (2022)
21. Ellouze, M., Hadrich Belguith, L.: Semantic analysis based on ontology and deep learning for a chatbot to assist persons with personality disorders on Twitter. Behav. Inf. Technol. 1–20 (2023)
22. Ellouze, M., Hadrich Belguith, L.: A deep learning architecture based on advanced textual language models for detecting disease through its symptoms associated with a reinforcement learning algorithm. In: International Conference on Software Technologies, pp. 207–229. Springer, Switzerland, Cham (2022)
23. Ellouze, M., Belguith, L.H.: Approach based on Bayesian network and ontology for identifying factors impacting the states of people with psychological problems from data on social media. In: International Conference on Model and Data Engineering, pp. 128–141. Springer, Switzerland, Cham (2023)
24. Ellouze, M., Mechti, S., Belguith, L.H.: A hybrid approach based on linguistic analysis and fuzzy logic to ensure the surveillance of people having paranoid personality disorder towards COVID-19 on social media. Int. J. Gen. Syst. **52**(3), 251–274 (2023)

The Impact of AI on Knowledge Management

Hassene Ben El Arem[1(✉)] and Anis Jarboui[2]

[1] Department of Management, University of Sfax, Sfax, Tunisia
hassene.elarem@gmail.com
[2] Department of Finance, University of Sfax, Sfax, Tunisia

Abstract. This article theoretically studies the theoretical influence of artificial intelligence (AI) on knowledge management within companies. By combining essential concepts of knowledge management with advances in artificial intelligence, this study examines how new technologies are changing knowledge management methods and impacting human resource dynamics. Internal and external knowledge, as well as new challenges and opportunities posed by AI, are comprehensively examined to provide an in-depth view of the implications of AI for strategic knowledge management.

Keywords: Artificial intelligence · knowledge management · internal knowledge · external knowledge · human resources · organizational transformation

1 Introduction

In a context where economic competitiveness is essential, knowledge management occupies a central position in the strategy of companies and organizations. Previous work, such as that of Frenay and Maroy (2003) and De Ketele (1996), highlights the crucial importance of internal and external knowledge in organizational development and performance. Internal knowledge, acquired through experience and learning within the organization, as well as external knowledge, derived from interactions with the external environment such as partnerships and collaborations, are essential drivers of value creation and the sustainability of businesses.

However, the advent of artificial intelligence (AI) brings a radical upheaval in the management and valorization of knowledge within organizations. While traditional knowledge management approaches focus on the assessment and development of human knowledge, AI integration introduces new paradigms and challenges [4,9,10,13,15,17]. AI, defined as the ability of machines to simulate human intelligence to accomplish various tasks, is revolutionizing knowledge management processes by providing innovative tools and techniques for analyzing, forecasting, and optimizing organizational knowledge.

L. H. Belguith and K. Shaalan (Eds.): LPKM 2024, LNNS 1303, pp. 271–278, 2025.
https://doi.org/10.1007/978-3-031-85067-7_27

This introduction aims to explore the theoretical impact of AI on knowledge management, examining how technological advances in this area are transforming knowledge management practices and influencing the dynamics of human resources within companies and organizations. By integrating key concepts of knowledge management with advances in AI, this study aims to provide in-depth insight into the opportunities and challenges of using AI in strategic knowledge management.

2 Theoretical Foundation and Research Hypothesis

In recent years, artificial intelligence has received a significant amount of theoretical and empirical attention in an effort to understand its effects on business. In the same vein, we can study the concept of artificial intelligence as "the simulation of human cognitive processes by machines, particularly computers, in order to perform tasks that generally require human intelligence" [12]. Likewise, Artificial Intelligence encompasses "a set of technologies that enable machines to perform tasks that traditionally require human intelligence, such as speech recognition, computer vision, decision-making, and machine learning" [7].

Artificial intelligence is one of the subjects that has attracted the attention of researchers, especially in the field of management. Indeed, artificial intelligence refers to the ability of machines to simulate human intelligence in order to accomplish various tasks [10]. This definition encompasses a wide range of techniques and approaches, ranging from machine learning algorithms to deep neural networks. The work of [13] highlights that AI allows machines to analyse large amounts of data, identify complex patterns, and make autonomous decisions based on that data. Similarly, [5] explain that AI systems are able to learn and adapt from the data they receive, allowing them to improve their performance over time without intervention direct human. By addition, [9] points out that the application areas of AI are numerous and varied, ranging from voice and facial recognition to machine translation, autonomous driving and the prediction of financial trends. On the other hand, [15] note that AI is also used in more complex applications such as knowledge management, where it can help identify knowledge gaps, recommend training programs personalized training and predict future knowledge needs. In the same vein, [17] highlight that AI has significant benefits for organizations, including greater operational efficiency, faster decision-making, and deeper data analyses. Combining these different perspectives, it can be concluded that artificial intelligence represents a rapidly evolving field that offers vast opportunities to improve business performance and capabilities.

For knowledge management, it is "the process by which an organization collects, organizes, stores and shares internal and external knowledge in order to promote organizational learning, innovation and informed decision making" [1].

Knowledge management can also be defined as "the systematic process of creating, acquiring, capturing, disseminating and using knowledge within an organization to improve its performance and competitiveness" [8].

Several researchers address the subject of knowledge management, such as:

According to [6], knowledge management encompasses both tacit knowledge, which resides in the minds of individuals, and explicit knowledge, which is formalized and documented. These authors developed the SECI model which illustrates the conversion between tacit and explicit knowledge.

The work of [16] highlights the importance of knowledge management to improve decision-making, promote innovation and stimulate organizational performance. [16] has helped illuminate the organizational benefits of knowledge management in his research.

Similarly, according to [1], knowledge management refers to the processes and practices aimed at identifying, creating, storing, sharing and effectively using knowledge within an organization. Indeed, knowledge management includes several key processes, including identifying critical knowledge, capturing and documenting knowledge, sharing and disseminating knowledge, and finally using and applying knowledge. These processes are often described in the knowledge management literature, with researchers such as Dalkir (2013) providing models and frameworks for organizations.

On the other hand, according to [2] who examined the role of information technologies in knowledge management, knowledge management is based on information technologies and data management systems that allow storing, to organize and efficiently retrieve knowledge within the organization. Likewise, knowledge management is also closely linked to organizational culture and leadership practices, as highlighted by [3], which highlight the importance of creating an environment favorable to learning and development innovation.

After having studied and defined the concepts of AI and knowledge management, we will now try to theoretically study the impact of artificial intelligence (AI) on knowledge management.

We find theoretically that artificial intelligence has a significant effect on knowledge management in several areas such as:

1. Improving access to knowledge: AI can play a crucial role in improving access to knowledge within organizations by enabling more efficient search and retrieval of relevant information. Work such as that of [11] showed how AI technologies, such as machine learning and natural language processing, can be used to develop smarter and more adaptive knowledge management systems.
2. Facilitating tacit knowledge management: AI can also help facilitate tacit knowledge management by identifying, extracting and capturing information from unstructured data such as documents, emails and conversations. Studies like that of [14] show how machine learning techniques can be used to automatically extract tacit knowledge from textual data.
3. Improved decision making: By integrating AI capabilities into knowledge management systems, organizations can improve decision making by providing relevant insights and recommendations based on data analysis. Research by [4] highlight the positive impact of the use of AI on the quality and speed of decisions in organizational environments.

4. Personalization of learning and knowledge development: AI can be used to personalize learning and knowledge development by identifying individual learner needs and recommending suitable educational resources. Recent work, such as that of [10], explore how AI systems can be used to create personalized and effective learning experiences.
5. Identifying Knowledge Gaps: AI techniques, such as predictive analytics and clustering, can help identify knowledge gaps within organizations by analyzing patterns and trends in data of knowledge. Research by [9] demonstrate how AI can be used to automatically detect knowledge gaps and suggest corrective actions.

In conclusion, AI has significant potential to transform knowledge management by improving access to knowledge, facilitating tacit knowledge management, improving decision making, personalizing learning and knowledge development, and identifying knowledge gaps. These points highlight the growing importance of AI in effective knowledge management within organizations. Based on all of the above, we postulate the following hypothesis to analyze its direct effect:

Hypothesis: Artificial intelligence has a positive effect on effective knowledge management?

3 Research Methodology

To study the effect of artificial intelligence (AI) on knowledge management, a mixed methodological approach will be adopted, combining both qualitative and quantitative methods. This approach will enable a deep understanding and rigorous analysis of the relationship between AI and knowledge management, taking into account both perceptions and objective data.

<u>Literature Reviews</u>

A systematic literature review will be conducted to identify relevant previous research on AI and knowledge management. This literature review will provide an understanding of the current state of knowledge, emerging trends, and gaps in existing research. In the same vein, knowledge management is a strategic area for organizations, involving the collection, storage, sharing and effective use of knowledge to support organizational objectives [1]. Integrating artificial intelligence (AI) into this process offers new insights and unique challenges. Work such as that of Maier et al. (2019) explored how AI can facilitate the automatic extraction of knowledge from large amounts of unstructured data, which can enrich organizational knowledge bases. Likewise, more recent research, such as that of [10], looked at the use of AI to improve knowledge management by identifying internal experts and facilitating knowledge sharing within organizations. By integrating machine learning and natural language processing techniques, AI systems can analyze employee interactions with knowledge bases and recommend relevant content, thereby promoting the effective dissemination of tacit expertise [13]. However, the use of AI in Knowledge management also raises concerns.

Studies such as those by [5] have examined data privacy and information security risks when using AI systems to manage sensitive knowledge. Additionally, algorithmic biases in AI systems, as highlighted by [9], could influence the selection and presentation of knowledge, raising questions of fairness and objectivity in organizational knowledge sharing.

Qualitative Interviews

Semi-structured interviews will be conducted with HR professionals, managers and AI experts to gather their perspectives and experiences on the use of AI in knowledge management. The interviews will provide in-depth qualitative insights into the challenges, opportunities, best practices, and ethical implications of integrating AI into this field.

To deepen our understanding of the impact of artificial intelligence (AI) on knowledge management, qualitative interviews were conducted with senior executives and human resources professionals in various organizations. These interviews gathered varied perspectives on the challenges, opportunities and best practices associated with integrating AI into knowledge management.

Participants highlighted several potential benefits of using AI in knowledge management, including its ability to automate time-consuming administrative tasks related to knowledge assessment and professional development planning. Some also noted that AI could facilitate a more personalized approach to knowledge development, recommending training programs and resources tailored to each employee's individual needs.

However, the interviews also revealed significant concerns regarding the use of AI in knowledge management. Participants raised concerns about the reliability of AI algorithms used to assess employee knowledge, highlighting the need for increased transparency and rigorous oversight to avoid potential bias. Additionally, some highlighted the organizational challenges associated with adopting new technologies, such as resistance to change and the need to develop new knowledge among employees and managers.

Quantitative Survey

An online survey will be developed and administered to a representative sample of companies and organizations from different sectors of activity. The survey will quantify organizations' perceptions, attitudes and practices regarding the use of AI in knowledge management. The quantitative data collected will identify trends, correlations and relevant statistical patterns To conduct a quantitative investigation into the impact of AI on knowledge management, here is an example of questions that could be asked, along with an explanation of the methodology used:

1. Questionnaire development: A structured questionnaire was developed taking into account the objectives of the study and key variables such as the use of AI in knowledge management, respondents' perceptions on benefits and challenges associated, as well as the organizational and human impact. Here is an example of the questions asked in the questionnaire survey:

  - To what extent is your organization currently using artificial intelligence technologies in the area of knowledge management?
  - What do you see as the main benefits of using AI in knowledge management within your organization?
  - What challenges or obstacles do you face in integrating AI into knowledge management?
  - How do you perceive the impact of the introduction of AI on organizational culture and relationships between employees and managers?
  - What are your recommendations for successful implementation of AI in knowledge management, based on your experience and observations?
2. Sample selection: A representative sample of organizations of different sizes and industries was selected to ensure diversity of perspectives. Specific criteria such as company size, industry, and maturity of AI adoption were considered in the selection process.
3. Questionnaire administration: The questionnaire was distributed to participants via methods such as online surveys, telephone interviews or paper questionnaires, depending on the respondents' preferences.
4. Quantitative data analysis: The responses were collected and processed using statistical data analysis software. Techniques such as descriptive analysis, correlation tests and regression analyses were used to examine the relationships between the variables studied.
5. Interpretation of results: The results of the survey were interpreted taking into account the research objectives and the questions asked. Conclusions were drawn by highlighting key trends, significant correlations and implications for knowledge management.
6. Analysis of qualitative data: Qualitative data from the interviews will be analysed using thematic analysis to identify recurring themes, patterns and emerging insights. Quantitative survey data will be analysed using statistical techniques such as regression and analysis of variance (ANOVA) to examine relationships between key variables and assess the impact of AI on business management knowledge.

4 Results

The results of the literature reviews, qualitative interviews and quantitative survey will be integrated to provide a holistic and in-depth analysis of the effect of AI on knowledge management. This integration of results will make it possible to draw robust conclusions and formulate practical recommendations for practitioners and decision-makers in the field of human resources and knowledge management. By combining these different methods, this research aims to make a significant contribution to understanding the impact of AI on knowledge management, while providing valuable insights to guide organizational practices and strategic decisions. Indeed, when studying this relationship between AI and knowledge management, here are some results that could be observed, validated or concluded by contributing to other work such as:

1. Improving the effectiveness of knowledge management: Recent research, such as that conducted by [10] and [13], have shown that integrating AI into knowledge management processes can significantly improve efficiency by facilitating the collection, organization and dissemination of knowledge within the organization.
2. Acceleration of organizational learning: According to the findings of the study by [5] and [9], the use of AI can accelerate the organizational learning process by quickly identifying trends, patterns, and relevant insights from available data.
3. Personalization of the learning experience: The work of [14] and [4] showed that AI allows training and knowledge development programs to be personalized based on individual employee needs, which can lead to better knowledge acquisition and retention.
4. Identification of key knowledge: Recent studies conducted by [15] and [17] demonstrated that AI tools can be used to identify key knowledge within the organization, thereby enabling better allocation of human resources.
5. Predicting knowledge needs: Research by [9] and [10] showed that AI-based predictive analytics can help organizations to anticipate future knowledge needs, identifying current gaps and forecasting talent requirements to achieve strategic objectives.
6. Optimization of recruitment and retention processes: Studies by [11] and [15] highlighted that AI can optimize recruitment processes by identifying the most qualified candidates and predicting employee retention factors.
7. Strengthening the culture of knowledge sharing: Research conducted by [5] and [10] showed that AI can strengthen knowledge sharing culture by facilitating collaboration and encouraging active employee contribution.

5 Conclusion

Artificial intelligence (AI) represents a major advancement in knowledge management, providing unprecedented opportunities to improve the collection, storage, sharing and use of knowledge within organizations. Recent theoretical work has highlighted the multiple ways in which AI can enrich knowledge management practices, including automating knowledge extraction, analysis, and recommendation processes, as well as facilitating increased personalization interactions with organizational knowledge.

However, despite its undeniable benefits, integrating AI into knowledge management also raises challenges and concerns. Issues related to data privacy, information security and algorithm transparency remain major concerns, requiring increased vigilance from organizations. Additionally, over-reliance on AI could compromise the role of individuals in creating and sharing knowledge, jeopardizing the human dimension of knowledge management.

Thus, to take full advantage of the opportunities offered by AI in knowledge management, it is essential to adopt a balanced and thoughtful approach, integrating the technical, ethical and social aspects of this transformation. Organizations must invest in training and knowledge development to ensure informed use of AI, while putting in place policies and practices that promote trust, transparency and fairness in knowledge management.

Ultimately, AI represents a powerful tool for transforming the way organizations manage their knowledge, paving the way for new opportunities for innovation and growth. However, its successful integration requires a holistic and ethical approach, placing humans at the center of the dynamics of knowledge management and ensuring that this transformation benefits the whole of society.

References

1. Alavi, M., Leidner, D.E.: Review: knowledge management and knowledge management systems: conceptual foundations and research issues. MIS Q. **25**(1), 107–136 (2001)
2. Alavi, M., Tiwana, A.: Knowledge management: the information technology dimension. In: Easterby-Smith, M., Lyles, M.A. (eds.) The Blackwell Handbook of Organizational Learning and Knowledge Management, pp. 104–121. Blackwell Publishing Ltd., Oxford (2003)
3. Cabrera, A., Crissman, J., Bernal, E., Nora, A., Terenzini, P., Pascarella, E.: Collaborative learning: its impact on college students' development and diversity. J. Coll. Stud. Dev. **43**, 20–34 (2002)
4. Chen, Y., Wang, H., Li, Z.: Leveraging artificial intelligence for knowledge management: a systematic literature review. J. Knowl. Manag. **25**(5), 993–1019 (2021)
5. Chen, Z., Wang, Y.: Exploring the impact of artificial intelligence on competency development: a comparative study. Int. J. Hum. Resour. Manag. **32**(4), 567–584 (2021)
6. Dyer, J.H., Nobeoka, K.: Creating and managing a high-performance knowledge-sharing network: the Toyota case. Strateg. Manag. J. **21**(3), 345–367 (2000)
7. Goodfellow, I., Bengio, Y., Courville, A.: Deep Learning. MIT Press, Cambridge (2016)
8. Jennex, M.E.: Knowledge Management in Modern Organizations. Idea Group Inc. (IGI) (2008)
9. Kim, H., Lee, S., Park, M.: Artificial intelligence and competence management: a case study of technology firms. J. Knowl. Manag. **25**(1), 87–104 (2023)
10. Li, Y., Zhang, Q., Wang, L., Chen, X.: Advancements in artificial intelligence: implications for competence management. J. Organ. Innov. **15**(2), 45–62 (2023)
11. Liu, J., et al.: Enhancing knowledge management systems with artificial intelligence: a review. Inf. Syst. Front. **23**(3), 587–607 (2021)
12. Russell, S.J., Norvig, P., Davis, E., Wirth, N.: Artificial Intelligence: A Modern Approach, 4 edn. Pearson (2021)
13. Smith, J., Jones, A.: AI and human resource management: a review of current trends and future directions. Hum. Resour. Manag. Rev. **28**(3), 301–318 (2022)
14. Wang, L., et al.: Knowledge management in the era of artificial intelligence: challenges and opportunities. J. Knowl. Manag. **24**(6), 1197–1219 (2020)
15. Wang, Q., Li, M., Zhang, H.: The role of artificial intelligence in shaping competence management practices. Int. J. Manag. Rev. **20**(2), 176–193 (2023)
16. Zack, M.H.: Developing a knowledge strategy. Calif. Manag. Rev. **41**(3), 125–145 (1999)
17. Zhang, L., Wang, J., Li, X.: Artificial intelligence and competency development: insights from the banking sector. J. Manag. Stud. **39**(4), 512–529 (2022)

A Rule-Based System for Translating Libyan Dialect Dual Forms to Modern Standard Arabic

Husien Alhammi[1(✉)] and Kais Haddar[2]

[1] Faculty of Economics and Management, University of Sfax, Sfax, Tunisia
h1974hami@gmail.com
[2] Laboratory MIRACL, University of Sfax, Sfax, Tunisia

Abstract. Machine translation from Arabic dialects to modern standard Arabic has recently gained attention. The quality of machine translation remains a challenge, and additional work is needed for improvement. This research presents an automatic machine translation system for translating the Libyan dialect short dual-form sentences to Modern Standard Arabic, which is the first machine translation effort to translate the Libyan dialect to modern standard Arabic. However, our work involves leveraging a pre-existing Libyan dialect resource, a bilingual dictionary, to build a machine translation system. The system uses a rule-based approach that relies on explicitly defined linguistic rules for the Libyan dialect and modern standard Arabic. In this paper, the translation process is divided into four phases: analysis, lexical transfer, lexical selection, and generation. In addition, to produce the fluent modern standard Arabic sentences, a 2-gram parallel corpus was used. The evaluation process is done by comparing the system results with the original human translation. The system was assessed using a test dataset of 100 short dual-form sentences. The BLEU, TER and ChrF scores of the system were 41.37%, 40.3% and 66.9%, respectively.

Keywords: Libyan dialect · MSA · machine translation · dual forms

1 Introduction

Throughout the Arab world, modern standard Arabic (MSA) is the standardized form of Arabic and is used in formal contexts such as education, literature, government, and the media, while Arabic dialects are often the spoken language by Arabs in informal contexts such as everyday conversation and informal interactions. Obviously, due to dialectal variations among Arabic dialects, speakers of different Arabic dialects can generally comprehend each other to a certain degree. Therefore, MSA can be used as an intermediary language between Arabic speakers to enable them to understand different Arabic dialects. However, translating Arabic dialects into MSA would be a challenging task due to their oral nature, lack of standard orthographies, complex morphology, and the scarcity of

L. H. Belguith and K. Shaalan (Eds.): LPKM 2024, LNNS 1303, pp. 279–290, 2025.
https://doi.org/10.1007/978-3-031-85067-7_28

resources [1,2]. The dual-form system presents one of these challenges due to the differences between MSA and dialectal Arabic. In detail, the dual-form system in Maghrebi dialects, such as Libyan dialect (LD), is completely different from the MSA. A dual form is a grammatical number that expresses the quantity of two objects. Maghrebi dialects are mainly variations of MSA and strongly influenced by Berber languages and other languages such as French and Italian. Dual forms in Maghrebi dialects can be expressed by using similar words. For example, the word زوز is used in Libyan and Tunisian dialects, and the word جوج in Morocco. Maghrebi dialects dual forms differ from MSA and other Arabic dialects, such as Egyptian dialect (ED). To explain how the nouns are converted to dual forms, here is an example. The singular forms of the English word "a car" are كرهبة, عربية and سيارة in LD, ED, and MSA, respectively. To convert the singular noun كرهبة in LD to dual form, the word زوز is placed before the plural noun كراهب, which is the plural form of the singular noun كرهبة. The obtained phrase, which is زوز كراهب, is the dual form of the singular noun كرهبة. In contrast, in ED and MSA, the last letter ة in the singular nouns is converted to ت, and then the suffix ين is attached to the singular nouns to make their dual forms: عربيتين and سيارتين in ED and MSA, respectively. Due to a lack of resources, such as training data, a rule-based approach was used to build the system.

The main contribution of the paper lies in building a rule-based machine translation (MT) system that will be able to translate the short dual-form sentences in LD into MSA so that speakers of other Arabic dialects can understand the LD. The other objective of this work is to make LD understandable to people worldwide. This can be done by translating the MSA text that represents the LD into other international languages, such as English. In this situation, MSA serves as a bridge between LD and English.

2 Related Work

Recently, several researchers have shown interest in translating dialectal Arabic (DA) into MSA. These researchers have used diverse approaches to develop their DA-MSA MT systems. An overview of the most significant research contributions made in this domain will be provided in this section. Wael Salloum and Nizar Habash presented a rule-based system in 2012 that translates dialectal Arabic to MSA, which is called Elissa [3]. A hybrid approach, which combines a rule-based and a statistical approach to translating Moroccan Arabic dialect to MSA, was developed by Tachicart, Ridouane, and Karim Bouzoubaa in 2014 [4]. Hamdi et al. introduced a rule-based bidirectional translation system to translate between Tunisian dialect (TD) and MSA verbs in 2013 [5]. In addition, Sadat et al. worked on a rule-based approach to develop a TD-to-MSA word-based MT system [6]. The system achieved a score of 14.32 BLEU. In 2015, Meftouh et al. [7] built from scratch the PADIC, which is a multi-dialectal parallel Arabic corpus that initially includes MSA as well as five Arabic dialects. They also developed phrase-based MT systems with a trigram language model. Roqayah

Al-Ibrahim, in 2020 [8], used deep learning to build a neural machine translation (NMT) framework for translating the Jordanian dialect (JD) into MSA using a recurrent neural network (RNN) encoder-decoder model with multiple hidden layers. The results showed that sentence-to-sentence translation achieved a BLEU accuracy of 63.2% and word-to-word translation had a BLEU accuracy of 91.3%. Mohamed Sghaier and Mounir Zrigui, in 2020 [9], presented a MT system for translating TD to MSA using a rule-based approach. The system achieved 23.28%, 23.85%, and 40.98% in the WER, TER, and BLEU scores, respectively. In 2021, Roua Torjmen and Kais Haddar [10] used a linguistic approach based on the NooJ linguistic platform using finite-state transducers to develop a translator to translate the TD into MSA. Their system achieved 0.97 for recall, 0.91 for precision, 0.88 for F-measure, 80.22 for BLUE, 11.58% for WER, and 12% for TER. Amel Slim [11], in 2022, used a NMT approach to build a system to translate the Algerian Arabic dialect into MSA. They used a fine-tuning transfer learning strategy to train a sequence-to-sequence model. Moukafih, Youness, et al. presented in 2022 [12] a NMT system based on multitasking learning strategies to translate Arabic dialects into MSA and between one another.

3 LD Dual Forms

The LD is mostly composed of MSA and Berber languages, and it is also heavily influenced by Italian and Turkish. As a result, it differs significantly from MSA and other Arabic dialects in terms of vocabulary, grammar, and pronunciation. A dual form is one of the most common variations between LD and MSA. Two distinct dual form types are used in LD based on the area where it is spoken. In the east region, dual forms can be created by adding the suffix ـين to the singular noun, which is similar to the MSA dual form in accusative and genitive cases. For example, the suffix ـين is attached to the singular noun دار, meaning "room", to transfer it to the dual form دارين , which means "two rooms". In contrast, in the west region, the word زوز , which means "pair" or "two" in English, is typically placed before the plural nouns in order to convert them into dual forms. For example, the two words زوز ديار , which mean "two rooms", are the dual noun of the singular noun دار , which means "room", where the word ديار is the plural form of the singular noun دار . Generally, the word زوز comes with plural and singular nouns to transfer them to their dual forms. Furthermore, the word زوز can come with the plural demonstrative pronoun هضم , which means "these", to convert it to a dual form. Moreover, the preposition letter ب and conjunction letter و can be attached to the word زوز as a prefix.

4 MSA Dual Forms

The dual form in MSA is used for nouns, pronouns, adjectives, and verbs. For nouns, it is formed by adding certain suffixes to the singular form. For example, the suffix ين can be added to the singular noun طفل, which means "child", to generate its dual form طفلين, which means "two children". Furthermore, words in MSA dual sentences must agree in number, gender, and case. For example, in the number agreement, the noun قطتين and adjective جميلتين in the MSA phrase قطتين جميلتين , which means "two beautiful cats", agree in number.

4.1 Dual Forms of Nouns

Dual forms of nouns can be in three different cases: nominative, accusative, and genitive. The suffix ان is added to the singular noun to make it a dual form in the nominative case. Likewise, the singular noun can be converted to a dual form by adding the suffix ين to its singular form in the accusative and genitive cases. The nouns and adjectives in dual forms must agree in gender and case. For example, the singular form sentence جاء الطالب الجديد, which means "the new student came" can be transferred to dual form by adding the suffix ان to the singular noun الطالب and the singular adjective الجديد. The obtained sentence is جاء الطالبان الجديدان, which means 'the two new students came", in a dual form with the gender and nominative case agreement. In addition, dual forms for nouns in MSA can also be relative pronouns or demonstrative pronouns for both masculine and feminine. Furthermore, the second-person dual pronoun أنتما can be used as a subject for both masculine and feminine genders. While the third-person dual pronoun هما can be a subject, object, or possessive pronoun for masculine and feminine.

4.2 Dual Forms of Verbs

Depending on the sentence tense, dual suffixes can be attached to verbs in singular form to change them to dual form. Each tense has its own dual suffix. In the present tense, the verb can be in the nominative genitive or accusative cases. The suffix ان is attached to the present verbs in the nominative case. while the suffix ا is added to the present verbs for the genitive or accusative cases. For example, the singular form sentence الولد يأكل الخبز , meaning "the boy is eating bread", can be mapped to dual form by adding the suffix ان to the subject الولد and the verb يأكل. This results in the sentence الولدان يأكلان الخبز in a dual form with the nominative case, which means "the two boys are eating bread". Likewise, in the genitive and accusative cases, the suffix ا is added to the verb in order to convert it to dual form. For example, in the genitive case, the dual form sentence الولدان لن يأكلا الخبز, meaning "The two boys will not eat bread", is generated by adding the suffixا to the singular verb يأكل. The dual verbs in future tense can be generated by either adding the prefix س to the verbs or placing the word سوف before the dual verbs in the present tense. In the past and imperative tense, the dual suffix ا , which is called ألف الاثنين alif al-ithnayn, could be added to the verbs in singular form to change them to their dual forms.

5 MSA and LD Word Ordering

Mapping sentences from LD into MSA requires careful consideration of sentence structures (word ordering) in both languages. The MSA and LD share the same sentence types, which can be generally classified as either a nominal or a verbal sentence [13,14]. The verb-subject-object (VSO) is the default word ordering in verbal sentences in MSA [15,16], while the subject-verb-object (SVO) word ordering is commonly used in verbal sentences in LD. On the other hand, in nominal sentences, subject-predicate is the typical word ordering in both languages.

6 System Design and Architecture

Although the MSA and the LD are also closely related languages because of the similarities between them, there are also notable differences between them in terms of vocabulary, grammar, pronunciation, and idiomatic expressions. However, the LD dual system is completely different from the MSA dual system. Consequently, translating the dual forms of the LD into the MSA is a challenging task. Therefore, the author, who is a native speaker of the LD with good linguistic knowledge, made a great effort to create the hand-written linguistic rules for building the system. Our system works under the hypothesis that words in LD that are not found in the bilingual dictionary are tagged as unknown words and remain as they are in the MSA. Unknown words are usually those that are shared and used in both languages. For example, the LD phrase فيه زوز كواشيك علي الطاولة, which means "there are two spoons on the table", can also be translated into هناك ملعقتان على الطاولة in MSA. Obviously, the word الطاولة in the LD phrase, which means "the table", is not translated and left as i0s in the MSA. It is clear that both sentences have the same sentence structure (word reordering). The LD sentence can be translated to MSA through four phases: analysis, lexical transfer, lexical selection, and generation phase. Figure 1 illustrates the main phases in our system.

6.1 Tokenization

In this step, the system basically uses white space to split the sentence into words. Each word represents a token.

6.2 Morphological Analysis

Morphological analysis plays a significant role in MT by enhancing the understanding of context [17]. Therefore, morphological features, such as part-of-speech (POS), gender, and number, must be assigned to each token. In this work, a bilingual dictionary that contains LD-MSA words, along with their linguistic features, is used to assign POS, gender, and number to each token [18]. A light rule-based morphological segmentation system was also developed for LD to divide agglutinative words into their constituent morphemes. LD words that do not match any word in the bilingual dictionary are marked as unknown words. Ambiguity could occur when a word in the LD has multiple meanings or POS tags. Word ambiguity poses a challenge because it can lead to inaccurate or misleading translations. For instance, the word باهي, which means "good" or "ok", might be an adjective or adverb depending on the context.

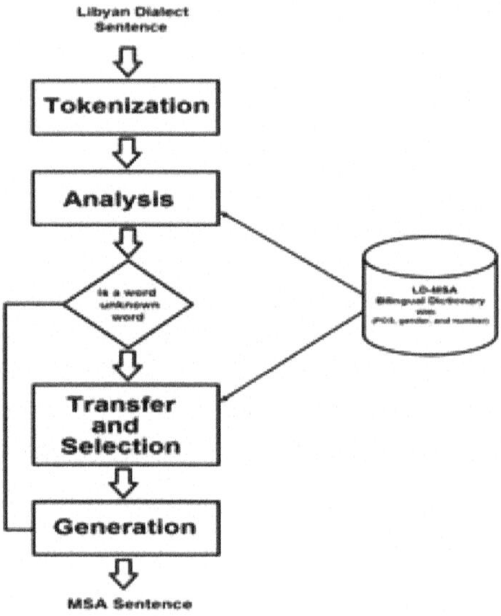

Fig. 1. Illustrates the system architecture

6.3 Lexical Transfer and Selection

Lexical transfer refers to the process of transferring or mapping words from LD into MSA. In this system, the bilingual dictionary is used to map LD words to their corresponding words in the MSA. The system uses two types of transfers: lemma-to-lemma and features-to-features transfers. The features-to-features transfers are used to convert clitics and affixes to their equivalents in MSA. However, words that are labeled as unknown in the analysis phase cannot be translated and are kept unchanged in MSA. With regard to the word زوز, the word could be completely omitted from the text in some cases. In other situations, the word زوز can be transferred to several MSA equivalents based on the words that surround it. Table 1 lists the different MSA equivalents for the word زوز when it is surrounded by nouns, verbs, adjectives, and pronouns.

Table 1. Lists some of the MSA equivalents for the word "زوز".

Accusative/Genitive	Nominative	English Meaning	LD Dual Forms
هذين (masculine) / هاتين (feminine)	هذان (masculine) / هاتان (feminine)	These two	هضوما الزوز
كوبين	كوبان	Two cups	زوز طيس
احضرهما	احضرهما	Bring both	هاتهم الزوز
الاثنين سيئين (masculine) / الاثنتين سيئتين (feminine)	الاثنان سيئان (masculine) / الاثنتان سيئتان (feminine)	Both are bad	الزوز بلهون
الطالبين كليهما (masculine) / الطالبتين كلتيهما (feminine)	الطالبان كلاهما (masculine) / الطالبتان كلتاهما (feminine)	Two students	الطلبلة الزوز
كلينا (both genders)	كلانا (both genders)	We both	حني الزوز
كليكما (both genders)	كلاكما (both genders)	You both	انتم الزوز

In lexical selection, an ambiguous word is a word that can have multiple POS tags. Depending on the POSs of the words that surround it, the lexical selection process selects an appropriate word from multiple words for the ambiguous word. For example, if the ambiguous word comes after the preposition, the word that has a noun tag can be assigned to the ambiguous word. Furthermore, depending on the words that immediately surround them, many words can have different meanings. For instance, depending on the word that comes before it, the word حقها could mean either "a right" or "to cost". It means "how much" when it comes after the word قداش to form the phrase of two words قداش حقها. In contrast, it means "it has a right" when it follows the word من to create the phrase of two words من حقها. To mitigate this problem, a 2-gram parallel corpus that contains more than 1700 pairs of two words in the LD and their corresponding words in MSA was created. These two words are essentially collocation words that often go together naturally in a language to convey a specific meaning. This 2-gram parallel corpus was used to improve the fluency and disambiguation of the translation. For example, the LD words من النو, which means "Because of the extreme heat", can be translated using their MSA equivalents in the 2-gram corpus, which is من شدة الحر.

6.4 Generation

The generation phase is the final step in the MT process. The MSA sentences should be correct in terms of grammar, structure, and meaning. The generation phase contains two steps: morphological and syntactic generation. The morphological generation focuses on generating the correct inflected or derived word

Table 2. Demonstrates how the system works.

Process	Description
I saw two pretty cats in this car شبحت زوز قطاطيس قنينات فالسيارة هدي	LD sentence
Splitting the sentence into individual words <شبحت> <زوز> <قطاطيس> <قنينات> <فالسيارة> <هدي> <this> < in the car > <pretty> <cats> < two> <I saw> Morphological segmentation is applied to break down words into their morphemes. <شبحـت> <زوز> <قطاطيس> <قنينـات> <فـالـسيارة><هدي>	Tokenization
<شبح:v+ت:pron> <زوز> <قطاطيس:P.F.N> <قنين:S.M.Aj+ات:F.P.Suf> <ف:Prep+ال:Dart+سيارة:unk.S.F.N> <هدي:F.Dpron>	Morphological Analysis
< رأي + ت > <omitted> <قطط> <جميلات> <في+ السيارة> <هذه>	Lexical transfer
Creating the surface forms for all words. < رأيت> <omitted> <قطط> <جميلات> <في> <السيارة> <هذه> Changing the plural forms of the words "قطط" and "جميلات" to their singular forms "قطة" and "جميلة" in order to change them to dual forms. < رأيت> <omitted> <قطة> <جميلة> <في> <السيارة> <هذه> For word ordering, the word "هذه" is repositioned before the word "السيارة". رأيت قطة جميلة في هذه السيارة Depending on the word case agreement, the singular form of the word "قطة" is changed to its dual form "قطتين". رأيت قطتين جميلة في هذه السيارة A word agreement between the noun "قطتين" and the adjective "جميلة" is applied to change the adjective form from "جميلة" to "جميلتين". رأيت قطتين جميلتين في هذه السيارة	Generation
Finally, a well-structured MSA sentence is generated. رأيت قطتين جميلتين في هذه السيارة I saw two pretty cats in this car	MSA sentence

forms based on their features, such as prefixes and suffixes. The purpose of syntactic generation is to construct a grammatically correct sentence structure in the MSA. This includes choosing the correct word ordering, verb tense, and agreements between different words. In this phase, in order to convert nouns to dual forms, broken plural nouns are converted to singular nouns using a list of nouns. The list contains 12,249 singular nouns with their corresponding broken plural nouns [19]. The system steps for translating LD dual sentences to MSA are shown in Table 2.

7 Results and Discussion

In order to create the test dataset, we asked native speakers of the LD to write 100 short sentences in dual forms. We also asked them to avoid writing negated words and idioms because these issues require particular handling. Furthermore, we collected 100 sentences from social media to increase the dataset size. These sentences are then translated into MSA by two human experts. The total dataset of 200 sentences was equally divided into two datasets. The first dataset was used to extract dual form grammatical rules. The second dataset was used as a reference translation to evaluate our system. Our evaluation methodology is based on making comparisons between the system outputs and human translations (reference translations). The online tools were used to calculate translation scores [20,21]. The TER, BLEU and ChrF scores that our system achieved were 40.3, 41.37 and 66.9, respectively. Table 3 illustrates the BLEU scores. A demo version of our system is available online [22].

Although the system was able to successfully generate the correct translation, it failed to generate the correct translation in many cases. There are some mistakes related to the word that can be translated to two or more words in MSA.

For example, to translate the LD words زوز كاويشات, which means "two sports shoes", to MSA, the plural word كاويشات is translated to its equivalent "أحذية رياضية" in MSA, and then it is converted to the singular form حذاء رياضي. After that, convert the singular form to the dual form حذاءان رياضيان. The system fails to translate the word زوز كاويشات and produces the wrong translation, which is حذاء رياضيان instead of حذاءان رياضيان. See Table 4 (row 3). Furthermore, because the LD contains a large number of homographs, the system makes many mistakes. For instance, the word دخان in Table 4 (row 4) has two meanings: "smoke" and "cigarettes". The two meanings of the word دخان have the same POS tag (noun). The system wrongly chose the word دخان rather than سجائر, which mean "smoke" and "cigarettes" respectively. Some mistakes occur when the morphological segmentation fails to correctly segment words. The wrong segmentation of words led to an incorrect translation. For example, the word وبوي in Table 4 (row 2), which means "my dad" was wrongly segmented into the word وي and the prefixes و + ب. As a result, the word وي did not match any words in the dictionary and was tagged as an unknown word. Furthermore, the system generated some mistakes related to verb conjugation. The dictionary only provides the verb forms for the third-person masculine singular in the perfect, imperfect, and imperative tenses. For example, the first-person masculine singular for perfect tense needs to be modified to match the third-person masculine singular verb form in the dictionary. Additionally, the system produced mistakes related to the nouns that end with the letters ات. The problem appears when plural nouns whose last two letters are ات are changed to their singular forms. The system generated the incorrect word ليمونين instead of the correct word ليمونتين when it added the wrong suffix ين to the word ليمون rather than the suffix تين. See Table 4 (row 1). Additionally, the generation of verb surface forms requires lots of improvement. For instance, in Table 4 (row 4), the system was unable to generate the correct surface form of the verb قال, which means "said". It can be noticed that the system output and reference translation both employ synonyms to express the same concept. For example, the words المنزل and البيت can be translated to the English word "a house". The word المنزل was generated by the system, and the word البيت is found in the reference translation. See Table 4 (row 1). Likewise, the words بنت and فتاة can be translated to "girl" in English. It is acknowledged that synonymous wording can negatively impact translation scores. In order to precisely calculate the system's accuracy scores, we adopted a two-step approach. First, both the system outputs and reference translations in MSA were translated into English using Google Translate. Subsequently, the scores were calculated based on these English translations. Obviously, translating the system output and reference translations to English resulted in a significant improvement in the translation scores, which were 68.42, 24.1, 79.2 and 87.2

for BLEU, TER, ChrF and COMET, respectively. BLEU scores are shown in Table 5.

Table 3. Shows BLEU scores of the system.

Type	1-gram	2-gram	3-gram	4-gram
Individual	65.76	46.04	35.96	30.45
Cumulative	63.76	53.35	46.30	41.37

Table 4. Shows sentences from the test set that were generated incorrectly by the system.

Our System	Reference Translation	English Meaning	Test Dataset	No
خذي معك ليمونين و تفاحة من المنزل	خذي معك برتقالتين وتفاحة من البيت	Take two oranges and an apple from home with you	خودي معاك زوز ليمات و تفاحة من الحوش	1
أمي وبوي الحبهمان	أمي وأبي كلاهما أحبه	My mom and dad I love them both.	امي وبوي الزوز انحبهم	2
لديه حذاء رياضيان	لديه حذاءان رياضيان	He has two sport shoes	عنده زوز كاويشات	3
اسمع قوله أحضر معه علبتين دخان	اسمع، قل له أن يحضر معه علبتين من السجائر	Listen, tell him to bring two packs of cigarettes.	اسمع قوله اجيب معاه زوز باكوات دخان	4

Table 5. BLEU scores for system output and reference translations in English.

Type	1-gram	2-gram	3-gram	4-gram
Individual	83.58	71.93	64.77	58.53
Cumulative	82.76	76.78	72.31	68.42

8 Conclusion

This research was to design and implement a rule-based MT system that is able to translate dual forms from the LD into MSA. The system used the existing LD resource, a LD-MSA bilingual dictionary. Additionally, this work also provides the 2-gram parallel corpus, which other MT systems can leverage to enhance their translation. A reference translation dataset of 100 dual-form sentences in the MSA was used to test the system. The system achieved a score of 41.37% in BLEU, 40.3% in TER and 66.9% in ChrF. In contrast, these scores improved greatly when they were calculated using the English translation of the system output and the reference translation. These results demonstrate that a high rate of synonyms might lead to low translation scores. In our future work, we endeavor to further improve the translation quality of our system by developing and implementing new rules, provided by linguist experts, and enriching a dictionary with more words. The translation quality can also be enhanced by creating a hybrid system that merges our system with the statistical approach.

References

1. Slim, A., Melouah, A.: Low resource Arabic dialects transformer neural machine translation improvement through incremental transfer of shared linguistic features. Arab. J. Sci. Eng. 1–17 (2024)
2. Younes, J., et al.: Language resources for Maghrebi Arabic dialects' NLP: a survey. Lang. Resour. Eval. **54**, 1079–1142 (2020)
3. Salloum, W., Habash, N.: Elissa: a dialectal to standard Arabic machine translation system. In: Proceedings of COLING 2012, Demonstration Papers (2012)
4. Tachicart, R., Bouzoubaa, K.: A hybrid approach to translate Moroccan Arabic dialect. In: 2014 9th International Conference on Intelligent Systems, Theories and Applications (SITA-14). IEEE (2014)
5. Hamdi, A., et al.: The effects of factorizing root and pattern mapping in bidirectional Tunisian-standard Arabic machine translation. In: Proceedings of Machine Translation Summit XIV, Papers (2013)
6. Sadat, F., et al.: Collaboratively constructed linguistic resources for language variants and their exploitation in NLP application-the case of Tunisian Arabic and the social media. In: Proceedings of Workshop on Lexical and Grammatical Resources for Language Processing (2014)
7. Meftouh, K., et al.: Machine translation experiments on PADIC: a parallel Arabic dialect corpus. In: Proceedings of the 29th Pacific Asia Conference on Language, Information and Computation (2015)
8. Al-Ibrahim, R., Duwairi, R.: Neural machine translation from Jordanian Dialect to modern standard Arabic. In: 11th International Conference on Information and Communication Systems (ICICS). IEEE (2020)
9. Sghaier, M., Zrigui, M.: Rule-based machine translation from Tunisian dialect to modern standard Arabic. Procedia Comput. Sci. **176**, 310–319 (2020)
10. Torjmen, R., Haddar, K.: Translation system from Tunisian Dialect to Modern Standard Arabic. Concurr. Comput. Pract. Exp. **34**(6), e6788 (2022)
11. Slim, A., et al.: Improving neural machine translation for low resource Algerian Dialect by transductive transfer learning strategy. Arab. J. Sci. Eng. **47**(8), 10411–10418 (2022)
12. Moukafih, Y., et al.: Improving machine translation of Arabic dialects through multi-task learning. In: AIxIA 2021-Advances in Artificial Intelligence. 20th International Conference of the Italian Association for Artificial Intelligence, Virtual Event, 1–3 December 2021, Revised Selected Papers. Springer, Cham (2022)
13. Abdelrazaq, D., Abu-Soud, S., Awajan, A.: A machine learning system for distinguishing nominal and verbal Arabic sentences. System **1**(2), 16–17 (2018)
14. Homeidi, M.: Does Arabic have one or two main basic sentence types? A descriptive pedagogic account. Educ. Res. Appl. **6**(01), 1–6 (2021)
15. AlShammiry, K.: Subject verb agreement in Modern Standard Arabic (MSA) and Saudi Dialect of Arabic (SA): a new minimalist account. Int. J. Lang. Lit. **4**(2), 83–92 (2016)
16. Boudelaa, S., Marslen-Wilson, W.: Morphological structure in the Arabic mental lexicon: parallels between standard and dialectal Arabic. Lang. Cognit. Process. **28**(10), 1453–1473 (2013)
17. Paulo, M.: Analysis of context-aware translation memories: part-of-speech pattern distribution and gender neutral translation memories. Dissertation (2022)
18. Alhammi, H., Haddar, K.: Toward building a bilingual dictionary for LD-modern standard Arabic machine translation. In: KDIR (2022)

19. Ouamer, M., Tajmout, R., Bouzoubaa, K.: Arabic broken plural model based on the broken pattern. In: International Conference on Digital Technologies and Applications. Springer, Cham (2022)
20. Vanroy, B., Tezcan, A., Macken, L.: MATEO: MAchine Translation Evaluation Online. In: The 24th Annual Conference of the European Association for Machine Translation, European Association for Machine Translation (EAMT) (2023)
21. Interactive BLEU score evaluator. https://www.letsmt.eu/Bleu.aspx. Accessed 2 Apr 2024
22. Online system. http://dra.net.ly/translation/duals/. Accessed 1 Apr 2024

Towards an Approach for Extending BPMN with Quality Dimension

Zohra Alyani[1(✉)], Mohamed Turki[2,3], Karima Dhouib[3], and Faiez Gargouri[3]

[1] FSEG Sfax, MIRACL Laboratory, Sfax, Tunisia
Zohra.alyani@gmail.com
[2] College of Computer Science and Engineering, University of Hail, Hail 81451, Kingdom of Saudi Arabia
moh.alturki@uoh.edu.sa
[3] ISIM Sfax, MIRACL Laboratory, Sfax, Tunisia
{karima.dhouib,faiez.gargouri}@isims.usf.tn

Abstract. Building a quality management system (QMS) relies heavily on two key steps: mapping and modeling business processes. The process approach, a core principle of QMS, emphasizes identifying and managing processes and their interactions systematically. This ensures achieving desired outcomes aligned with the organization's policies and strategic goals. Following this approach, organizations need to establish, implement, maintain, and continuously improve their quality management system (SQM), encompassing all relevant processes and their connections, to meet regulatory requirements. This article aims to present our approach for extending the BPMN formalism by a quality dimension based on the MDA approach and on an ontological approach for the definition of a meta-model for the quality management which covers the ISO 9001 Norm.

Keywords: Quality Management · Business process modeling · Ontology Core quality · ISO 9001 · Ontological engineering · BPMN2.0

1 Introduction

The Quality Management System (QMS) is based on an in-depth understanding of an organisation's business processes to ensure the consistent quality of its products and services. In fact, it is the business processes that define the key stages in producing a product or providing a service, transforming inputs into outputs that meet customer expectations. To improve their performance and increase their responsiveness, organisations need to identify and manage all the sensitive processes that represent a source of value to be developed [1]. In this context, the modelling and automation of these processes is a crucial stage which requires the use of standard modelling language. However, ISO standards do not require the use of any specific formalism for modelling processes. The main objective of this paper is to propose an extension to the standard BPMN2.0

L. H. Belguith and K. Shaalan (Eds.): LPKM 2024, LNNS 1303, pp. 291–299, 2025.
https://doi.org/10.1007/978-3-031-85067-7_29

formalism. This extension must include the key concepts of the Quality Management domain and must enable them to be taken into account during the modelling phases of process execution.

The remainder of the paper is organized as follows: Sect. 2 depicts the contribution of BPM to quality management systems. Section 3, illustrates the Role of ontologies in BPM and MQ. Section 4 present the different methods for designing BPMN extensions that are the most considerable in the literature. While, Sect. 5, discusses the review of the existing extensions BPMN. Section 6, details our approach. Eventually, Sect. 7 wraps up the conclusion, displays certain pertinent concluding remarks and provides new perspectives for future works.

2 The Contribution of BPM to Quality Management Systems

Organisations are now adopting a 'process-centric' approach, placing business processes (BPs) at the heart of overall integration thinking. In practice, this strategy promotes the operational efficiency of processes and enables them to adapt easily to potential changes within the organisation [La Rosa et al., 2013]. It also improves organisational agility and optimises the value chain. To facilitate this process vision of organisations, many theorists and practitioners have proposed frameworks for analysing and modelling BP. Additionally, the richness and liveliness of the concept of "Process" or "Business Process" (BP) in an organisational context has given rise to a variety of definitions, some of which are complementary. Following a review of the literature, BP *"A process is a set of correlated or interactive activities that transform input elements into output elements"* [2]. Then, Business Process Management is defined in [3] as: "the use of methods, techniques and software systems to design, execute, control and analyse business processes involving people, applications, documents and other information sources".

The authors in, [4], *"business process management includes the concepts, methods and techniques that enable the modelling, administration, configuration, implementation and analysis of business processes".*

BP management is essentially based on an explicit representation of the BP and its activities, as well as on execution constraints between these activities. Once modelled, a BP can be analysed and improved before being implemented. In the following, we present the different aspects of BP management.

The BPM approach makes it possible to monitor and manage the evolution of a BPM from its conception to its operation. In the literature, there is no uniform view of the number of phases in the BPM lifecycle, which can vary depending on the granularity chosen for identifying the phases [5]. Generally speaking, the lifecycle of a BPM is essentially made up of six stages Fig. 1.

- **Identification of the BP**: this stage firstly sets out the problem to be solved and secondly identifies the needs that the BP must satisfy.

- **BP Modelling/Discovery**: the requirements specified in the identification stage can be realised by modelling a BP schema that reflects these requirements. The BP modelling step can be replaced by a discovery step if there are previously modelled BPs that meet the needs identified above.
  BP Analysis: this stage is devoted to identifying and the adoption of an ontological approach in this step can significantly influence the modeling of the process documenting the problems relating to the modelled BP. Indeed, the adoption of an ontological approach in this step can significantly influence the modeling of the process.
  BP Redesign: the aim of this stage is to take into account the various problems identified in the analysis phase. This phase consists of making the necessary adjustments to the BP definition, in order to obtain a more appropriate definition. The result of this stage is a BP definition that is ready to be instantiated.
  BP implementation: his stage allows the instantiation of a BP schema, then the deployment of this instance in a BP management system for execution. The execution of the BP is ensured by an execution engine and can call upon an actor and/or external applications.
  BP monitoring and control: once the BP instance is running, relevant data from this execution is collected and analysed. This data is used to check whether the PO is complying with its objectives and expectations. More specifically, this stage involves detecting recurring errors and deviations from planned processing, so that corrective action can be taken. In this way, new problems can be identified that require the cycle to be repeated. In addition, the adoption of a quality management system (QMS) is a strategic decision for an organization. It can significantly improve overall performance and provide a strong foundation for sustainable development initiatives. Business Process Management (BPM) plays an essential role in improving organisations' quality management systems (QMS). By merging BPM principles with QMS, companies can optimise their processes, strengthen consistency and compliance, and achieve superior product and service quality. BPM offers a structured methodology for analysing, modelling, optimising and implementing business processes,

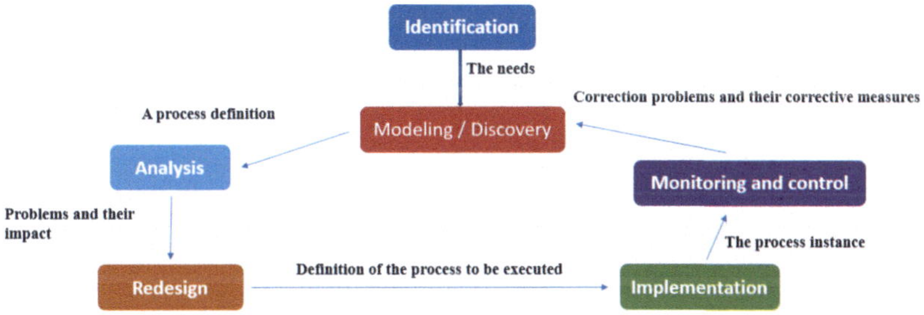

Fig. 1. Life cycle of a BP [5]

enabling inefficiencies to be identified, redundant tasks to be eliminated and manual quality-related steps to be automated. This approach reduces costs, improves productivity and accelerates delivery times. In addition, BPM standardises quality processes, ensuring consistent execution in line with regulatory requirements and industry standards, reducing errors, improving traceability and optimising risk management. By encouraging a collaborative, process-centred approach, BPM involves all stakeholders in the continuous improvement of quality-related processes, fostering a quality-focused corporate culture where quality is a top priority rather than a mere administrative formality. By integrating BPM tools and methodologies into existing QMSs, organisations benefit from a centralised, global view of quality processes, making it easier to manage change, measure performance and identify opportunities for continuous improvement.

In summary, BPM offers a valuable framework for optimising quality management systems and achieving operational excellence. By integrating BPM principles into QMSs, companies can improve the efficiency of quality-related processes, strengthen consistency and compliance, foster a quality culture and achieve tangible business results.

3 Role of Ontologies in BPM and MQ

The term 'ontology' is borrowed from philosophy and refers to [6]: *"that part of metaphysics which applies to being as being, independently of its determinations"*. For the disciplines of artificial intelligence and knowledge engineering, a consensual definition of the notion of ontology is that of [7]: *"an ontology is an explicit specification of a conceptualisation"*. An application ontology is an ontology that brings together all the domain concepts of a given application. these concepts can belong to different levels of abstraction, so ontologies can be composed of high-level foundational ontologies, domain core ontologies and domain ontologies.

- **High-level foundational ontology:** at the most abstract level, high-level foundational ontologies provide a set of abstract concepts and abstract relationships to structure any domain. Examples include BFO (Basic Formal Ontology), GFO (General Formal Ontology) and DOLCE.
- **Core domain ontologies:** at an intermediate level, "core" domain ontologies: The authors in [8] define a minimal set of generic and central concepts for each domain concerned.
- **Domain ontology:** at the most specific level, core domain ontologies are in turn refined to define domain-specific concepts through specialisation.

Additionally, many business process management (BPM), Business Process Quality Management (BPQM) and process frameworks, methods, or approaches (e.g., Lean, Six Sigma, Business Process Reengineering (BPR), Total Quality Management (TQM), Zero Defect, Business Process Modeling Notation

(BPMN), Business Process Execution Language (BPEL) have their own vocabulary. Each of these vocabularies has its own definition of terms, such as business process, process step, process activity, events, process role, process owner, process measure, and process rule. It provides fundamental process concepts that can be used to document corporate knowledge and structure process knowledge by defining relation process concepts. The BPM ontology is presented as a shared vocabulary that structures knowledge in two ways. First, it allows practitioners to structure their business knowledge by adding meaningful relationships between the vocabulary terms. Second, it organizes concepts in hierarchic "is-a" relationships that allow a polymorphic inheritance of properties.

It is evident that, ontological engineering plays a crucial role in the field of Business Process Management (BPM) and Quality Management (MQ) by providing a rigorous and shared conceptual framework for modeling, analyzing. This is why we are interested in extending BPMN to include the quality dimension by building a core ontology. Before presenting our approach to extending the BPMN, we will give an overview of the various works relating to extensions that have been proposed for the BPMN format.

4 Related Works: BPMN Extension

Today, the execution of business processes generates a large amount of data representing events that can be used for process monitoring and analysis. Authors, in [9] presented a structured and rigorous method for developing extensions to the BPMN 2.0 meta-model based on Model-Driven Architecture (MDA). This approach enables extensions to be conceptually modelled using UML, graphically represented using the BPMN extension mechanism and transformed into XML Schema documents compatible with BPMN tools.

The authors in [10], reported a rigorous BPMN4SBP extension that adheres to the existing BPMN extension mechanism. This extension facilitates the multi-dimensional modeling of Sensitive Business Processes (SBPs), enabling the creation of expressive, complete, unambiguous, and comprehensible models. It achieves this by enriching and improving the functional, organizational, behavioral, and informational dimensions of the BPMN 2.0.2 specification. Additionally, the extension integrates new aspects related to the knowledge dimension and the intentional dimension of SBPs.

In [11], the authors proposed a BPMN4CP extension, developed using the BPMN extension mechanism and the extension method of [9]. This extension marks the first systematic approach to adapting the BPMN process modeling language to Clinical Pathways (CPs). To ensure a sound design of the extension elements, the authors performed an equivalence check between domain concepts and BPMN elements. They found that BPMN covers a significant number of requirements related to CPs but falls short in representing specific constraints and concepts within the CP context.

The authors, in [12] enacted an extension BPMN4VC to the BPMN (Business Process Model and Notation) standard that enables versions of collaborative processes to be modelled. These processes are represented in the form of collaborations in the BPMN language. BPMN4VC has been developed to address the issue of flexibility in collaborative processes, a major challenge facing Business Process Management (BPM) systems before widespread adoption in enterprises. The BPMN4VC metamodel supports both static and dynamic aspects of collaboration versioning. As such, BPMN4VC is the meta-model that provides an effective solution for managing the flexibility and complexity of collaborative processes, thereby contributing to better management of enterprise business processes.

The authors, in [13,14] defined a scientific approach based on Model-Driven Architecture (MDA) to ensure the transformation of the extended BPMN 2.0 meta-model into an executable business process model. Their work focused on extending the functional dimension of BPMN 2.0 to better model sensitive business processes (SBPs). To specify the extensions and define the rules for transforming conceptual models, they used the QVT (Query-View-Transformation) standard. They then developed an Eclipse Activiti plugin for modelling and executing SBPs using the extended BPMN 2.0 standard. This plugin implements the extensions made to the functional dimension of BPMN 2.0, enabling effective execution of SBPs.

Authors, in [15] eleborated BPMN-L, a new and effective language for modeling process landscapes. BPMN-L maintains visual consistency with existing landscape modeling approaches, particularly value chains, while also aligning structurally with the BPMN standard. The authors employed a design science research methodology to develop BPMN-L. This involved characterizing, designing, and implementing the solution. The development process included analysis of existing process landscape languages and approaches. Following implementation, the authors demonstrated BPMN-L based diagrams and validated them through quantitative analysis and a "proof-of-concept" approach. The results confirm that BPMN-L offers both visual compatibility with current landscape modeling approaches, especially value chains, and suitability for multi-abstraction layered modeling based on BPMN. Analysis of sample landscape diagrams using BPMN-L demonstrates the practical applicability of all its specified concepts. Overall, the proposed BPMN-L solution has the potential to significantly impact process landscape.

Authors in [16] introduced an approach to the specification and management of the resources that are associated to each BP task and that are required to support the execution of processes and related complex systems. The extension proposed for Performability-enabled BPMN (PyBPMN) builds upon an approach that leverages principles and standards outlined in the Model Driven Architecture (MDA). This approach offers significant benefits in terms of customizable features and enhanced automation, particularly for performance and reliability analysis. Furthermore, in this study, the extension has been expanded to encompass intricate resource modeling and management aspects.

Authors, in [17] introduced social business process meta-models (SBP2Ms), a concept that enriches existing business process meta-models by incorporating social aspects. Additionally, it proposes a domain-specific language (DSLs) for modeling these SBP2Ms. These DSLs are defined based on the SBP2M framework and are designed as extensions to the standard BPMN notation. The proposed DSLs are both simple and generic due to the limited number of extensions needed and their ability to model SBP2Ms regardless of the specific Web 2.0 technology being used. To demonstrate the applicability of these proposed notations, the authors implemented them within a BPMN2 modeler editor.

In what follows, authors in [18,19] propose a new decision-making method for outsourcing business processes (BPs) that are aware of the Internet of Things (IoT). The researchers developed both the MEREC and ELECTRE IS methods using the Java programming environment to implement their approach. To test their proposal, they created an IoRT-aware BP specific to the agriculture sector. This BP was developed within the Eclipse Modeling Framework (EMF) using the open-source BPMN 2.0 modeler plug-in for Eclipse, which is an integrated development environment (IDE).

5 Discussion

In the literature review, we found that existing extensions to BPMN support several dimensions, such as the functional, organisational, behavioural, informational, intentional and knowledge dimensions. However, the quality dimension is absent from the standard BPMN model. To address this gap, we propose a BPMN extension specifically designed to integrate quality considerations.

6 BPMN Extension Approach Through the Quality Dimension

BPMN provides perspectives to support notation extensions. These perspectives allow researchers to extend BPMN with additional modelling features such as: performance measures, cost and reliability, security requirements, data modelling. In this paper, we propose a rigorous approach to extend BPMN through the quality dimension, to do this, we thought of using the MDA model-driven approach to have an executable BP specification.

The OMG (Object Management Group) [20] defined the MDA approach as a response to the problems associated with the continuous evolution of technologies. It is at the origin of model-driven engineering. Everything that makes up the MDA approach: modelling, transformations and implementation technologies. Thus, several definitions of MDA have been proposed in the literature, we present the most common, *"MDA is an approach to specifying computer systems which separates the specification of the functions of a system from that of the implementation of these functions on a particular platform* [21]. This method is composed of three successive pmodels "CIM: Computation Independent Model", "PIM: platform independent model" and "PSM: platform specific model": a model of analysis and conceptualization of the BP domain, a model of preparation of the BPMN extension, and a model of definition of the extension meta-model.

In addition, the first step is to build the ontology of the quality domain. This involves analyzing the domain and designing the ontology modules we propose for our approach. These ontology modules are based on ISO9000 documentation. We propose the following ontology modules: “Performance Indicator”, “Risk”, “Objective” “Document” and “Resource”... Besides, the second step “PIM” illustaretes the Extended BPMN2.0, Finally, the PSM model describes the execution of extended BMN. The following figure describes the different steps (Fig. 2).

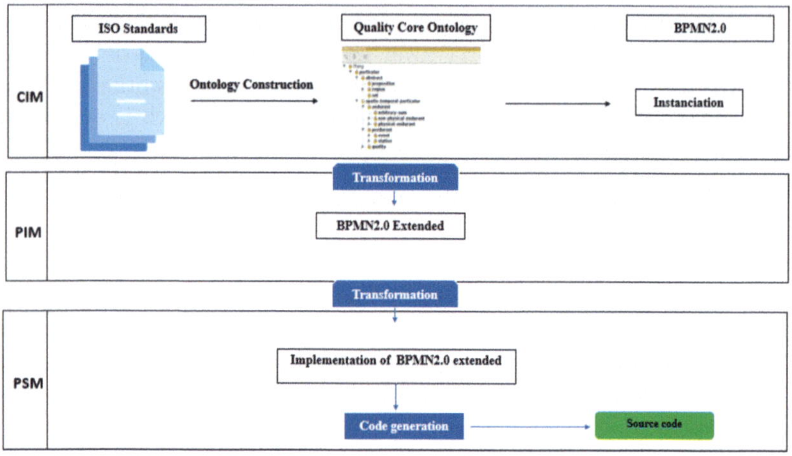

Fig. 2. The Proposed MDA Approach

7 Conclusion and Perspectives

This article proposes a rigorous new extension that adheres to the BPMN extension mechanism. This extension enables multi-dimensional modeling of business processes (BPs), resulting in more expressive, complete, unambiguous, and understandable models. It enriches and improves the functional, organizational, behavioral, and informational dimensions of the BPMN 2.0 specification, while also integrating new aspects related to the quality dimension.

In the future, we also plan to refine our proposal by specifying a meta-model for the quality management domain based on a kernel ontology. This approach will then enable us to define all the model transformation rules needed to implement the extensions to be made to the BPMN 2.0 standard. The ultimate goal is to propose a complete approach for modelling and executing business processes. Finally, it will be essential to test and validate the proposed approach by applying it to a concrete case study.

References

1. Turki, M.: 1M.Turki proposition d'une méthode multicritère et d'une ontologie noyau des processus d'organisation pour l'aide à l'identification des processus sensibles. Thèse (2012)
2. ISO: ISO 9000: - Systèmes de management de la qualité (2000)
3. Van Der Aalst, W.M.P., et al.: Business Process Management & Workflow (2003)
4. Smirnov, S., Reijers, H.A., Weske, M.: Business process model abstraction: a definition, catalog, and survey (2012)
5. Dumas, M., La Rosa, M., Mendling, J., Reijers, A.H.: Fundamentals of Business Process Management (2013)
6. Rey, A., Delesalle, S.: Problèmes et conflits lexicographiques. Langue française (1979)
7. Feliu, J., Vivaldi, J., Cabré, M.T.: Ontologies: a review (2002)
8. Gangemi, A., Fisseha, F., Keizer, J., Lehmann, J., Liang, A.: A core ontology of fishery and its use in the fishery ontology service project (2004)
9. Stroppi, L.J.R., Chiotti, O., Villarreal, P.D.: Extending BPMN 2.0: method and tool support. In: Business Process Model and Notation (2011)
10. Ben Hassen, M., Turki, M., Gargouri, F.: Extending BPMN models with sensitive business process aspects. In: 26th International Conference on Knowledge-Based and Intelligent Information & Engineering Systems (KES 2022) (2022)
11. Braun, R., Schlieter, H., Burwitz, M., Esswein, W.: BPMN4CP: design and implementation of a BPMN extension for clinical pathways. In: IEEE International Conference on Bioinformatics and Biomedicine, pp. 9–16 (2014)
12. Said, I.B., Chaâbane, M.A., Andonoff, E., Bouaziz, R.: BPMN4VC-modeller: easy-handling of versions of collaborative processes using adaptation patterns. Int. J. Inf. Syst. Change Manag. **10**(2), 140–189 (2018)
13. Alyani, Z., Turki, M.: An MDA approach for extending functional dimension for sensitive business processes execution. In: International Conference on Intelligent Systems Design and Applications (2022)
14. Alyani, Z.: Extension of the functional dimension of BPMN based on MDA approach for sensitive business processes execution. In: ICIKS 2023, pp. 243–254 (2023)
15. Polančič, G.: BPMN-L: a BPMN extension for modeling of process landscapes. Comput. Ind. **121**, 103276 (2020)
16. Bocciarelli, P., D'Ambrogio, A., Giglio, A., Paglia, E.: A BPMN extension to enable the explicit modeling of task resources (2016)
17. Yahya, F., Boukadi, K., Maamar, Z.: Social business process modeling. In: Encyclopedia of Information Science and Technology. igi-global.com (2018)
18. Fattouch, N., Rekik, M., Wakrime, A., Boukadi, K.: A model-driven engineering approach for business process based SaaS services composition. In: AICCSA 2019 (2019)
19. Fattouch, N., Ben Lahmar, I., Boukadi, K.: Towards a meta-modeling approach for an IoRT-aware business process. In: ECMS 2022 (2022)
20. OMG. MDA guide version. 2.0. OMG Document: ormsc (2014)
21. Bézivin, J.: Sur les principes de base de l'ingénierie des modèles (2004)

Author Index

L. H. Belguith and K. Shaalan (Eds.): LPKM 2024, LNNS 1303, pp. 301–302, 2025.
https://doi.org/10.1007/978-3-031-85067-7

The manufacturer's authorised representative in the EU is Springer Nature Customer Service Centre GmbH, Europaplatz 3, 69115 Heidelberg, Germany. If you have any concerns regarding our products, please contact ProductSafety@springernature.com

Printed and bound by CPI Group (UK) Ltd, Croydon, CR0 4YY

02/10/2026

02227132-0009